THE TEACHING OF GEOMETRY
AT THE PRE-COLLEGE LEVEL

THE TEACHING OF GEOMETRY AT THE PRE-COLLEGE LEVEL

PROCEEDINGS OF THE SECOND
CSMP INTERNATIONAL CONFERENCE CO-SPONSORED BY
SOUTHERN ILLINOIS UNIVERSITY AND CENTRAL
MIDWESTERN REGIONAL EDUCATIONAL LABORATORY

Edited by

HANS-GEORG STEINER

University of Erlangen-Nürnberg, PH Bayreuth, West Germany

First published in

EDUCATIONAL STUDIES IN MATHEMATICS, Vol. 3, No. 3/4; Vol. 4, No. 1

ISBN 978-94-017-5638-9   ISBN 978-94-017-5896-3 (eBook)
DOI 10.1007/978-94-017-5896-3

EDITOR'S PREFACE

It has been a privilege for me to collaborate with the Comprehensive School Mathematics Program (CSMP), Carbondale, Illinois (U.S.A.), as European co-director, mathematician-in-residence, and consultant since 1965. It was during this time that what was once a dream became a reality: children, teachers, mathematics educators, mathematicians, educational measurement and evaluation specialists, media and systems experts are working together under one roof for the purpose of finding new ways of viewing and organizing, of learning and teaching mathematics.

Significant features of CSMP are its strong emphasis on mathematical content, the active involvement of mathematicians at every stage of the development of a highly individualized mathematics curriculum, and the collaboration with other mathematics projects in the U.S.A. and abroad. The CSMP International Conferences on the Teaching of Mathematics at the Pre-college Level bring together pioneers from other projects and mathematicians from various countries who are specialists in their fields and who are deeply concerned with the teaching of mathematics and the mathematical learning process.

Each conference consists of a variety of activities. Like the first conference on the teaching of probability and statistics (Carbondale, March 18–27, 1969)*, this second CSMP international conference was not restricted to giving lectures. Films and learning materials were presented, discussions and workshops among the participants and the CSMP staff were held, classes in all phases of the CSMP curriculum development observed, and demonstration lessons arranged.

This publication is a collection of the papers which were presented or prepared by the participants. The spectrum ranges from basic reflections on the goals of geometry teaching, historical perspective and psychological investigations on the learning of mathematics, to didactical analysis of particular areas of geometry and proposals for the incorporation of more advanced and recent mathematical ideas in the curriculum. The Conference

---

* Proceedings of this conference have been published: see *The Teaching of Probability and Statistics. Proceedings of the First CSMP International Conference*, (edited by Lennart Rade), Almqvist and Wiksell, Stockholm, 1970; John Wiley and Sons, Inc., New York, London, Sydney, 1970.

Recommendations express a consensus which grew out of extensive and lively discussions.

The conference chairman and editor of this book wishes to express his gratitude to Southern Illinois University Extension Services, especially to Chancellor R. W. MacVicar and Mr. Andrew Marcec for their cooperation in providing conference rooms and in arranging for various social activities, to the CSMP staff for their great help in organizing the conference and preparing the papers in their final form by translating, editing and typing, and to Professor Freudenthal for his support in making this publication possible.

HANS-GEORG STEINER

Bayreuth, West Germany

CONTENTS

BACKGROUND

Mathematics teaching is in a state of transition in all parts of the world. Numerous experiments aimed at modernizing the mathematics curriculum have been undertaken.

The Comprehensive School Mathematics Program (CSMP) was envisioned in 1963 as a response to a research impetus in mathematics education manifested in the reports of several prominent committees of mathematicians and mathematics educators from the United States and abroad.

CSMP, located in Carbondale, Illinois, U.S.A., is a major effort to modernize both the content of the pre-college mathematics curriculum and the methods of teaching.

In 1966, CSMP was established and became affiliated with Southern Illinois University. A proposal for a long-range curriculum development project was written during this year. The proposal was presented to the Central Midwestern Regional Educational Laboratory (CEMREL), and CSMP was incorporated as one of its major programs in the spring of 1967.

The Central Midwestern Regional Educational Laboratory was established under Title IV of the Elementary and Secondary Education Act of 1965, and began operation in June, 1966. One of 15 educational laboratories in the United States, whose purpose is to improve education for the children of the nation, CEMREL's mission is the development, CEMREL has set as its mission the development of individualized curricula in mathematics and the arts and humanities for all students in kindergarten through twelfth grade; it is also concerned with the development of effective instructional techniques and materials for children who have learning disabilities. Foremost authorities from across the nation compose the advisory committees to these programs. CEMREL is also concerned with developing instructional management systems, making use of the latest technology, to be used in conjunction with the curriculum development programs. Various related projects as well are a part of CEMREL's activities.

CEMREL is governed by a Board of Directors made up of outstanding educators, civic leaders, businessmen and labor officials from Illinois, Kentucky, Missouri and Tennessee. Headquarters for the laboratory are located in St. Ann, Missouri at 10646 St. Charles Rock Road, and programs offices are maintained in Carbondale, Illinois, Bowling Green, Kentucky, and in

Chattanooga and Nashville, Tennessee. Wade M. Robinson is the executive director.

CEMREL works cooperatively with the State Departments of Education in the four states, as well as with colleges, universities and, of course, the elementary and secondary schools.

The primary goal of CSMP is the development of individualized mathematics curricula for students of ages 5–18 which provide for each student a program sound in content, enjoyable, and most appropriate for his needs and abilities. Each activity package will provide a teaching program which may involve individual study, teacher instruction, small group interaction, reading a text, watching a demonstration film, listening to a tape, playing mathematical games, or a combination of these and other procedures.

In considering strategies to achieve this ultimate goal, namely the development of mathematics curricula that are sound in content and appropriate for the needs of future adults in a changing society, the developers of CSMP decided the program must be discipline-oriented. By this is meant that, while all pedagogical aspects of mathematics education are of deep concern, priority is given to the selection of *mathematical content* that is sound, relevant, and enjoyable. The implications of this decision are that the mathematical community must be deeply involved in the program, that there be mathematicians physically in residence, and that mathematicians must guide the program. To date, this has been the case, and it is a principle of CSMP procedure that every phase of future development of the program will continue to enjoy the strong involvement of mathematicians.

The major role of mathematicians in CSMP is the selection and analysis of content. The CSMP attitude is that content selection should, as far as is possible, be unfettered by traditional notions of "what children can do" or "what teachers can teach"; it should instead be guided by what is important in mathematics, by what the outcomes of mathematics instruction should be, and by what students actually demonstrate they can do. For these reasons, content selection within CSMP is based on empirical data gathered from probes made with students in the trying-out of ideas generated by mathematicians. Serious content suggestions are accepted or rejected only after adequate trials in the classroom.

To have a truly individualized curriculum, one must take into account the different views of mathematics held by various people. The creative research mathematicians view it in various ways, depending on their fields of research. There are various users of mathematics (scientists, engineers, social scientists, businessmen, etc.) who view mathematics in other ways; mathematics educators are concerned with curriculum and methods, while school administrators, parents, students, and the man in the street view it in terms of the exigencies of their situations and in terms consistent with their backgrounds. Each has needs that some form of mathematics can fulfill. Each can enhance his services to society by bringing to his vocation an "appropriate" background in mathematics.

An assumption accepted by CSMP is that all these views of mathematics are viable and valid. One goal for the content of CSMP is a curriculum so designed that none of these views is excluded; students at each stage of their schooling have, therefore, a maximum number of avenues open for their adult uses of mathematics.

In order to obtain such a curriculum, several alternative approaches to one and the same topic will need to be developed. It is, therefore, essential that a detailed analysis of the content selected for the CSMP curriculum be made by the mathematicians involved. CSMP has to be open to any suggestion which might affect its goals, its choices of content, its techniques of presentation. This means that the program should have channels of information about trends in mathematics as a developing science, as well as about research in mathematics education, both here and abroad.

One such useful channel has been opened with the launching, by CEMREL, of a series of CSMP international conferences on the teaching of mathematics at the pre-college level. The first conference, which attracted scholars from every continent, was held in Carbondale, Illinois, from March 18 to March 27, 1969, and dealt with the teaching of probability and statistics.*

This book is a report of the second international conference held in Carbondale, March 19–28, 1970, to investigate the teaching of geometry in the schools. Future conferences will examine the teaching of algebra, of logic and foundations, and of analysis.

The last century has witnessed a burgeoning of geometrical thinking. The fields of projective and affine geometry have taken their place alongside classical Euclidean geometry, and their interrelations have been studied through the ideas of transformation geometry. Historically, modern axiomatic theory was first exemplified in the context of geometry. With the development of the theory of metric spaces, of topology and of linear algebra, geometrical structures could be viewed as part of a wider and more general system, while geometrical ideas and terminology penetrated to a great advantage into general theories and their applications.

Since the turn of the century many scholars and educators have urged the incorporation of such ideas and terminology in the secondary school curricula. Under Felix Klein's influence, the use of transformations and vectors

---

* *The Teaching of Probability and Statistics: Proceedings of the First CSMP International Conference on the Teaching of Mathematics at the Pre-College Level*, Almqvist & Wiksell, Stockholm, 1970; John Wiley & Sons, Inc., New York.

became a significant aspect of German school reform. Many countries are now engaged in similar reform programs. The treatment of geometry totally within the framework of linear algebra was strongly advocated in several recent conferences, and at the 1960 Aarhus Conference, in particular.*

When long-range objectives are to be set for a K-12 program the problems are numerous indeed. That is why, in the original invitation letter to the prospective Geometry Conference participants, we were forced to draw up a rather lengthy list of topics for lectures and for discussion:

(1) Introduction of geometry in the early grades: topological aspects; spacial geometry before plane geometry; experimental attitude towards geometry; motion geometry; symmetries.

(2) Geometry for the intermediate grades: role of affine geometry; transition from an experimental to a theoretical-deductive attitude; the role of mathematization; order and orientation; concept of angle; geometry and numbers (algebraization); group theoretical aspects of geometry; congruence; similarity; geometrical constructions; measures, such as length, angle measure, area, volume.

(3) Geometry for the upper grades: geometry and axiomatics; linear algebra approach; coordinate geometry; linear mappings; bilinear forms and conics; orthogonal group; the place of trigonometry; projective geometry, non-euclidean geometry; convex sets; combinatorial geometry; topological aspects of geometry; applications of geometry and the use of geometrical language in other fields.

The papers presented at the Conference could not be expected to cover the entire field of problems in need of attention. Several topics of interest were brought up and clarified during the discussion periods and at several working sessions.

We believe that these proceedings do offer a broad background for future curriculum decisions.

CEMREL-CSMP, Carbondale, Ill., U.S.A.

---

* Aarhus Universitet Matematisk Institut: Elementaer Afdeling No. 7, *Lectures on Modern Teaching of Geometry and Related Topics*, ICMI Seminar, 1960.

THE CSMP DEVELOPMENT OF GEOMETRY

Of all the decisions one must make in a curriculum development project with respect to choice of content usually the most controversial and least defensible is the decision about geometry. Pressures shaping such decisions in our time range from "Euclid is dead; teach them linear algebra", to "Teach them a rigorous Euclidean geometry with Euclid's deficiencies corrected".

The approaches to geometry that will finally be adopted by CSMP should be designed to give varied and rich experiences to students at several levels, rather than a single approach at one level. It is not a coincidence that this International Conference on the Teaching of Geometry is convened at a time when CSMP is making its final decisions about the kinds of geometries it will teach at three levels of its program. The following is a description of CSMP's current planning; certainly this conference will prove a deciding factor in revisions of these plans.

ELEMENTARY SCHOOL: FIRST EXPERIENCES

The CSMP elementary geometry program provides initial geometric experiences and will eventually dove-tail with the geometry of the secondary program. Such a program must not only satisfy the prerequisites for an anticipated secondary program, but also miss no opportunity to exploit each child's intuition with respect to the geometry of the world of two and three dimensions. Because of the non-standard nature of the content of the elementary program, it is described here in some detail.

In K through 2, the average child does not read fluently, and so all his geometrical experiences must necessarily be gained via concrete situations. These take the form of recognizing and naming simple planar and space figures, making distinctions between the inside and the outside of simple closed curves, and experimenting with tesselations and reflective symmetries. The vehicles for these experiences consist of teacher-directed play and games involving attribute blocks (both flat and solid), mazes, figure fitting, mirror cards, paper folding, etc.

When the child can read, he is allowed to proceed independently at his own pace and at his own depth of comprehension. At this stage most of his initial observations of geometrical situations are over and he begins to explore the relations between the things he observes. He considers congruence of planar figures by deciding whether or not they "match." He is introduced to translations as slide mappings of the plane, and reflections are introduced as fold mappings of the plane. He decides whether certain planar figures are reflective symmetric by means of folding or tracing. Rotations are seen as turn mappings of the plane, and the child is able to find images of points and figures under any of these transformations, or under a composition of such transformations.

This knowledge of the isometries of the plane is then exploited in the definitions of parallel and perpendicular, and in describing various planar figures such as parallelogram, square, rhombus. He is introduced to parallel projection as a first experience of a mapping of the plane which is not one-to-one. His knowledge of parallel projections is also extended to three dimensions by means of activities dealing with shadows of planar figures as well as three-dimensional shapes. His first experience of similarity of planar figures is in terms of homotheties, which are seen as combinations of translations and dilatations.

After the child is led by experimentation to discover Euler's formula for a polyhedron, he is introduced to solids with holes in them; he works out their Euler characteristics and sees a pattern. Thus, he will be able to predict the Euler characteristic of a solid, given the number of holes.

The earlier notions of symmetry are now described in terms of invariant sets of isometries of the plane. He is introduced to the properties of a group by consideration of the group of symmetries of an equilateral triangle. The group concept is then further developed when the set of translations is shown to be a group. A central symmetry or halfturn is introduced, and it is shown that the set of central symmetries generates the group of all central symmetries and translations.

Vectors are introduced in terms of translations and are used to show some standard results in the plane.

The area of a rectangle is found by means of counting unit squares. Using the fact that translations preserve area, it is shown that parallelograms between the same parallels and on the same base are equal in area. As a corollary, the child can find the area of any triangle. Volume is introduced by an extension of the techniques used in finding areas.

The child completes his intuitive knowledge of similarity by considering similarities which involve rotations and/or reflections as well as translations and dilatations. He is introduced to the idea of orientation in the plane, and he reconsiders the isometries of the plane in terms of their orientation-preserving or reversing properties. Having done this, the child has a brief look at orientation in space.

The elementary program culminates in the introduction of angle measure and various construction methods.

Having completed this program, students will have covered, in an intuitive and experimental way, the background necessary for an appropriate secondary program in which such first experiences are formalized. Every student will have experienced geometric activities commensurate with his ability and interest.

MIDDLE SCHOOL: EXPLOITATION OF GEOMETRIC INTUITION

It is planned that the CSMP curriculum will some day extend from K through 12. Currently, the primary and secondary programs are being developed disjointly. The present need is for a geometry course preceding the secondary program which provides the intuitive experiences in geometry on which the more formal treatments in the secondary program can be built.

This intermediate course in geometry serves two purposes. For the time being it will be a substitute for the geometric background anticipated in the projected CSMP primary program. In the future it will serve as a bridge for entry into the CSMP secondary program by students who have not experienced the CSMP primary program.

The course, called *Intuitive Geometry*, begins with descriptions of incidence properties of points, lines, and planes that are suggested by physical experiences. When intuition fails to support the necessary notions of completeness of points in a line, agreements and definitions are made which replace intuition and which lead to a naive introduction to real numbers and distance mappings.

As vehicles for the development of congruence, symmetry, similarity and classification of figures, the primitives *line reflection* and *dilatation* are selected.

After descriptions are given of mappings and composites of mappings in general, the intuition concerning motion of figures is exploited to introduce mappings of the plane and space. From the primitive (mirror) reflection mapping, all the isometries of the plane are constructed as composites of reflections. Certain properties extend to all composites of reflections, giving characterizing properties of all mappings in the group of isometries.

Congruence of figures is then defined in terms of isometries, and properties of congruence stem from properties of isometries. Much of the usual classification of figures is then done in terms of reflective symmetries. Similarity of figures is described and studied in terms of composites of dilatations and isometries. In particular, right triangles are sorted into similarity classes, and the trigonometric ratios are defined for each class. Occasional extensions are made of reflections and their composites from the plane to space.

Vectors are introduced in terms of translations, and the course terminates with some elementary consequences of the properties of a plane vector space. Coordinate frames are used occasionally to provide an algebraic representation.

This course is in an intermediate position in several ways. It lies between the experimental probing of geometric experiences in the primary school and the formal deduction of the secondary program. As such it utilizes all the paraphernalia of tracings, mirrors, paper folding to arrive at geometric conjectures, and then makes heuristic arguments about the conjectures. It also begins to perfect the enormously powerful device of mappings in searching for geometric truths, a process to be formalized in the secondary program.

SECONDARY SCHOOL: FORMAL GEOMETRY

Two books in the Elements of Mathematics series, CSMP's secondary program, are devoted to Geometry. Book 10, with the tentative title *Elements of Geometry*, develops the synthetic approach to geometry, while Book 12, *Linear Geometry with Trigonometry*, is planned to utilize the linear algebra approach.

Book 10 will not develop all of Euclidean geometry. On the contrary, Book 10 is planned as a short book that will develop only a fragment of geometry, but that fragment rigorously.

Since no history has as yet been included in the *Elements of Mathematics* series, and since the history of geometry is very rich, introductory and concluding historical chapters will be included in Book 10. The first chapter will investigate the origins of mathematics (not just geometry), both in the East and in the West, and will culminate with Euclid.

The middle chapters of the book will develop the formal theory. Affine geometry in 2- and 3-space will be studied with considerable emphasis given to characterizing the finite affine planes. Hilbert's system will be mentioned, but Euclidean geometry will not formally be developed in Book 10 beyond ordered affine geometry. Rather, projective planes will be investigated, particularly finite projective planes. This will be an easy task, as the usual correspondence between affine and projective planes will be established. Also, of course, duality will be stressed: given the logical background of the students, duality should be an extremely meaningful and interesting topic.

The concluding chapter of Book 10 will pick up the history of the parallel postulate and it should serve as an informal introduction to non-Euclidean geometry for any student who wishes to read further in that subject.

Book 12 is in a very preliminary planning stage; thus, only a rough outline can be given. The main objective of the book should be a thorough investigation of plane and space geometry using the language and tools of linear algebra. Thus, emphasis will be on the development of two- and three-dimensional vector spaces over the field R of real numbers and the use of these vector spaces to study geometry. CSMP students have studied fields other than the field of real numbers; therefore vector spaces can be defined over arbitrary fields with no restriction on the dimension. Examples and exercises might well be given about vector spaces over the field of complex numbers, the field of rational numbers and finite fields; nevertheless, the main investigation should be restricted to two- and three-dimensional vector spaces over the field of real numbers. If students obtain a thorough knowledge and working facility in these dimensions, the transition to higher dimensions is almost trivial. Similar remarks hold for the study of higher dimensional geometry.

The proposed content of Book 12 is divided into the six 'chapters' listed below.

1. Vector Spaces and Linear Transformations
2. Affine Spaces and Affine Transformations
3. Bilinear Functions (Inner Products)
4. Euclidean Geometry and Euclidean Transformations
5. Trigonometry
6. Further Topics from Linear Algebra and Geometry.

After the conference these topics will be developed into a more detailed outline.

GEOMETRY CONFERENCE
RECOMMENDATIONS

(1) Geometry has evolved as an important means of understanding and organizing spatial phenomena. It is widely and multifariously applied in everyday life, whether in a highly mathematized form or not. Geometrical concepts, including their aesthetic aspects, have pervaded and elucidated other fields of mathematics. The richness of geometry, which lies in the wide variety of its concepts and problems, should be carefully exploited to give the student an anticipation of other branches of mathematics, which have either developed as powerful tools to study space (such as linear algebra, group theory) or as answers to questions that could be posed simply in geometric terms, but which could not be solved, or solved simply, within the framework of geometry (such as topology, measure theory).

(2) The teaching of geometry should take into account the points stressed in the previous paragraph; in particular, geometry should play a part in the teaching of the ideas of conceptual model building. Moreover, the teaching of geometry should respect obvious facts of developmental psychology: it can start at an early age with concrete material in concrete situations; rather than being imposed according to a preconceived system, deductive structures should arise to the degree to which children are able to analyze situations. It depends on the maturity and on the gifts of students, as well as on the part attributed to geometry, to what extent geometry is deductively organized and axiomatized. As an organizing and classifying element in geometry, groups are essential and should be used extensively. As to the question of what geometrical subjects should be taught, one should free oneself from the traditional view of geometry, and carefully weigh subjects and aspects that have not been given sufficient attention in the past; for instance: incidence structures, convexity, transformations, metrics, topological ideas, order, orientation, etc.

(3) Geometry should not be used primarily as a vehicle to teach logic; to do so would give a distorted view of both subjects. Correct mathematics should be a goal of every activity, whether implications are drawn from informal exploration or from formal reasoning. The degree of rigor in the teaching of mathematics may vary according to circumstances, but that should never be an excuse to misinform or to mislead the student.

(4) Integrated studies of mathematical curricula and teaching methods should be the concern of permanent centers, such as CSMP, employing the

services of a full-time staff, regularly consulting with mathematicians of high standing, and organizing conferences that will bring together teachers and mathematicians from home and abroad.

The effective teaching of geometry requires the use and the further development of suitable materials, such as apparatus for practical work, books and packages for individual and group use, and a wide range of visuals, including films. The conference commends CSMP for the admirable efforts which it has already made in this direction.

(5) The mathematics taught in teacher training courses (and its geometric content, in particular) should not be restricted to background knowledge: teachers should also be confronted with mathematical topics in the context in which they are to be taught. This means that courses in Advanced Mathematics from an Elementary Viewpoint should be designed with a view to the communication of elementary and advanced aspects of the subject matter and of its teaching. Insofar as the interrelation of geometry with other subjects has been neglected, a more balanced approach should be adopted.

n-GONS

INTRODUCTION

The starting point of my little theory of n-gons was the well-known theorem

(i) *In any quadrangle the midpoints of the sides form a parallelogram* (Figure 1).

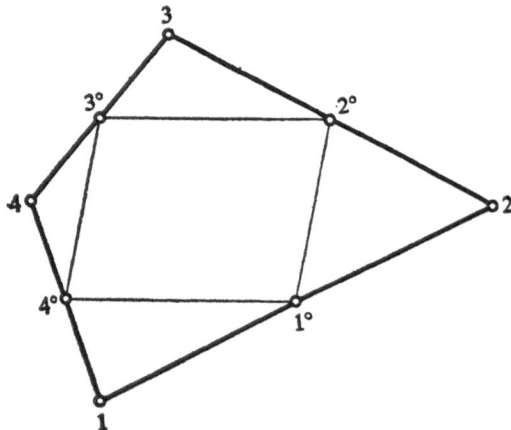

Fig. 1.

This theorem we interpret as follows. To any quadrangle A we assign *the quadrangle $A°$ formed by the midpoints of the sides*. Then we have a mapping of the set of all quadrangles into itself. Furthermore we have a special class of quadrangles, the set of parallelograms. The theorem says: The mapping maps the set of all quadrangles into this special class. It "specializes" the set of quadrangles. This specialization diminishes the maximum dimension:

the given quadrangle is not necessarily a plane figure, but a parallelogram is at most two dimensional.

The initial theorem "If A is a quadrangle, then $A°$ is a parallelogram" raises the question: does the construction of the *n-gon $A°$ formed by the midpoints of the sides* of a given *n*-gon A specialize for any *n*? We see immediately that it does not specialize for $n=3$. In general, it does not specialize for *n* odd. But for $n=2m$ we have:

(ii) *If A is a 2m-gon, then $A°$ is a 2m-gon, in which the omitting sides are closed vector polygons.*

More precisely: If we mark the sides of $A°$ alternately by a and b, then the a-sides form a closed vector polygon and the same holds for the b-sides. Figures 2 and 3 show a 6-gon and a 10-gon with this property.

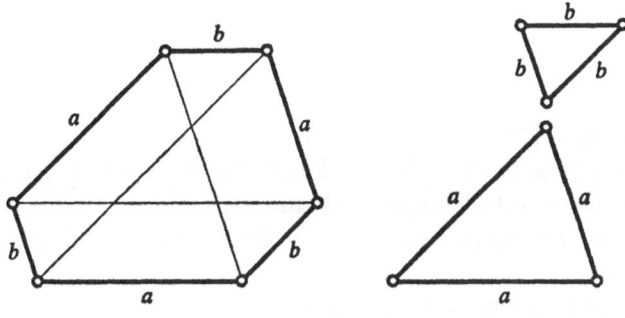
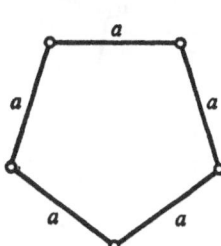

Fig. 2.

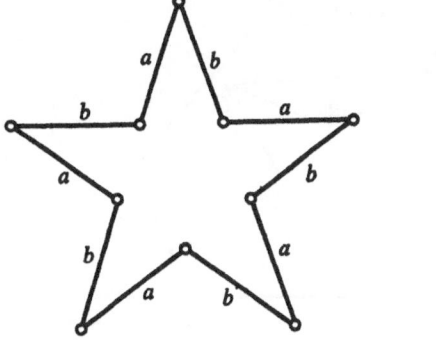

Fig. 3.

If for any three consecutive vertices of a given n-gon A we construct the fourth parallelogram point (for the vertices 1, 2, 3 the fourth parallelogram point 1', for the vertices 2, 3, 4 the fourth parallelogram point 2', etc.), we get an n-gon A'. Figure 4 shows a triangle A and its triangle A'.

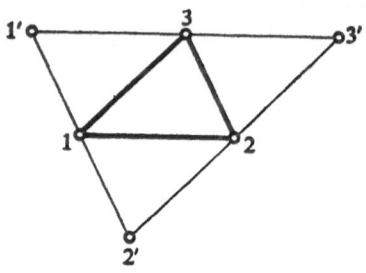

Fig. 4.

The following holds:

(iii) *If A is a 6-gon, then the fourth parallelogram points of the triples of consecutive vertices of A form a prism* (Figure 5).

Figures 6 and 7 show how prisms are to be defined (note the enumeration of the vertices).

Theorem (iii) seems to be rather surprising, since the mapping of A onto A' does not specialize for $n = 1, 2, 3, 4, 5, 7, 8, 9, 10, 11$.

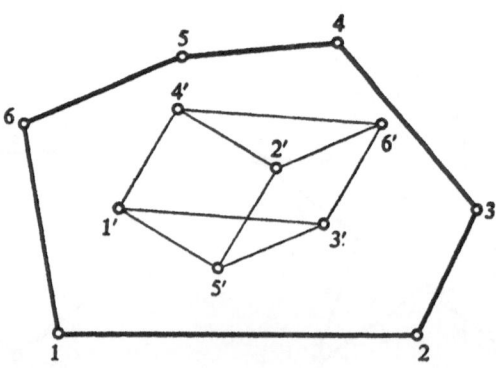

Fig. 5.

From these and many other examples we derive a notion of a "cyclic" class of n-gons and of a "cyclic" mapping of the set of all n-gons into itself. Working in a vector space over a field K (with $n \cdot 1 \neq 0$) we define an n-gon as an ordered n-tuple of vectors:

$$(a_1, a_2, ..., a_n).$$

To any n-tuple $(c_0, c_1, ..., c_{n-1})$ of elements of K we define a cyclic class and a cyclic mapping, each by a cyclic system of linear equations. A Main Theorem says that the number of cyclic classes is finite for any n and any K.

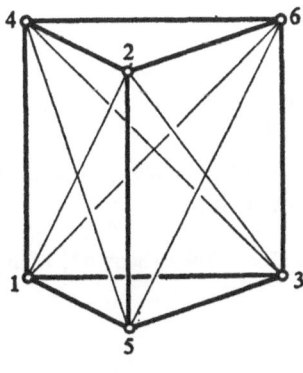

Fig. 6.

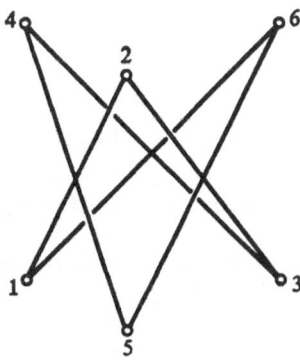

Fig. 7.

More precisely, the cyclic classes of n-gons form a finite Boolean algebra. Therefore any n-gon has a certain atomic structure: any n-gon has a unique decomposition into n-gons of the atomic cyclic classes. The components of an n-gon (the n-gons of the atomic cyclic classes) are regular in some sense. How regular they are depends on K.

I do not venture to assert that the Main Theorem may be proved in school classes, but the theory contains a lot of quite elementary items:

(1) There are many theorems connecting special cyclic classes with special cyclic mappings, as the examples (i), (ii), (iii) did.

(2) It is interesting to investigate for a given n (e.g., $n=6$) cyclic classes and their geometric properties and to order cyclic classes as a Boolean algebra.

(3) We have, I think, nice special decomposition-theorems. Examples, such as the decomposition of a parallelogram into two squares, are given at the end of this paper.

In such an elementary manner, the subject has been treated successfully in two schools at Kiel. I would like to add a few methodological remarks. I would say that n-gons are rather primitive objects (some examples are known to everyone) and that the theorems I would propose can be made visual, especially by drawing figures and diagrams. Within the theory we have many problems of any degree of difficulty, and among them sufficiently numerous little questions which may be used as exercises. (There is a general theory valid for any n, but nevertheless it remains true that pentagons and hexagons are different things.)

We use a vector space. Substantially, all we need of a vector space is the definition. Even linear dependence may remain unknown and we do not need the scalar or vector products. (It seems to me that in the school texts there is a certain lack of geometrical applications of the pure vector space of arbitrary dimension.)

On the whole, I regard the theory as an instructive connection between geometry and usual algebra and an elementary treatment of the n-gons may be continued by a treatment of such topics as Boolean algebra, linear mappings, polynomials (especially cyclotomic polynomials which occur in our school curriculum), rings of polynomials, quotient rings.

Reference: F. Bachmann und E. Schmidt, n-Ecke. BI-Hochschultaschenbuch, Mannheim/ Wien/Zürich, 1970.

1. DATA, n-GONS

Let n be a natural number, K a (commutative) field with $n \cdot 1 \neq 0$, V a vector space over K which contains an element $\neq o$. The dimension of V is arbitrary (finite or infinite), independent of the given number n.

The elements of V are called *points*, in particular the zero vector o, the zero point (origin). The n-tuples

$$(a_1, a_2, ..., a_n)$$

of elements of V are called *n-gons*. n-gons are denoted by $A, B, ...$, in particular the *zero-n-gon* $(o, o, ..., o)$ by O, the set of all n-gons by $\mathscr{A}_n$.

For n-gons we define *addition* and *multiplication with elements* $c \in K$ by

$$(a_1, a_2, ..., a_n) + (b_1, b_2, ..., b_n)$$
$$= (a_1 + b_1, a_2 + b_2, ..., a_n + b_n)$$
$$c(a_1, a_2, ..., a_n) = (ca_1, ca_2, ..., ca_n).$$

Then the set $\mathscr{A}_n$ of all n-gons is a vector space $V^n = V \oplus ... \oplus V$. This vector space is called the *vector space of the n-gons*.

Some of the vertices $a_1, a_2, ..., a_n$ of an n-gon may be equal. The n-gons

$(a, a, ..., a)$ are called *trivial n-gons*. Since a trivial *n*-gon may be regarded as a 1-gon counted *n* times, we denote the set of all trivial *n*-gons by $\mathscr{A}_{1,n}$.

2. CYCLIC CLASSES OF *n*-GONS

Closely connected with an *n*-gon $(a_1, a_2, ..., a_n)$ are the *n*-gons

$$(*) \qquad (a_2, ..., a_n, a_1), (a_3, ..., a_1, a_2), ..., (a_n, a_1, ..., a_{n-1}),$$

resulting from the given *n*-gon by a cyclic permutation of the vertices.

According to the geometric idea of an *n*-gon we study primarily sets of *n*-gons which contain all *n*-gons (*) if they contain an *n*-gon $(a_1, a_2, ..., a_n)$.

The set of all *n*-gons $(a_1, a_2, ..., a_n)$ which satisfy a linear equation

$$c_0 a_1 + c_1 a_2 + \cdots + c_{n-1} a_n = o$$

with given coefficients $c_0, c_1, ..., c_{n-1} \in K$, does not, in general, have the desired property.

We consider *cyclic systems of equations*, i.e., homogeneous systems of linear equations

$$c_0 a_1 + c_1 a_2 + \cdots + c_{n-1} a_n = o$$
$$c_0 a_2 + c_1 a_3 + \cdots + c_{n-1} a_1 = o$$
$$\cdots$$
$$c_0 a_n + c_1 a_1 + \cdots + c_{n-1} a_{n-1} = o$$

with given coefficients $c_0, c_1, ..., c_{n-1} \in K$. The set of all *n*-gons satisfying such a system contains all *n*-gons (*) if it contains an *n*-gon $(a_1, a_2, ..., a_n)$.

A *cyclic class of n-gons* is defined as the set of all solutions of a cyclic system of equations. Cyclic classes are special subspaces of the vector space of the *n*-gons.

Any *n*-tuple $(c_0, c_1, ..., c_{n-1})$ of elements of K defines a cyclic class. Different *n*-tuples may define the same cyclic class.

Three basic examples. The *n*-tuple $(0, 0, ..., 0)$ defines the set of all *n*-gons. $a_1 - a_2 = o, ...$ describes the set of trivial *n*-gons. $a_1 = o, ...$ has only the solution $(o, o, ..., o) = O$. Therefore $\mathscr{A}_n, \mathscr{A}_{1,n}, \{O\}$ are cyclic classes.

The periodic classes. Let d be a divisor of n and $n = dd$. $a_1 = a_{d+1}, ...$ describes the cyclic class of the periodic *n*-gons

$$(**) \qquad (a_1, ..., a_d, a_1, ..., a_d, ..., a_1, ..., a_d).$$

Such an *n*-gon may be interpreted as a *d*-gon counted d times (but it is a

dd-gon), and we denote the class of these n-gons by $\mathscr{A}_{d,\bar{d}}$. Extreme cases: $\mathscr{A}_{1,n}$ is the class of the trivial n-gons, $\mathscr{A}_{n,1} = \mathscr{A}_n$ is the set of all n-gons.

If we write the vertices of an n-gon $(a_1, a_2, ..., a_n)$ as the array

$$
\begin{array}{cccc}
a_1 & a_{d+1} & \cdots & a_{n-d+1} \\
a_2 & a_{d+2} & \cdots & a_{n-d+2} \\
& \cdots & & \\
a_d & a_{2d} & \cdots & a_n ,
\end{array}
$$

then the rows are d-tuples and may be called the *omitting sub-d-gons of* $(a_1, a_2, ..., a_n)$. One method to define cyclic classes consists in postulating something for the omitting sub-polygons. The periodic n-gons (**) are the n-gons with the property that the omitting sub-d-gons are trivial.

Degree. We say that a cyclic class $\mathscr{C}$ has degree f if f is the maximum number of arbitrary points $a_1, a_2, ..., a_f$ which can be extended to an n-gon

$$(a_1, a_2, ..., a_f, a_{f+1}, ..., a_n)$$

of $\mathscr{C}$. If degree $\mathscr{C} = f$, then for the n-gons of $\mathscr{C}$ there exists a representation with $a_1, a_2, ..., a_f$ as "parameters".

Remark. Any cyclic class $\mathscr{C}$ is invariant under all endomorphisms α of V: $(a_1, a_2, ..., a_n) \in \mathscr{C}$ implies $(\alpha a_1, \alpha a_2, ..., \alpha a_n) \in \mathscr{C}$.

3. EXAMPLES OF CYCLIC CLASSES FOR $n=4$ AND $n=6$

(a) $n=4$. There are three periodic classes of quadrangles: $\mathscr{A}_4$, $\mathscr{A}_{2,2}$ (the class of doubly counted 2-gons), $\mathscr{A}_{1,4}$. Furthermore we define the class of *parallelograms* by

$$a_1 - a_2 + a_3 - a_4 = o, ... \quad \text{or} \quad \tfrac{1}{2}(a_1 + a_3) = \tfrac{1}{2}(a_2 + a_4), ...$$

The diagram (Figure 8) shows the inclusions among these four classes of quadrangles and contains the degrees of the classes.

(b) $n=2m$. The $2m$-gons satisfying

$$a_1 - a_2 + a_{m+1} - a_{m+2} = o, ... \quad \text{or}$$
$$\tfrac{1}{2}(a_1 + a_{m+1}) = \tfrac{1}{2}(a_2 + a_{m+2}), ...$$

(the second system says that all diagonals have the same midpoint) are called *2m-parallelograms*. Figure 9 shows a special 12-parallelogram. The $2m$-parallelograms form a cyclic class of degree $m+1$.

In a 4-parallelogram the alternating sum of the vertices is o. In general, the cyclic system of equations

$$a_1 - a_2 + a_3 - a_4 + \cdots + a_{2m-1} - a_{2m} = 0, \ldots$$

characterizes the $2m$-gons with alternating sum o as a cyclic class of degree $2m-1$, the ASO-*class of* $2m$-*gons*. Examples are known from Figures

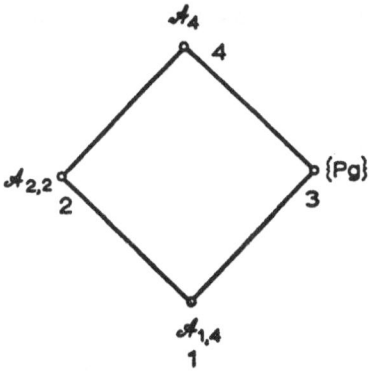

Fig. 8.

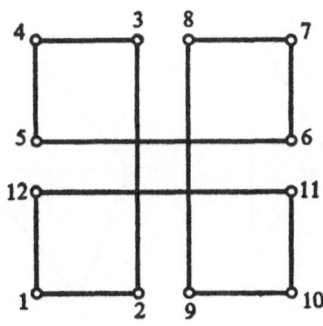

Fig. 9.

2 and 3. By comparing the degrees, we see that the ASO-class coincides with the class of the $2m$-parallelograms only for $n=4$. For $n=4m$ we have an inclusion: In any $4m$-parallelogram the alternating sum of the vertices is o.

(c) $n=6$. There are four periodic classes: $\mathscr{A}_6$, $\mathscr{A}_{3,2}$ (the class of doubly counted triangles), $\mathscr{A}_{2,3}$ (the class of 2-gons counted three times), $\mathscr{A}_{1,6}$.

Furthermore we have the class of 6-parallelograms, the 6-ASO-class, the class of *prisms* defined by

$$a_1 - a_4 = a_3 - a_6 \dots,$$

and the class $\mathscr{R}_6$ of *affinely regular 6-gons*, i.e. 6-gons with the property that all triples of consecutive vertices have the same fourth parallelogram point:

$$a_1 - a_2 + a_3 = a_2 - a_3 + a_4, \dots. \text{ (Figures 10–13.)}$$

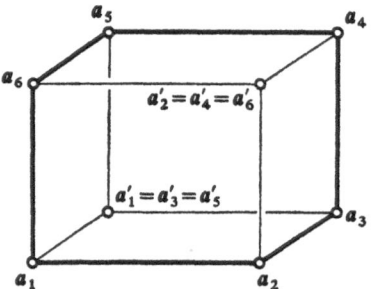

Fig. 10. 6-parallelogram.

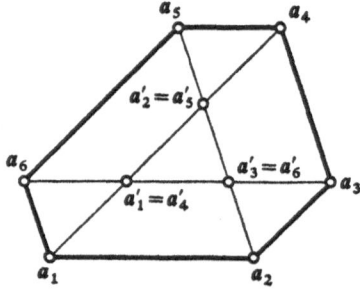

Fig. 11. ASO-6-gon.

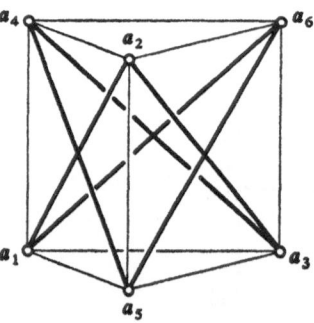

Fig. 12. Prism.

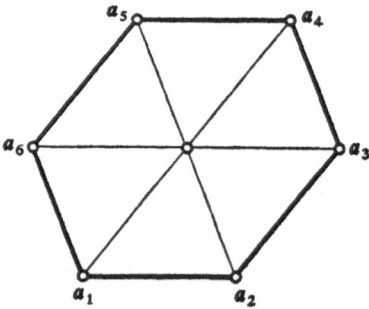

Fig. 13. Affinely regular 6-gon.

The diagram (Figure 14) shows the inclusions that hold among these eight cyclic classes of 6-gons and contains the degrees. If two segments slope up from a class, one may ask if this class is the full intersection of the two upper classes. And if we have two segments sloping down from a class, we may ask if the class is the sum of the two lower classes.

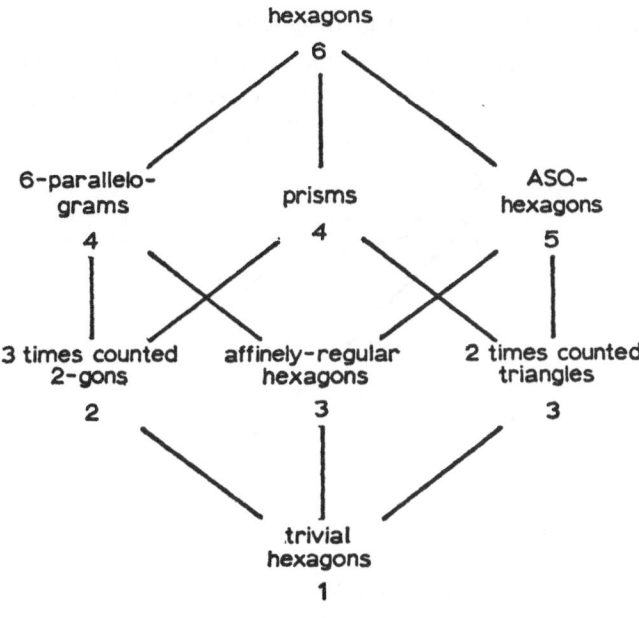

Fig. 14.

By considering the fourth parallelogram points $a'_i = a_i - a_{i+1} + a_{i+2}$ we see, for instance:

(i) *The affinely regular 6-gons are precisely the 6-parallelograms with alternating sum o.*

A simple computation shows:

(ii) *The prisms are precisely the sums of the 2-gons counted three times and the doubly counted triangles* (Figure 15).

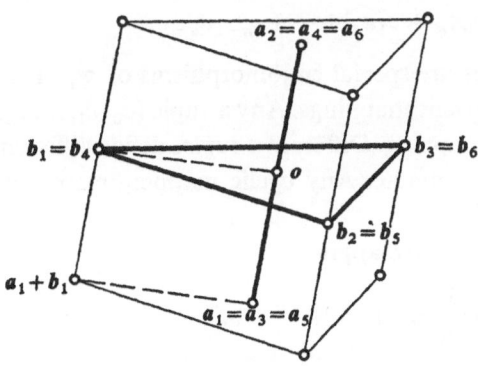

Fig. 15.

(d) $n=8$. Again it is easy to give eight cyclic classes: The four periodic classes, the class of doubly counted parallelograms, the class of 8-parallelograms, the 8-ASO-class, and the class of 8-gons having the property that the two omitting sub-quadrangles are parallelograms (Figure 16).

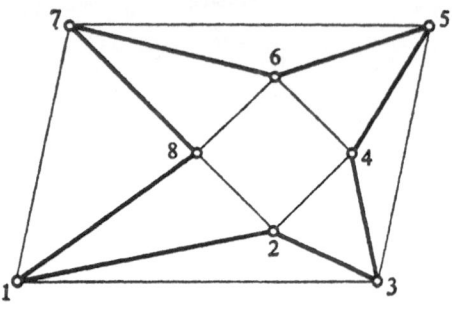

Fig. 16.

The diagram of these classes differs in an interesting way from the diagram of the eight classes of 6-gons given above. For instance the periodic classes form a chain, and the class of 8-parallelograms is contained in the 8-ASO-class.

4. CYCLIC MAPPINGS

A mapping of the vector space $\mathscr{A}_n$ of all n-gons into itself is called *cyclic* if there exist elements $c_0, c_1, \ldots, c_{n-1} \in K$ such that any n-gon $(a_1, a_2, \ldots, a_n)$ and its image $(b_1, b_2, \ldots, b_n)$ are related by the equations:

$$b_1 = c_0 a_1 + c_1 a_2 + \cdots + c_{n-1} a_n$$
$$b_2 = c_0 a_2 + c_1 a_3 + \cdots + c_{n-1} a_1$$
$$\cdots$$
$$b_n = c_0 a_n + c_1 a_1 + \cdots + c_{n-1} a_{n-1}.$$

Cyclic mappings are special endomorphisms of $\mathscr{A}_n$. The cyclic classes are the kernels of the cyclic mappings. Any n-tuple $(c_0, c_1, \ldots, c_{n-1})$ of elements of K defines a cyclic mapping. Different n-tuples define different cyclic mappings. Cyclic mappings commute. Any cyclic mapping maps any cyclic class into itself.

Example. The cyclic mapping

$$\sigma: b_1 = \frac{1}{n}(a_1 + a_2 + \cdots + a_n), \cdots$$

assigns to any n-gon $(a_1, a_2, \ldots, a_n)$ the trivial n-gon $(a, a, \ldots, a)$, a being the center of gravity of $(a_1, a_2, \ldots, a_n)$. σ is a cyclic projection and "trivializes" all *n*-gons. (The kernel of σ consists of all *n*-gons with o as center of gravity. These *n*-gons form a cyclic class.)

5. EXAMPLES OF CYCLIC MAPPINGS FOR *n*=4 AND *n*=6

(a) $n=4$. We denote the cyclic mapping

$$(a_1, a_2, a_3, a_4) \rightarrow (b_1, b_2, b_3, b_4),$$

which assigns to any quadrangle the quadrangle formed by the midpoints of the sides by κ_2:

$$\kappa_2: b_1 = \tfrac{1}{2}(a_1 + a_2), \dots.$$

The image quadrangle

$$\kappa_2(a_1, a_2, a_3, a_4) \\ = \left(\tfrac{1}{2}(a_1 + a_2), \tfrac{1}{2}(a_2 + a_3), \tfrac{1}{2}(a_3 + a_4), \tfrac{1}{2}(a_4 + a_1)\right)$$

has alternating sum o, and is therefore a parallelogram:

κ_2 maps $\mathscr{A}_4$ into the class of parallelograms. κ_2 trivializes exactly the twice counted 2-gons (a_1, a_2, a_1, a_2); it maps $\mathscr{A}_{2,2}$ onto $\mathscr{A}_{1,4}$.

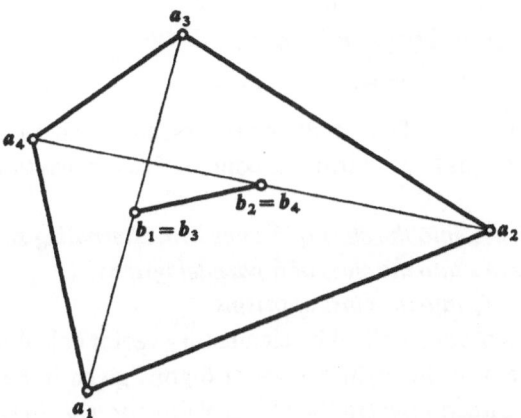

Fig. 17.

We ask for a cyclic mapping that maps $\mathscr{A}_4$ into $\mathscr{A}_{2,2}$. Obviously, the mapping

$$\mu_2: b_1 = \tfrac{1}{2}(a_1 + a_3), \dots$$

which assigns to any quadrangle the midpoints of its diagonals (Figure 17), has the desired property. μ_2 trivializes precisely the parallelograms.

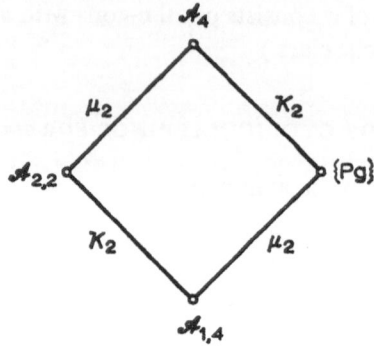

Fig. 18.

By now, for any segment of the diagram of the four cyclic classes of quadrangles given in 3, we have a cyclic mapping which maps the upper class into
the lower one (Figure 18). We have $\kappa_2\mu_2 = \sigma$: $\kappa_2\mu_2$ trivializes all quadrangles.

(b) $n=6$. By κ_2, κ_3, α_3 we denote the cyclic mappings

$$(a_1, a_2, ..., a_6) \rightarrow (b_1, b_2, ..., b_6)$$

given by the following cyclic systems of equations:

$$\kappa_2: b_1 = \tfrac{1}{2}(a_1 + a_2), ...$$
$$\kappa_3: b_1 = \tfrac{1}{3}(a_1 + a_2 + a_3), ...$$
$$\alpha_3: b_1 = a_1 - a_2 + a_3, ...$$

κ_2 assigns to each pair of consecutive vertices, and κ_3 to each triple, the center
of gravity; α_3 assigns to each triple of consecutive vertices the fourth parallelogram point.

(i) κ_2 maps $\mathscr{A}_6$ into the class of 6-gons with alternating sum o.
(ii) κ_3 maps $\mathscr{A}_6$ into the class of 6-parallelograms.
(iii) α_3 maps $\mathscr{A}_6$ into the class of prisms.

These statements are verified by elementary vector calculus.

In the diagram of the eight classes of 6-gons given in Section 3, κ_2 maps
in the NW–SE direction, κ_3 in the NE–SW direction, α_3 in the N–S direction,
but always the upper class into the lower one (Figure 19). If, for instance,
A is a 6-parallelogram, then $\kappa_2 A$ is affinely regular (Figure 20).

If to an arbitrary 6-gon we apply the mappings κ_2, κ_3, α_3 successively, in
any order, then the 6-gon will be specialized step by step and finally it is
killed. The final result is the trivial 6-gon consisting only of the center of
gravity counted 6 times ($\kappa_2 \kappa_3 \alpha_3 = \sigma$).

In all these examples, a certain cyclic mapping φ maps a certain class $\mathscr{A}$
into a certain subclass $\mathscr{B}$. What can be said about the φ pre-images of a

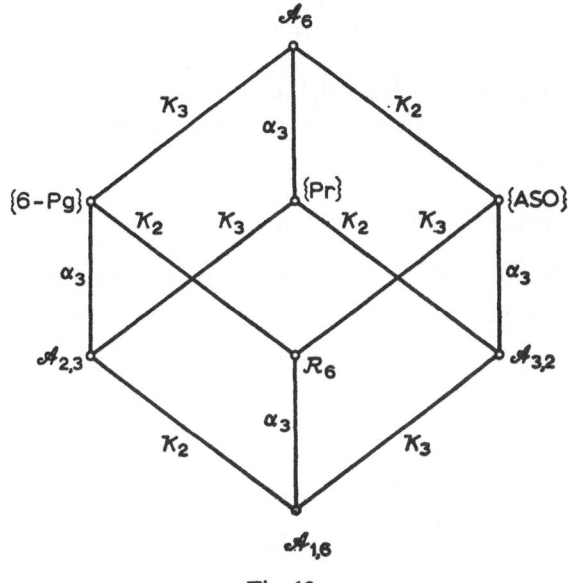

Fig. 19.

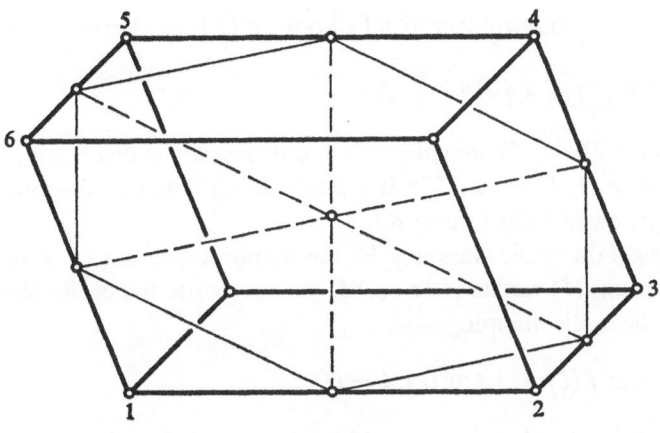

Fig. 20.

given *n*-gon $B \in \mathscr{B}$? In the case $n=4$, $\varphi = \kappa_2$, we have to search out the quadrangles A which are circumscribed about a given parallelogram B (the A's with $\kappa_2 A = B$). There exists exactly one circumscribed quadrangle with an arbitrarily chosen point as the first vertex, and among all these circumscribed quadrangles there is precisely one parallelogram; the construction of this parallelogram is obvious. Analogous questions may be studied in all other examples.

6. MAIN THEOREM

We denote the algebra of polynomials over K by $K[x]$.

In order to get a convenient representation of the cyclic mappings, we use the special cyclic mapping

$$\zeta: (a_1, a_2, ..., a_n) \rightarrow (a_2, ..., a_n, a_1).$$

Then in the algebra of the endomorphisms of $\mathscr{A}_n$, the cyclic mapping given by the n-tuple $(c_0, c_1, ..., c_{n-1})$ of elements of K may be written as

$$c_0 + c_1\zeta + \cdots + c_{n-1}\zeta^{n-1},$$

carrying $(a_1, a_2, ..., a_n)$ to

$$(c_0 + c_1\zeta + \cdots + c_{n-1}\zeta^{n-1})(a_1, a_2, ..., a_n).$$

Let $K[\zeta]$ be the set of cyclic mappings (a subset of the algebra of endomorphisms of $\mathscr{A}_n$).

(+) $\qquad f(x) \rightarrow f(\zeta)$

is an algebra-homomorphism of $K[x]$ onto $K[\zeta]$; the kernel is $(x^n - 1)$ and therefore

$$K[\zeta] \cong K[x]/(x^n - 1).$$

Theorem: *The cyclic mappings form a commutative algebra $K[\zeta]$ over K, of dimension n. $1, \zeta, \zeta^2, ..., \zeta^{n-1}$ is a basis. $K[\zeta]$ is the group-algebra of the cyclic group generated by ζ (over K).*

Since we get the cyclic mappings by the homomorphism (+), and since the cyclic mapping $f(\zeta)$ carries A to $f(\zeta)A$, we may write the cyclic classes – the kernels of the cyclic mappings – as

(+ +) $\qquad \mathrm{Ker}\, f(\zeta) := \{A : f(\zeta) A = O\}.$

For an $f(x) \in K[x]$, (+ +) is said to be the *cyclic class defined by the polynomial f(x)*. If $f(x)$ is a polynomial of degree $< n$ (including the zero polynomial):

$$f(x) = c_0 + c_1 x + \cdots + c_{n-1} x^{n-1},$$

then the cyclic class defined by $f(x)$ is merely the class defined by the cyclic system of equations with $(c_0, c_1, ..., c_{n-1})$ as the n-tuple of coefficients.

Now let

$$x^n - 1 = p_1(x) p_2(x) ... p_k(x)$$

be the decomposition of $x^n - 1$ into prime polynomials in $K[x]$. Since $n \cdot 1 \neq 0$ in K, $x^n - 1$ is squarefree. $x^n - 1$ has at least the decomposition

$$x^n - 1 = \prod_{d \mid n} F_d(x)$$

($F_d(x)$ the d-th cyclotomic polynomial over K). If the cyclotomic polynomials are prime polynomials over K, then the number k is the number of divisors of n which we denote by $\tau(n)$. We have

$$\tau(n) \leqslant k \leqslant n.$$

Now consider the *lattice* $L(x^n - 1)$ *of the* (classes of associated) *divisors of* $x^n - 1$ *in* $K[x]$. (sup and inf are the L.C.M. and the G.C.D..) $L(x^n - 1)$ is a Boolean algebra of 2^k elements (since $K[x]$ is a principal ideal domain, it is distributive; since $x^n - 1$ is squarefree, it is complemented).

Main Theorem:

(*) $t(x) \rightarrow \mathrm{Ker}\, t(\zeta)$ $(t(x) \in K[x]$ and $t(x) \mid x^n - 1)$

defines an isomorphism of the lattice $L(x^n - 1)$ onto a sublattice of the lattice of the subspaces of $\mathscr{A}_n$ and this sublattice consists precisely of all cyclic classes.

The Main Theorem implies:

(i) *The cyclic classes defined by the divisors of $x^n - 1$ are all the cyclic classes. Non-associated divisors define different cyclic classes.*

(ii) *Two cyclic systems of equations with n-tuples $(c_0, c_1, ..., c_{n-1})$, $(d_0, d_1, ..., d_{n-1})$ of coefficients define the same cyclic class if and only if the polynomials $\sum c_i x^i$, $\sum d_i x^i$ have the same G.C.D. with $x^n - 1$ in $K[x]$.*

Further fundamental consequences of the Main Theorem:

(iii) *The cyclic classes form a finite Boolean algebra. The number of cyclic classes is 2^k (k being the number of prime factors of $x^n - 1$ in $K[x]$). The number of cyclic classes is independent of the dimension of the given vector space V.*

(iv) *The Boolean algebra of the cyclic classes is a sublattice of the lattice of subspaces of $\mathscr{A}_n$. Sum and intersection of cyclic classes are cyclic classes.*

(v) *$\mathscr{A}_n$ is the direct sum of the atomic cyclic classes* (the atoms of the Boolean algebra of cyclic classes). *Any n-gon has a unique decomposition as a sum of n-gons of the atomic cyclic classes.*

(vi) *Any cyclic class consists of all n-gons with the property that certain atomic components are the zero n-gon.*

In addition to the Main Theorem we have

Theorem: *The isomorphism (*) transfers the degree:*

$$\text{degree } t(x) = \text{degree Ker } t(\zeta) \qquad (t(x) \mid x^n - 1).$$

Corollary: *The degrees of the cyclic classes satisfy the formula*

$$\text{degree } \mathscr{B} + \text{degree } \mathscr{C} = \text{degree}(\mathscr{B} + \mathscr{C}) + \text{degree}(\mathscr{B} \cap \mathscr{C}).$$

Remark. Over any allowable field there exist at least the $2^{\tau(n)}$ cyclic classes defined by the products of different cyclotomic polynomials $F_d(x)$ with $d \mid n$. All examples of cyclic classes given hitherto are such "ever existing classes".

7. CYCLIC CLASSES AS IMAGES

Among the cyclic mappings the idempotent ones, called *cyclic projections*, are especially important. It is an elementary fact (valid for any commutative zing with unity) that the idempotent cyclic mappings form a Boolean algebra with respect to "Boolean addition" (sum minus product) and multiplication; the complement of ε is $1 - \varepsilon$.

Since $K[\zeta] \cong K[x]/(x^n - 1)$ and $x^n - 1$ is squarefree, $K[\zeta]$ is a direct sum of k fields (k being the number of the prime factors of $x^n - 1$ in $K[x]$). It follows that $K[\zeta]$ contains exactly 2^k idempotent elements and that any cyclic mapping is associated with exactly one idempotent element, i.e., one cyclic projection.

The cyclic classes occur in the Main Theorem as the kernels of the cyclic mappings. But are they also the images of $\mathscr{A}_n$ under the cyclic mappings? This question is answered by the following

Theorem: *The following sets coincide:*

(1) *The cyclic classes of n-gons,*
(2) *The kernels of the cyclic mappings,*
(3) *The kernels of the cyclic projections,*
(4) *The images of $\mathscr{A}_n$ under the cyclic projections,*
(5) *The images of $\mathscr{A}_n$ under the cyclic mappings.*

(1)=(2) follows from the definitions, (3)=(4) from the simple fact that for an idempotent endomorphism ε of a vector space $\text{Ker } \varepsilon = \text{Im}(1 - \varepsilon)$. For (2)=(3) and (4)=(5) we use the fact that any cyclic mapping is associated with a cyclic projection and that associated cyclic mappings have the same kernel and map $\mathscr{A}_n$ onto the same set.

One general method for determining the cyclic projection which projects $\mathscr{A}_n$ onto a given cyclic class $\mathscr{C}$, is the following. $\mathscr{C}$ may be written as $\text{Ker } t(\zeta)$, with $t(x) \mid x^n - 1$. Consider in $K[x]$ the congruences

$$e(x) \equiv 1 \mod t(x), \qquad e(x) \equiv 0 \mod \frac{x^n - 1}{t(x)}$$

and call a solution $e_t(x)$. Then $e_t(\zeta)$ is the cyclic projection with the desired property:

$$\text{Ker } t(\zeta) = \text{Im } e_t(\zeta).$$

In particular

$$A = e_{p_1}(\zeta) A + e_{p_2}(\zeta) A + \cdots + e_{p_k}(\zeta) A$$

is the decomposition of an *n*-gon A into its atomic components.

Remark 1. As interesting cyclic mappings I would like to mention the involutory ones, the *cyclic reflections*. They form a group.

Remark 2. Cyclic mappings with $c_0 + c_1 + \cdots + c_{n-1} = 1$ preserve the center of gravity. The special cyclic mappings σ, κ_2, κ_3, μ_2, α_3, ζ, all have this property. If $f(x)$ is a polynomial of $K[x]$ and $f(1)$, the sum of its coefficients, is different from zero, we may define $f^*(x) := f(x)/f(1)$; then $f^*(1) = 1$, and $f^*(\zeta)$ is a cyclic mapping which maps any *n*-gon onto an *n*-gon with the same center of gravity.

Example for $n = 6$. The special cyclic mappings considered in 5, can be expressed in terms of cyclotomic polynomials

$$\kappa_2 = F_2^*(\zeta), \quad \kappa_3 = F_3^*(\zeta), \quad \alpha_3 = F_6^*(\zeta) = F_6(\zeta),$$
$$\sigma = F_2^*(\zeta) F_3^*(\zeta) F_6(\zeta).$$

8. REGULAR *n*-GONS

Roughly, if K is the field C of complex numbers, then the *n*-gons of the atomic cyclic classes are regular *n*-gons (in the usual sense) of a complex plane; if K is the field R of reals, then the *n*-gons of the atomic cyclic classes are affinely regular; if K is the field Q of the rational numbers, the "regularity" of the *n*-gons of the atomic cyclic classes is, in general, something new, and we speak of *Q*-regularity. (These statements need some refinement.)

An *n*-gon is called *Q-regular* if all omitting sub-*d*-gons (for all $d \mid n$, $d \neq 1$) have the same center of gravity. The *Q*-regular *n*-gons are for $n = 4$ the parallelograms, for $n = 6$ the affinely regular hexagons, and generally for $n = 2p$ (p prime) the 2*p*-parallelograms with alternating sum *o*. Figure 21 shows a

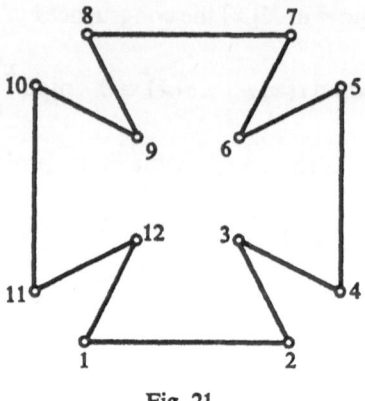

Fig. 21.

special Q-regular 12-gon. The reader is recommended to sketch other amusing examples, say for $n=12$.

Among the atomic cyclic classes, there always occurs the class of trivial n-gons (defined by the prime polynomial $x-1$). All other atomic classes are "zero point classes", i.e., cyclic classes which consist only of n-gons with o as center of gravity. (In any cyclic class $\mathscr{C}$, the n-gons of $\mathscr{C}$ which have o as center of gravity form a cyclic class.)

The class defined by the n-th cyclotomic polynomial $F_n(x)$ is the Q-regular zero point class. If $K=Q$, then this class is an atomic class and all other atomic zero point classes consist of Q-regular d-gons counted several times, with $d|n$.

For an arbitrary K the interesting problem is to study the atomic classes contained in the Q-regular zero point class (the classes defined by the prime factors of $F_n(x)$).

If $K=C$, these classes are defined by the linear polynomials $x-w$, w being a primitive n-th root of unity. Such a class consists of all n-gons

$$(a, wa, w^2a, ..., w^{n-1}a).$$

For $w=\exp(2\pi i/n)$, these n-gons are regular n-gons with o as center of gravity; each of them lies in a complex plane determined by its first vertex a. By running through the vertices in jumps relatively prime to n (i.e., by applying the mappings $(a_1, a_2, ..., a_n) \rightarrow (a_1, a_{1+j}, ..., a_{1+(n-1)j})$ with a j relatively prime to n), we get the n-gons for the other primitive w's. E.g., for $n=5$ we have $\varphi(5)=4$ such classes; if $K=C=V$, they are: the usual regular pentagons oriented counterclockwise, the same oriented clockwise, the regular pentagrams oriented counterclockwise, the same oriented clockwise – all with o as center of gravity (Figures 22 and 23).

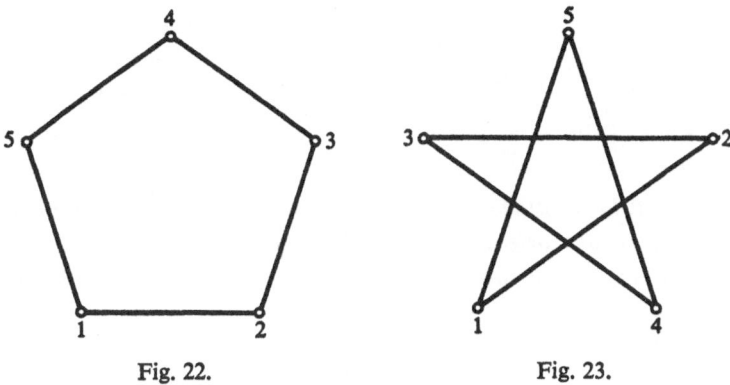

Fig. 22.   Fig. 23.

For $K = R$ (reals) the interesting classes are given by the cyclic systems

$$a_1 - ca_2 + a_3 = o,\ldots \quad \text{with} \quad c = 2\cos(2\pi j/n), \quad (j, n) = 1$$

(the sum of a_1, a_3 is c times a_2, etc.). For $n = 5$ we have two such classes: the affinely regular pentagons and the affinely regular pentagrams – both with o as center of gravity (Figures 24 and 25).

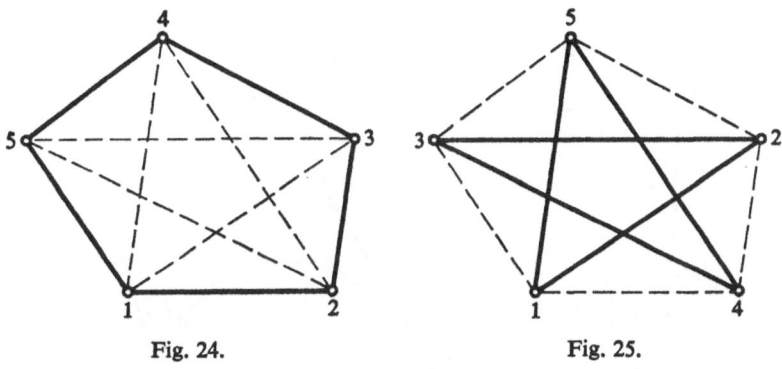

Fig. 24.   Fig. 25.

9. DECOMPOSITION OF n-GONS

We know that any n-gon has a unique decomposition into n-gons of the atomic cyclic classes. Coarsenings of this decomposition may be studied also. In the Boolean algebra of cyclic classes, to any class $\mathscr{B}$ we have a complement $\mathscr{C}$ with $\mathscr{A}_n = \mathscr{B} \oplus \mathscr{C}$ and therefore a unique decomposition of all n-gons.

Especially interesting are decompositions of the Q-regular n-gons, and the decompositions of the affinely regular n-gons into regular n-gons. We give only a few examples, which may be verified in a quite elementary manner. In the examples (i)–(iv), let all n-gons have o as center of gravity.

(i) *A quadrangle has a unique decomposition into a parallelogram and a doubly counted 2-gon.* (The parallelogram is the unique parallelogram having the same midpoint-parallelogram as the given quadrangle (Figure 26).)

(ii) *A hexagon has a unique decomposition into a prism and an affinely regular hexagon* (Figure 27).

(iii) *In a Euclidean plane any parallelogram has a unique decomposition into two squares, one oriented counterclockwise and one oriented clockwise* (Figure 28). (The squares may be constructed easily by ruler and compass.)

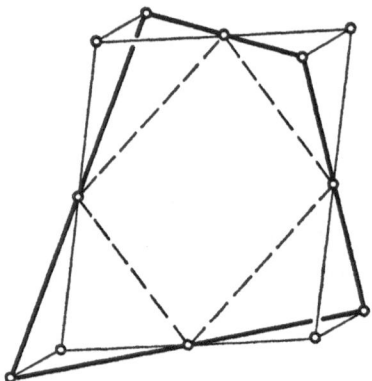

Fig. 26.

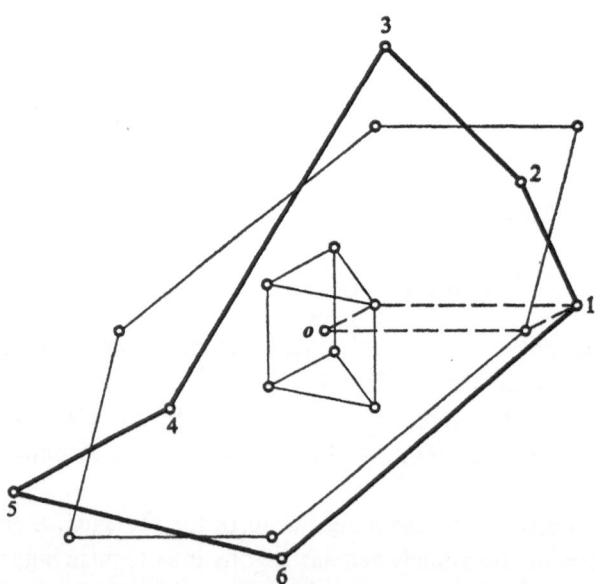

Fig. 27.

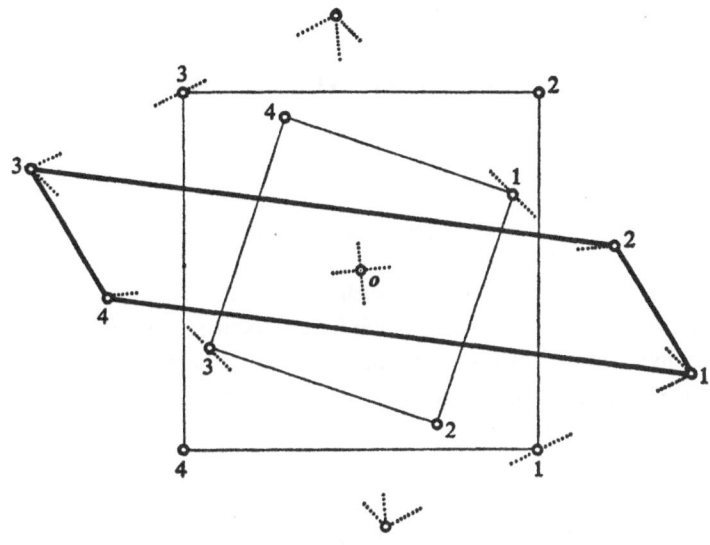

Fig. 28.

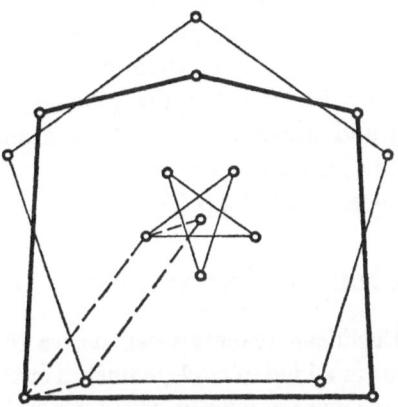

Fig. 29.

Analogously, we can decompose a triangle into two regular triangles, and an affinely regular hexagon into two regular hexagons.

(iv) *In the real affine plane a pentagon has a unique decomposition into an affinely regular pentagon and an affinely regular pentagram* (Figure 29).

Example (iii) may be motivated by the following question. In a Euclidean plane all quadrangles form a group under addition. (For the addition we need an origin, but the quadrangles need not have the same center of gravity.) All parallelograms form a subgroup. But do all squares form a subgroup?

Mathematisches Seminar, Christian-Albrechts-Universität, Kiel, G.F.R.

H. S. M. COXETER

INVERSIVE GEOMETRY

1. INTRODUCTION

Euclidean geometry deals mainly with points and straight lines. Its basic transformation is the *reflection,* which leaves fixed all the points on one line and interchanges certain pairs of points on opposite sides of this "mirror". All other *isometries* (or "congruent transformations" or "motions") are expressible in terms of reflections. (For simplicity, we are describing only *plane* geometry. In space we would reflect in a plane instead of a line.)

Analogously, *inversive* geometry deals with points and circles. Its basic transformation (invented by L. J. Magnus in 1831) is the *inversion,* which leaves fixed all the points on one circle and interchanges the inside and outside of this "circle of inversion". All other circle-preserving transformations (including similarities as a special case) are expressible in terms of inversions.

This kind of geometry is worthy of attention not only for the sake of its intrinsic beauty but also because it is the geometry of complex numbers [10; 6, pp. 145–147] and because the point pairs and circles of the real inversive plane provide an isomorphic model for the lines and planes of hyperbolic (non-Euclidean) space [4, p. 221].

2. AN OUTLINE OF THE AXIOMATIC APPROACH

From the ordinary Euclidean plane we can derive the inversive plane by regarding a straight line as a kind of circle, namely a circle that passes through a special point called *the point at infinity.* This extra point, which is added to the Euclidean plane to make the inversive plane, enables us to declare, without any exception, that

$$\text{Any three distinct points lie on just one circle.} \qquad (2.1)$$

If the three given points happen to be collinear, the "circle" is a straight line.

The idea of adding an ideal line to the Euclidean plane to make a projective plane has been ascribed to Desargues (1639). For more than two centuries this procedure was believed to be unique, as when Cayley remarked in 1859: "Metrical geometry is a part of descriptive geometry and descriptive geometry is all geometry". It was M. Bôcher, in 1914, who first saw clearly that the idea of adding a single ideal point to the Euclidean plane (to make an inversive plane) is entirely analogous to what Desargues did, and equally useful.

Mario Pieri, who gave the first satisfactory set of axioms for projective geometry (Cayley's "descriptive geometry") in 1899, did the same for inversive geometry in 1912. Subsequent authors have improved the details, so that now most of the familiar properties of points and circles can be deduced from just four axioms: (2.1) and the following three:

> *There exist four points not on a circle.*   (2.2)

> *If a point P is on a circle α while Q is not on α, there is just one circle through Q whose only common point with α is P.*   (2.3)

> *If each cyclically adjacent pair of four circles have a pair of common points, forming altogether eight distinct points, and if four of these points, one from each pair, lie on a circle, then the remaining four lie on a circle.*   (2.4)

Two circles are said to be *intersecting, tangent,* or *non-intersecting* according as they have two common points, one common point, or no common point. Thus the circles considered in (2.3) are tangent, with P as their point of contact.

Notice that this treatment makes no mention of "distance": the only relation used is that of *incidence*, which allows us to use such phrases as "a point P is on a circle α" or "two circles have a pair of common points".

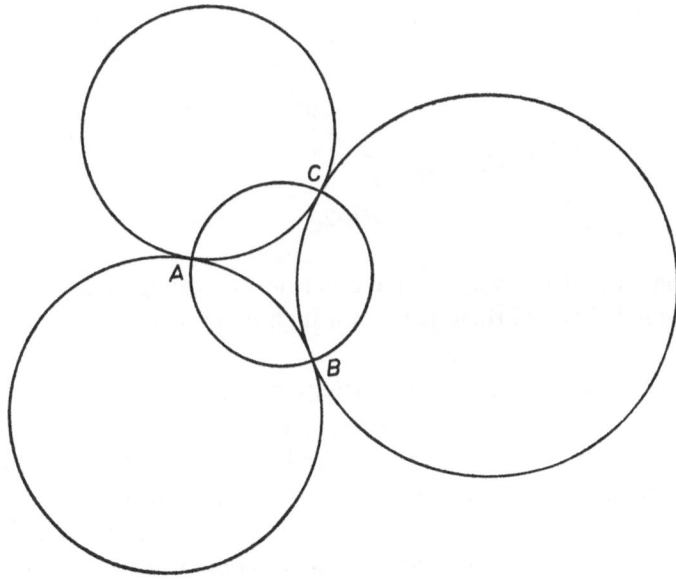

Fig. 1.

(Just as some geometers investigate "non-Desarguesian" projective planes that do not satisfy the theorem of Desargues, so they also investigate "non-Miquelian" inversive planes that do not satisfy Axiom (2.4), which is the theorem of Miquel.)

We usually say that two intersecting circles are orthogonal if they have perpendicular tangents at either of their points of interaction. It is an amusing problem to ask how the same notion can be described without using the word "perpendicular" or any of its synonyms. The solution is surprisingly simple (see Figure 1).

Two intersecting circles are *orthogonal* if one of them belongs to a triad of mutually tangent circles, touching one another at three distinct points A, B, C, while the other is the circle ABC.

For any point P not on a circle ω, the *inverse* point P' can be defined as the second intersection of any two circles through P orthogonal to ω. (In Figure 2, one of the two has been drawn as a straight line, and the other with its center on this line.)

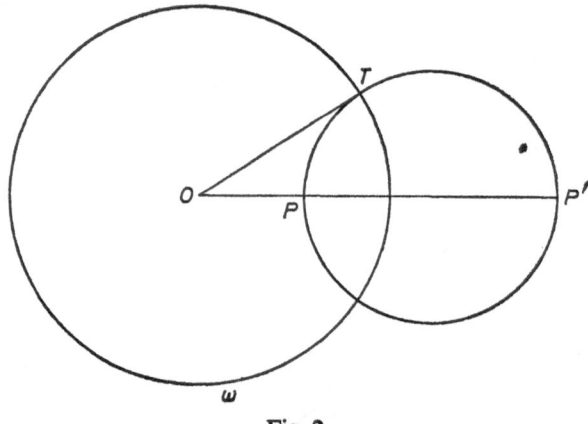

Fig. 2.

Inversion provides a very simple criterion for orthogonality: two circles are orthogonal if one of them passes through two points that are inverses in the other.

Any two distinct circles, α and β, determine a *pencil* of circles orthogonal to them, consisting of all the circles PP_1P_2 where P is a variable point while P_1 and P_2 are its inverses in α and β. If γ and δ are two such circles, the pencil of circles orthogonal to them includes α and β; it is the pencil "spanned" by α and β, and is conveniently denoted by $\alpha\beta$. One finds that the circles through two distinct points, L and L', form an *intersecting pencil*, orthogonal to the *non-intersecting pencil* for which L and L' are the *limiting points* (inverses in each member of the non-intersecting pencil). If α and β are tangent, $\alpha\beta$ is a *tangent pencil*, and the orthogonal circles through the point of contact form another tangent pencil. Three or more circles that belong to one pencil are said to be *coaxal*.

Any three non-coaxal circles, α, β, γ, span a *bundle* of circles, say $\alpha\beta\gamma$, consisting of the smallest set of circles which includes α, β, γ and, with every two of its members, the whole of the pencil spanned by them. If α, β, γ have a common point, $\alpha\beta\gamma$ consists of all the circles through this point; it is called a *parabolic* bundle. If α, β, γ are all orthogonal to one circle, $\alpha\beta\gamma$ consists of all the circles orthogonal to this one circle; it is called a *hyperbolic* bundle (because its members represent the lines of the hyperbolic plane). Finally, if α, β, γ have neither a common point nor a common orthogonal circle (e.g., if they are mutually orthogonal), $\alpha\beta\gamma$ is called an *elliptic* bundle (because its members represent the lines of the elliptic plane).

3. THE EUCLIDEAN APPROACH

Unlike projective geometry, which is easily developed from its axioms [see, for instance, 2], the details of pure inversive geometry require so much effort that the best practical procedure is to specialize one point, call it the point at infinity, and then use the powerful tool of Euclidean geometry. In other words, to solve a particular problem concerning circles, we first simplify the figure by inverting in a circle whose center lies on one or more of the circles (so as to replace these circles by straight lines) and then deal with the corresponding Euclidean problem. Afterwards we can perform the same inversion again so as to restore the circles.

Applying one of Euclid's theorems to Figure 2, we see that, if ω has center O and radius OT, $OP \times OP' = OT^2$. This remark brings us back to the classical definition:

Inversion in a circle ω, with center O and radius k, is the transformation that interchanges pairs of points, P and P', such that

$$OP \times OP' = k^2,$$

P and P' being on the same side of O. Thus each point on ω is self-inverse, every point outside ω has an inverse inside ω, each line through O is self-inverse, and we can regard the point at infinity as being the inverse of O.

Inversion is a *circle-preserving* transformation; more precisely, it transforms the set of all straight lines and circles into itself. Fejes Tóth [8, p. 85] proves this as follows.

Consider first the inverse of a circle γ not passing through O. Let the line of centers of ω and γ cut γ in A and B, and let P be any other point on γ.

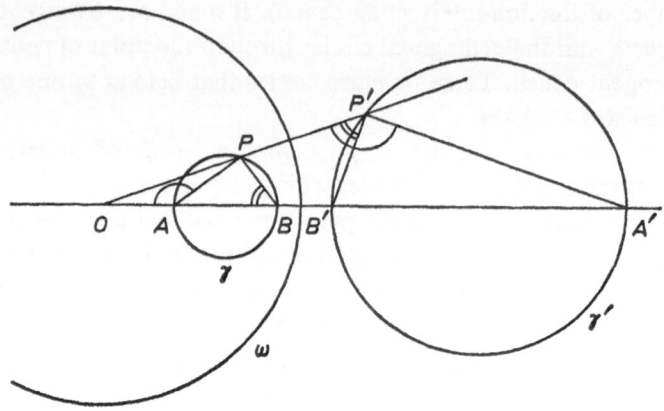

Fig. 3.

Since

$$OP \times OP' = OA \times OA' = OB \times OB',$$

we have pairs of oppositely similar triangles

$$OAP \sim OP'A', \qquad OBP \sim OP'B',$$

as in Figure 3. Thus

$$\begin{aligned}
\measuredangle B'P'A' &= \measuredangle OP'A' - \measuredangle OP'B' \\
&= \measuredangle OAP - \measuredangle OBP = \measuredangle APB = 90°,
\end{aligned}$$

showing that the inverse of γ is the circle with diameter $A'B'$. This proof is easily modified to cover the cases when γ is a circle passing through O, or a line not passing through O. The inverse of such a line γ is a circle γ' through O whose tangent at O is parallel to γ.

If follows that the two supplementary angles formed by two lines through

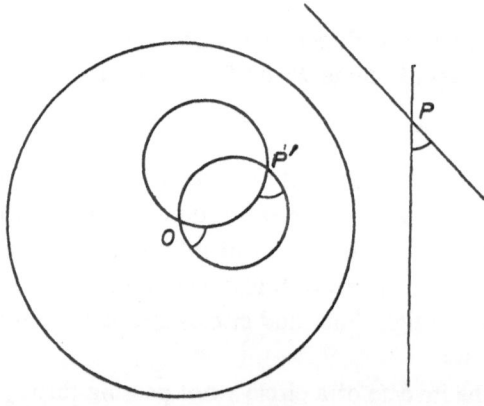

Fig. 4.

a point P are equal to the angles formed by their inverse circles, which intersect at O and again at P', as in Figure 4. In other words, inversion is an *angle-preserving* (or "conformal") transformation. In particular, *orthogonality* is preserved.

Let us see what our four axioms look like in Euclidean terms. Having already considered (2.1), we turn to (2.2). Taking one of the four points to be at infinity, we obtain the obvious statement that there exist three points not on a line.

When Q is at infinity, (2.3) says that there is just one tangent line at any point P on a circle α. More interestingly, when P (instead) is at infinity, (2.3) says that, *if Q is any point not on a line α, there is just one line through Q parallel to α.*

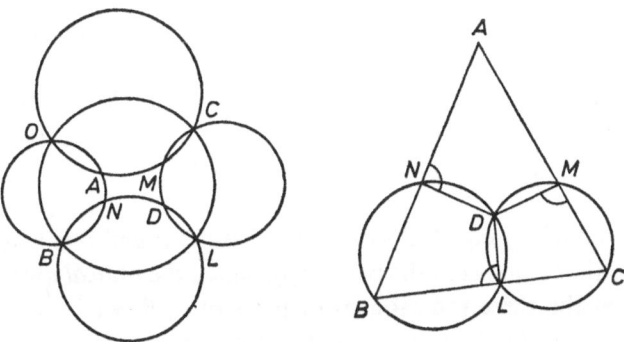

Fig. 5.

In (2.4), let OA, BN, LD, CM be the four pairs of common points of the four circles $AOBN$, $NBLD$, $DLCM$, $MCOA$, with the first of each pair on a circle, as in Figure 5. Can we assert the existence of a circle $ANDM$? Sending O to infinity, we obtain a triangle ABC with points L, M, N on its three sides, while circles NLB and CLM (having L in common) meet again in D. Now the question is whether D lies also on the circle AMN. The answer is Yes, by the Pivot Theorem [9, p. 17]. If D is inside the triangle, as in our figure, the proof is simply

$$\sphericalangle DMC = \sphericalangle DLB = \sphericalangle DNA.$$

In other cases it is just as easy.

4. PENCILS AND BUNDLES

Consider once more the intersecting pencil of circles through any two points L and L'. When L' is at infinity, this is simply the pencil of lines throgh L.

The orthogonal non-intersecting pencil, orthogonal to these lines, consists of all the circles with center L. In other words, any two intersecting circles can be inverted into intersecting lines, and any two non-intersecting circles can be inverted into concentric circles.

Similarly, two orthogonal tangent pencils can be inverted into two orthogonal "pencils" of parallel lines. (Parallel lines are tangent circles whose point of contact is the point at infinity.)

The sides of any triangle are circles having the point at infinity in common. The parabolic bundle spanned by them is simply the set of all lines in the Euclidean plane.

If α and β are two perpendicular diameters of a circle γ, so that α, β, γ are mutually orthogonal, the elliptic bundle $\alpha\beta\gamma$ consists of all the diameters of γ and all the circles that pass through pairs of diametrically opposite points of γ. In other words, $\alpha\beta\gamma$ consists of all the circles that arise from the great circles on a sphere by stereographic projection.

Another Euclidean approach to bundles uses Steiner's concept of power. With respect to a circle with center O and radius k, the *power* of a point P is

$$OP^2 - k^2\,.$$

The locus of points of equal power for two circles α and β is a straight line (the only line that belongs to the pencil $\alpha\beta$), called the *radical axis* of α and β. It follows that the three radical axes of pairs of α, β, γ (if not parallel) all pass through the *radical center*: a point of equal power for the three circles. The set of all circles for which a given point P has a given power p is a bundle: hyperbolic if $p>0$, parabolic if $p=0$, elliptic if $p<0$.

If α and β are intersecting circles, their radical axis obviously joins their two points of intersection. If α and β are non-intersecting, their radical axis can be constructed as in Figure 6, where γ is any genuine circle intersecting both of them. In such a case (and again when three circles are all non-intersecting but not coaxal), the radical center of α, β, γ is outside all of them, and the bundle $\alpha\beta\gamma$ is necessarily hyperbolic.

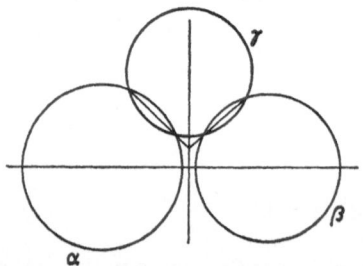 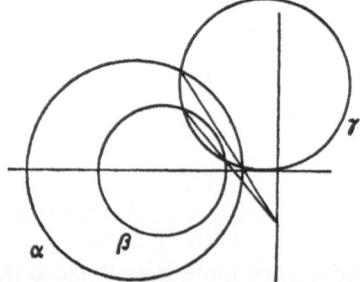

Fig. 6.

5. MID-CIRCLES AND ORTHOCYCLIC POINT PAIRS

We recall that two intersecting, tangent, or non-intersecting circles can be inverted into two intersecting lines, two parallel lines, or two concentric circles, respectively. When a circle is inverted into a line, inversion in the circle is transformed into reflection in the line.

Given any two distinct circles α and β, consider the locus of a point P such that two circles, tangent to both α and β, are tangent to each other at P. Inversion shows that this locus consists either of two orthogonal circles or of a single circle. More precisely, if α and β intersect, we have two circles bisecting their supplementary angles of intersection; if α and β are tangent, we have one circle tangent to both at the same point; and if α and β are non-intersecting we again have just one circle, belonging to the pencil $\alpha\beta$. Clearly, inversion in such a circle interchanges α and β. The classical name "circle of antisimilitude" [1, p. 28] is conveniently abbreviated to *mid-circle*.

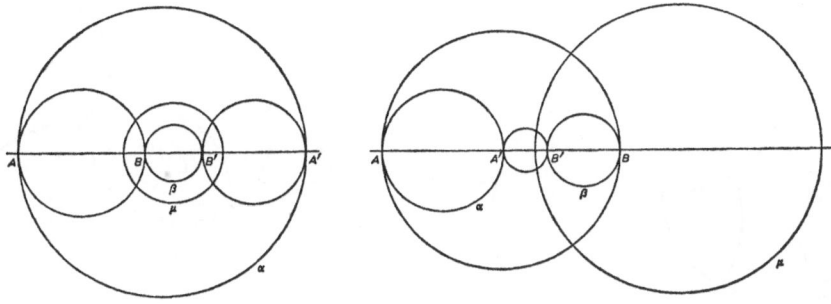

Fig. 7.

If α and β are congruent, their mid-circle (or one of their mid-circles) evidently coincides with their radical axis. In other cases the following constructions are available.

If α and β intersect, the centers of their mid-circles lie on the lines that bisect their angles of intersection.

If α and β are non-intersecting, let their diameters on their line of centers be AA' and BB', so named that A and B' separate B and A', as in Figure 7. Then a diameter of their mid-circle is formed by the limiting points of the two non-intersecting circles whose diameters are AB and $A'B'$. For, this circle, being orthogonal to the circles on AB and $A'B'$, inversts A into B, and A' into B'.

By making A' and B' coincide (at C, say) we obtain the limiting case when α and β are tangent circles. Then a diameter of their mid-circle is formed by C and its inverse in the circle on AB.

Notice that the definition of a mid-circle belongs to inversive geometry, although we have used Euclidean methods to develop its properties. The advantage of allowing ourselves such freedom is well illustrated by the following discussion of "orthocyclic point pairs", in which we first use a strictly inversive method and then, for comparison, the Euclidean method.

Two point pairs, LL' and MM', are said to be *orthocyclic* [1, p. 100] if there is a circle through L and L' that inverts M into M'. In this case, we can prove that the relation is symmetric: there is a circle through M and M' that inverts L into L'.

If all four points lie on one circle α, as in Figure 8, the circle λ through L and L' that inverts M into M' inverts α into itself and is therefore orthogonal to α. The circle μ, through M and M' orthogonal to α, is also orthogonal to λ (since λ inverts M into M'). Therefore the inversion in μ interchanges the two intersections of α and λ, which are L and L'.

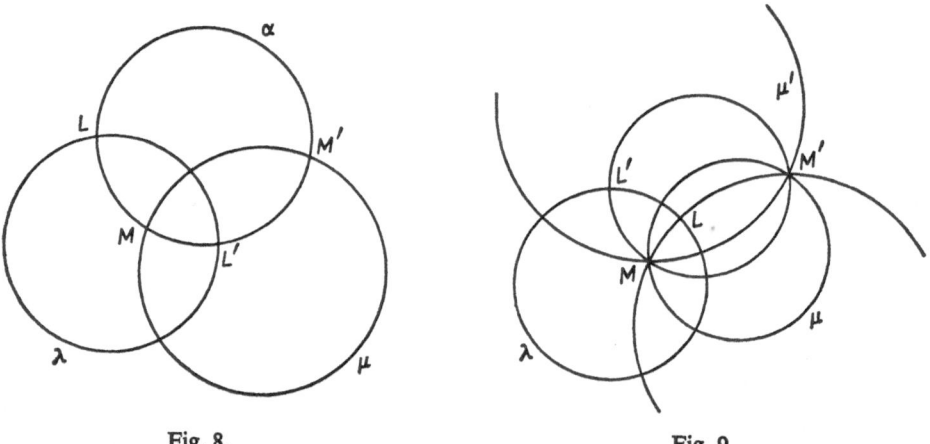

Fig. 8.   Fig. 9.

Figure 9 illustrates the remaining case, in which L, L', M, M' do not lie on one circle. The circle λ through L and L' that inverts M into M' is still orthogonal to every circle through M and M'. In particular, λ is orthogonal to the two intersecting circles LMM', $L'MM'$, and to each of their two mid-circles, say μ and μ'. Both μ and μ' invert the two intersecting circles λ and LMM' into the two intersecting circles λ and $L'MM'$. Since L is one of the intersections of the former pair, and L' of the latter, if μ does not invert L into L', μ does.

Euclidean geometry provides a far easier proof. Take L' to be the point at infinity. Then there is a line through L that reflects M into M'. Thus $LM = LM'$. The circle through M and M' with center L inverts L into L', as desired.

6. INVERSIVE DISTANCE

Any two intersecting circles have two supplementary angles of intersection, which are preserved when the circles are inverted into lines, and are bisected by the two mid-circles. Somewhat analogously, any two non-intersecting circles have an *inversive distance*, which is preserved when the circles are inverted into concentric circles, and is bisected by the mid-circle. For concentric circles, of radii a and b, the inversive distance is defined to be

$$\left| \log \frac{a}{b} \right|$$

so that, if three circles of a non-intersecting pencil are inverted into concentric circles whose radii satisfy $a > b > c$, we have the proper additive relation

$$\log \frac{a}{b} + \log \frac{b}{c} = \log \frac{a}{c}.$$

In particular, concentric circles of radii 1 and e have inversive distance 1.

In other words, the inversive distance between any two non-intersecting circles is defined to be *the logarithm of the ratio of the radii (larger to smaller) of two concentric circles into which the given circles can be inverted.*

It follows [7, p. 129] that if two non-intersecting circles α and β (neither surrounding the other) have centers A and B, radii a and b, and inversive distance v, the (Euclidean) distance between their centers is given by

$$AB^2 = a^2 + b^2 + 2ab \cosh v. \tag{6.1}$$

Moreover [7, pp. 130, 176 (Ex. 4)] if C is the foot of the perpendicular from A to a line γ outside α, and if λ is the inversive distance between α and γ,

$$AC = a \cosh \lambda. \tag{6.2}$$

As a limiting case we naturally regard tangent circles as having inversive distance 0.

Three non-intersecting circles α, β, γ are said to be *nested*, with γ *separating* α and β, if every circle that intersects α and β intersects γ too. In particular, for any three members of a non-intersecting pencil, one separates the other two in this sense. In the case of three concentric circles, this is the one whose radius is neither the greatest nor the least. It is interesting to observe that the mutual inversive distances of three nested circles satisfy a "non-triangle inequality":

Among the three inversive distances between pairs of three nested circles, one is greater than or equal to the sum of the other two. Equality holds only when the three circles are coaxal.

To prove this, let γ separate α and β, and let the inversive distances be

$$\lambda = (\alpha, \gamma), \quad \mu = (\beta, \gamma), \quad \nu = (\alpha, \beta).$$

Since α, β, γ are non-intersecting, there is at least one circle orthogonal to all of them. Taking one of the intersections of this circle with γ to be the point at infinity, we have a straight line γ perpendicular to the line AB joining the centers of α and β, as in Figure 10.

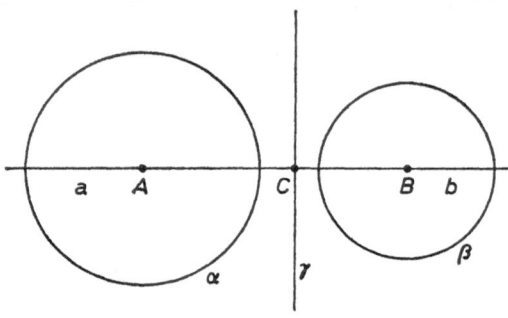

Fig. 10.

Equations (6.1) and (6.2) yield

$$a^2 + b^2 + 2ab \cosh \nu = AB^2 = (AC + CB)^2$$
$$= (a \cosh \lambda + b \cosh \mu)^2$$
$$= a^2 (1 + \sinh^2 \lambda) + 2ab \cosh \lambda \cosh \mu + b^2 (1 + \sinh^2 \mu),$$

whence

$$2ab \{\cosh \nu - \cosh (\lambda + \mu)\} = (a \sinh \lambda - b \sinh \mu)^2 \geqslant 0.$$

Thus

$$\nu \geqslant \lambda + \mu, \tag{6.3}$$

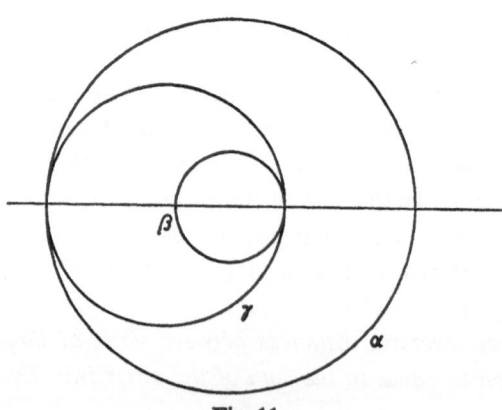

Fig. 11.

with equality only when $a \sinh \lambda = b \sinh \mu$. But

$$(a \sinh \lambda)^2 = (a \cosh \lambda)^2 - a^2 = AC^2 - a^2$$

is the power of C with respect to α, and similarly $(b \sinh \mu)^2$ is the power of C with respect to β. Hence equality occurs only when γ is the radical axis of α and β, that is, when α, β, γ are coaxal. (The extreme case when $\lambda = \mu = 0$ is illustrated in Figure 11.)

The inequality (6.3) is intimately related to Einstein's Twin Paradox (or Clock Paradox), which depends on the non-triangle inequality for time-like intervals [5, p. 11].

University of Toronto, Canada

REFERENCES

[1] J. L. Coolidge, *A treatise on the circle amd the sphere*, Oxford, 1916.
[2] H. S. M. Coxeter, *Projective geometry*, Waltham, Mass., 1964.
[3] H. S. M. Coxeter, *Non-Euclidean geometry*, Toronto, 1965.
[4] H. S. M. Coxeter, '*The inversive plane and hyperbolic space*', Abh. Math. Sem. Univ. Hamburg **29** (1966), 217–242.
[5] H. S. M. Coxeter, '*The problem of Apollonius*', Am. Math. Monthly **75** (1968), 5–15.
[6] H. S. M. Coxeter, *Introduction to geometry*, New York, 1969.
[7] H. S. M. Coxeter and S. L. Greitzer, *Geometry revisited*, New York, 1967.
[8] László Fejes Tóth, *Regular figures*, New York, 1964.
[9] H. G. Forder, *Geometry*, London, 1950.
[10] Hans Schwerdtfeger, *Geometry of complex numbers*, Toronto, 1962.

ROBERT B. DAVIS

THE PROBLEM OF RELATING MATHEMATICS TO THE POSSIBILITIES AND NEEDS OF SCHOOLS AND CHILDREN

I am clearly the worst geometer – perhaps one should say the only non-geometer – here. Conceding that I can add nothing of a geometric nature to this conference, I nonetheless do suggest there is one important matter to which I can speak.

We are talking about geometry in relation to *schools* and *children*. This involves some dangers for competent mathematicians, because they may erroneously assume that schools are run as they themselves would run them. They may assume that the sequential order of topics within the curriculum is rationally designed within some framework of pre-requisites, and nicely adjusted to the interests and abilities of different children. This, and many other natural assumptions, are usually not justified; school practice can be surprisingly crude and can produce hideous distortions of otherwise sensible ideas.[1]

School practice, in fact, is shaped by several traditions, four of which are:
(1) Mathematics as various people know it;
(2) Prevalent psychological theories, some explicit and some implicit;
(3) Educational theory;
(4) The accumulated conventional wisdom and practitioner's know-how of classroom teachers.

These traditions have gotten out of touch with one another, and often put further strain on already over-stressed schools by pulling in quite different directions.

My central thesis is that major improvement in the actual education of children is not possible until an effective reconciliation of these different traditions has been achieved. This means that at least a few of the mathematicians who are now getting interested in school mathematics must also get interested in typical school practice, and in the diverse intellectual traditions that have helped to shape school practice.

Suppose that a mathematician does pursue the study of practice and traditions in school mathematics; what sort of thing can he expect to find? Here are three examples that mathematicians *have* found:

Example 1. An instance of school practice in the teaching of exponents.

[1] Cf., e.g., Robert B. Davis, "Report from the States", *Mathematics Teaching*, Vol. 50, Spring, 1970, pp. 6–12.

The teacher spent two 45 minute periods, on consecutive days, conducting recitations on a single type of problem. She would write on the chalkboard

$$X^2 \cdot X^3 =$$

call on a student, and get the answer "X^5". She would then write a precisely analogous problem, say

$$p^{10} \cdot p^5 =$$

call on the next student, and so on. I repeat that two 45·minute periods were devoted to this activity.

What she did *not* do: she did not consider contrast cases, such as

$$X^2 + X^3 \, ;$$

she did not relate this fact to other parts of mathematics (for example, by the simple observation that we are adding the number of factors); she did not use an appropriate mathematical language to write this as a generalization

$$a^x \cdot a^y = a^{x+y} \, ;$$

and she did not consider implications or extensions of this idea (for example, to consider exponents that are not positive integers).

Since I do not know why she taught this lesson as she did, I am left to conjecture, but my guess is that she had in mind the stimulus-response conception of "knowledge", which says (roughly) that a teacher's job is to present a "stimulus" (e.g., "$X^2 \cdot X^3 =$"), to cause it to be paired up with a "correct response" (e.g., "X^5"), and then to cement this pairing as securely as possible by effective "bonding". (Among the discussions of stimulus-response, or "S–R" theory, a particularly interesting one for mathematicians is: Edward Chace Tolman, *Behavior and Psychological Man*, University of California Press, 1958, pp. 242–3.)

Personally, I doubt that many mathematicians would have handled two days' lessons (incidentally, for rather bright students!) in this way, but this report is a factual one, and presents an example of actual school practice (indeed, in a suburban high school that is rated well above average).

Example 2. An example of historical background.

Studies of the historical evolution of school mathematics nearly always make reference to the influence of a conceptualization of learning known as "faculty psychology".[2]

The idea is sometimes traced back at least to the ancient Greeks, and was particularly prevalent in the United States (sometimes under the name "phrenology") during the nineteenth century. Roughly, faculty psychology sought to identify specific mental functions ("memory", "reasoning", "will", "neatness" were among those suggested); where possible, to identify the part of the brain concerned with this function, and to develop each function by practice. This last notion was most influential on the selection of the school curriculum and on the style of classroom teaching. "Developing each function by practice" was regarded rather like "strengthening a muscle by exercise". Thus the selection of the curriculum content wasn't too important – all that was needed was good exercise material, as it were. The rote memorization of arithmetic facts, for example, was good practice in memorization, and was presumed to make your memory somehow work better, thereafter. Instruction was frequently harsh, in no way individualized, and simply expository, without explanations. If one was merely "exercising the faculty of reason", then this was essentially calisthenics, and all you needed was persistence, a good drill-master, and something hard to chew on. This view suggested that no effort at all be made to identify specific misconceptions in the minds of individual children. Nor was there any reason to relate one part of mathematics to another. Such questions didn't really arise.

Faculty psychology wasn't all wrong, and isn't entirely old-fashioned. In a much refined and much wiser form it is being revived, with significant additions, modifications and improvements, today. (Cf., for example, the use of electrodes implanted in the brain, to identify areas of specific brain function; and if one replaces the vastly oversimplified notion of "strengthening mental muscles by exercising them" with the far subtler notion of developing Polya-like heuristics, then one might possibly classify the heuristic theory of Seymour Pappert as belonging to this general intellectual tradition.)

The nineteenth-century version, however, as translated into typical school practice, did little to improve the learning of mathematics. Probably this original tradition is mostly dead (aside from the new, sophisticated revival), but some remnants show up occasionally in actual classroom settings.

Example 3. The Modern Use of Behavioral Objectives.

Some influential educational theorists nowadays advocate organizing the curriculum by specifying, in advance, and in rather narrowly-defined terms,

---

[2] See, for example, M. Vere De Vault and Thomas E. Kriewall, *Perspectives in Elementary School Mathematics*, Chas. Merrill Publishing Co., 1969, pp. 50–55.

just what objectives are supposed to be achieved by the instructional program – what, *exactly*, do you want the learner to *do*, once he has learned whatever it is that you're teaching.

Like faculty psychology, the idea isn't entirely without merit, but – again like faculty psychology – it can easily lead to undesirable classroom practice. There are recorded opinions of some educational theorists to the effect that visiting Westminster Abbey, or listening to Beethoven's last five quartets, cannot be defended as "educationally valuable" unless one states behavioral objectives for them. But to reduce these experiences, which are properly consummations in and of themselves, to the status of intermediate tasks (like picking up a pen or a pencil) en route to some other goal, is to invert the natural order of knowledge, wisdom, and emotion. One does not listen to Beethoven *in order to do something else*.

That this is relevant (and potentially harmful) to the learning of mathematics is demonstrated clearly by this actual occurrence:

A third grade boy named Kye created his own original algorithm for subtracting, making use of negative numbers (which Kye understood reasonably well):

$$\begin{array}{r} 64 \\ -28 \\ \hline -4 \\ 40 \\ \hline 36. \end{array}$$

Kye's teacher, never having seen this before, had not planned for it, and could not have done so. But she was alert, receptive, and respected children, so she listened to Kye, tried out his method on other problems, found that it seemed to work, was visibly impressed (to Kye's deep delight), and ultimately showed it to a mathematician (who, also, had never seen it before). How would Kye have felt if the teacher had merely rejected his method as *wrong*?

Now the mathematician had occasion to show this algorithm to an educational theorist who advocates a strict use of narrowly-defined behavioral objectives. One clearly could not state an objective for that lesson which said: "the child is to discover a new algorithm for subtraction". To do so would result in nearly every child failing, and behavioral objectives of that sort are not considered acceptable. But, if the discovery was *not* related to the behavioral objectives for the lesson, then it was just like visiting Westminster Abbey – it wasn't educationally valuable.

There may be educational theorists who accept this upside-down world – indeed, there *are* – and it is partly up to us as mathematicians to try to give them a more adequate view of what mathematics is really about. In part, these theorists fail to describe the teaching and learning of mathematics because they do not understand what mathematics is. Who will help them to understand?[3]

But do mathematicians understand what cognition is? I don't mean to suggest an arrogant approach to all of our colleagues in other disciplines – there seem to be things we can learn from them. This is especially true of the deeper diagnostic work of Piaget, Ginsburg, and others.

Some of the Piaget "conservation" experiments are highly suggestive, although they tend to get lost under incorrect denials that the phenomena exist (and the phenomena *are* surprising), and under unwise interpretations of what they mean. Here is an area where mathematicians and psychologists, working together, can accomplish more than either group can by themselves. Let me describe two actual interviews.

Interview I. Miriam, aged 4 years, 9 months; interviewed by David M. Clarkson.

Adult (putting down on the table two drinking straws of the same length, arranged like in Figure 1, so that they formed opposite sides of an incomplete rectangle): "Miriam, are these straws the same length?"

Fig. 1. Original placement of two drinking straws.

Miriam: "Yes."

Adult (after sliding one straw sideways, to get the configuration shown in Figure 2.) "Miriam, are they the same length now?"

Miriam: "No, this one (pointing to the one that had been moved) is longer."

Fig. 2. New placement of the two drinking straws.

Many adults who have not tried any of the Piaget tasks with children are unwilling to believe that such responses actually occur. It seems impossible

---

[3] There are also "non-mathematical" reasons for worrying about the tendency to circumscribe curricula by a narrow use of behavioral objectives. Particularly with younger children, there is the case of "massive readiness building" – i.e., many experiences do seem to add up to "enrichment" that manifests itself very significantly at later stages in the child's life, *although it has no apparent short-term effect*. We are thus faced with the quite real possibility of omitting valuable or essential experiences because we are not immediately aware of the damage that we are doing. Even with older children, there may be important side-effects to educational experiences that may need to be considered. Being initially unknown, these side effects (or their avoidance) cannot be specified as *a priori* objectives.

that children can say four things are more than four things, that drinking straws get longer if you move them, that a part is greater than a whole. Yet these are things that children do say, and with remarkable consistency, for the Piaget experiments replicate very reliably. In a moment I want to argue that these responses even make sense – provided you think about them correctly. Part of the clue is contained within the interviews themselves, and part in an experiment you can easily try for yourself, using yourself as a subject. (Unfortunately, sophisticated geometers may turn out to be about the only people in the world who, as subjects, will *not* replicate the usual results; that is also something you can test for yourselves.)

Continuing the interview with Miriam:
Adult (putting out 8 red cubical wooden blocks, spaced somewhat apart, in a linear array – like modern houses along a straight street):
"Miriam, can you get as many yellow blocks as there are red blocks?" Miriam could, and did. The array then looked like in Figure 3.

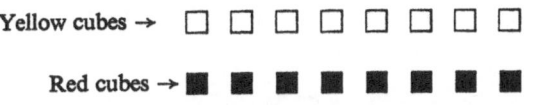

Fig. 3. Equally spaced rows of red and yellow blocks.

Adult: "Are there as many red cubes as yellow cubes?".
Miriam: "Yes."
Adult: "How do you know?"
Miriam answered this in two different ways, by counting (and getting 8 of each), and by pushing the two rows closer to one another (making the "street" between the "houses" narrower), so as to match the cubes one-to-one.
The adult now pushed the yellow cubes closer together, so that the array looked like in Figure 4.

Fig. 4. The yellow cubes have been moved closer together.

The row of red cubes was now longer, but there were still eight cubes of each color.
Adult: "Miriam, are there as many yellow ones as there are red ones?"
Miriam: "There are more red ones."

(At this point we again have one of those phenomena that adults often treat with disbelief, and which indeed seems, at first glance, incomprehensible. Miriam seems to be contradicting herself. Can any child be so foolish? Does she believe that pushing gently on the blocks has changed the *number* of blocks?)

With the short row of yellow cubes facing the long row of red cubes, the interview continues:

Adult: "How many red ones are there?"

Miriam (counts): "Eight."

Adult: "How many yellow ones are there?"

Miriam (counts) "Eight."

... and, at this point, a kind of Aha! smile appears on Miriam's face.

Before discussing this first interview, let's consider an excerpt from the interview which immediately followed Miriam's, again conducted by David M. Clarkson.

Interview II. Chris, age 6 years, 3 months. The adult had a jar[4] shaped like this:

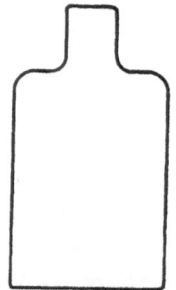

Fig. 5. Jar used in "gravity" experiment.

He had also prepared sheets of paper showing the bottle in the four different positions as shown in Figure 6, but these papers were out of sight at the beginning of the interview. He also had available a cover that screwed tightly onto the jar, and a brown paper bag in which the jar could be placed so that one could still *feel* the jar, and see its silhouette, but (since the paper was opaque) could not see the jar itself.

---

[4] This interview format was designed by Piaget's group in Switzerland for use by the British Mathematics Project, as a measure of a child's growth in the development of mathematical conceptualizations.

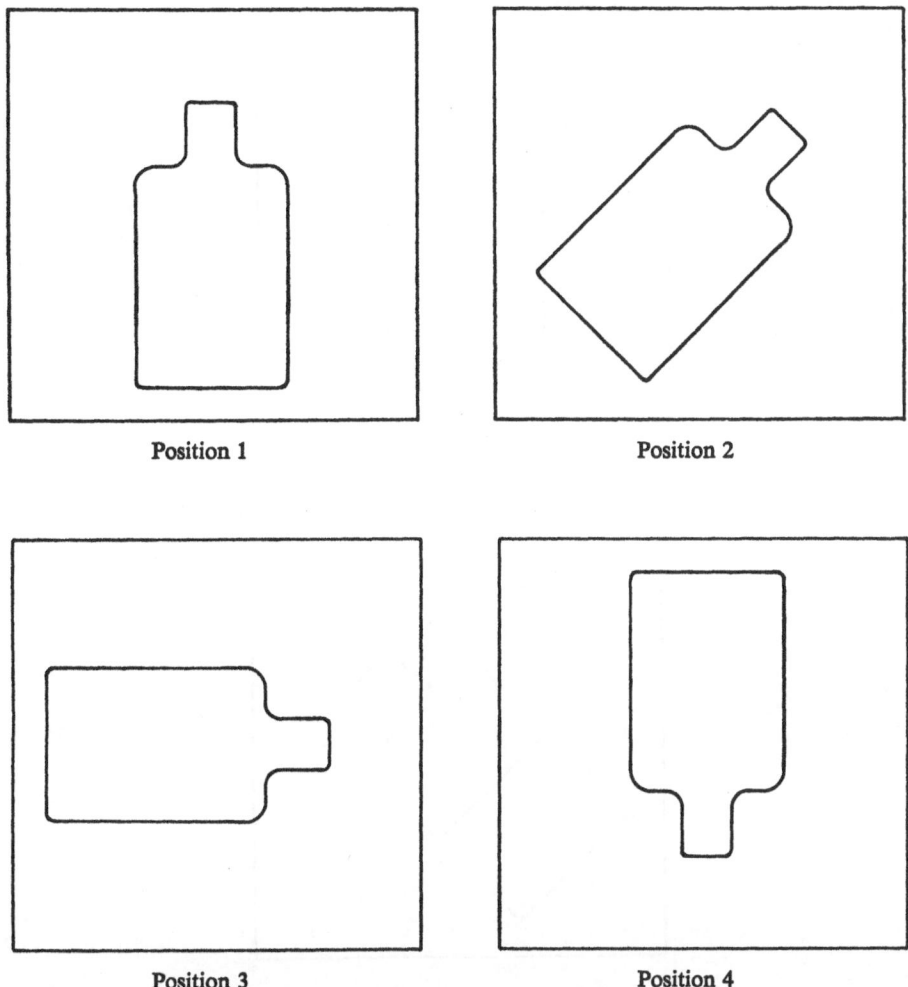

Position 1   Position 2

Position 3   Position 4

Fig. 6. Four form outlines of bottle drawn on paper in advance of the interview.

The interview now started with the adult filling the jar half-full with blue-dyed water, while Chris watched. The filling (and adjusting) stopped when Chris said he was satisfied that the jar was half full.

The adult now brought out one of his pictures of the bottle (Figure 6, Position 1) and drew a line on the picture to indicate the water level (Figure 7). Chris agreed this was essentially correct.

Now the adult screwed the cap tightly on, and put the bottle in the opaque paper bag; its outline was still visible, but the water level could not be seen, and had to be imagined.

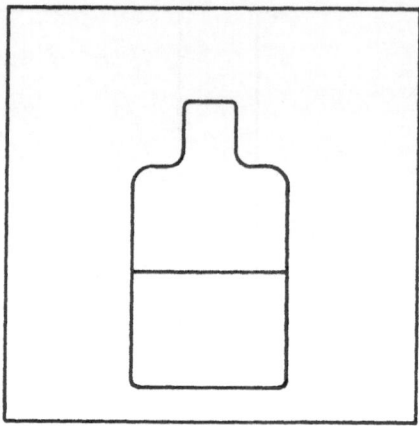

Fig. 7.   Line drawn on paper to indicate water level.

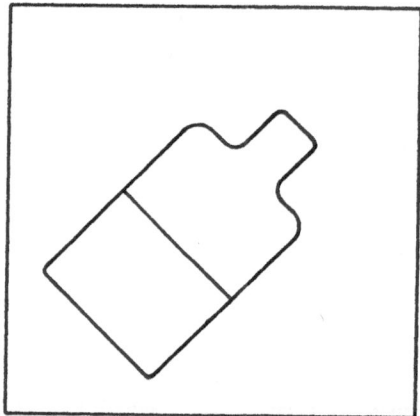

Fig. 8.   Line drawn by Chris to indicate water level.

At this point, the bag and bottle were tilted 45° (Position 2); Chris was asked to draw the water level. He drew the line as shown in Figure 8, and smiled confidently that he was right (although he had treated the water as if it were ice, frozen solidly in place; he had overlooked the effect of gravity). The bag and bottle were then rotated an additional 45°, and Chris was again asked to draw in the water level. He drew the level like in Figure 9 and was again confident that he was right.

Finally, this procedure was repeated with the bag and bottle in Position 4, and Chris drew the water level as shown in Figure 10.

Once again, you might think children could hardly be so foolish. Skeptical

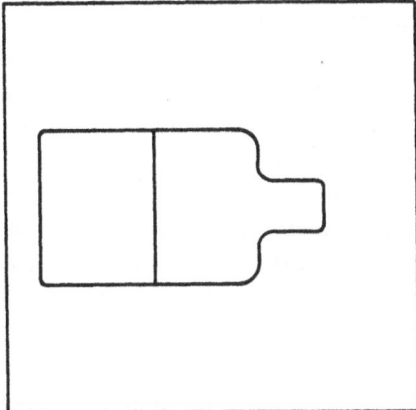

Fig. 9.   Line drawn by Chris to indicate water level.

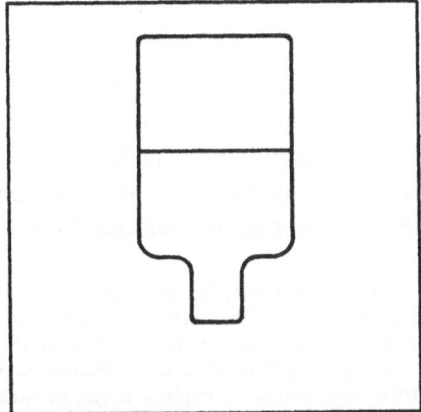

Fig. 10.   Line drawn by Chris to indicate water level.

adults (I was one myself) should try some of these interviews with children. The results reported here are typical.

Discussion of the Interviews: The possible relevance of such "Piagetian" interviews (there are many variants) for the creation of a mathematics curriculum has been a moot point for some time. I think a very strong case can be made that they do, in fact, provide the soundest possible basis for curriculum design, but developing a convincing argument requires consideration of many different interviews of different types. Taken together, a larger array of interviews offers a surprisingly adequate view of the gradual growth of mathematical concepts in children.

Besides their role in curriculum design, these interviews provide a much better basis for keeping track of the progress of individual children. They reveal clearly both understandings and misunderstandings, and thus provide a foundation for much better diagnosis of learning problems. Traditional "testing" and "evaluation" techniques (for example on standardized multiple-choice tests) have been very effective in assessing superficial verbal behavior, and notoriously ineffective at revealing deeper understandings and misunderstandings, especially in the case of idiosyncratic misconceptions (which in totality are very commonplace, though each specific misunderstanding can be quite rare[5]).

There are many discussions of Piagetian interviews now in the literature[6], with various interpretations of what the children's responses mean and what they imply for the curriculum. I think these discussions would read quite differently if mathematicians had been participants in the dialogue. I want to present here a somewhat different interpretation of cognition in children, as revealed by these interviews.

First, the conventional wisdom (I do not necessarily attribute it to Piaget, but it is quite clearly implicit in many conversations) attaches too many implications to Piaget's notion of four developmental stages in children, identifiable with chronological age: the so-called "sensory-motor" or "preverbal" stage (identified with very young children, from birth to 2 years old),

[5] Videotaped interviews with children now being collected by the Cornell University Center for Research in Education (by Professor Herbert Ginsburg and others) show clearly how exceedingly diverse misconceptions can be. Do you believe the perimeter of a plane figure determines its area? Many people unconsciously assume that it does, and are not even able to formulate their assumption in explicit terms in order to test it. Do you confuse the surface area of a solid with its volume? Many people consider them identical and cannot imagine that they could be different.

Of course, these are common (but previously unsuspected) misconceptions. (Notice how severely damaging they are; for example, imagine adding them as axioms in geometry.) If most misconceptions were common, one could easily design multiple-choice tests to diagnose them. But many children have their own, quite personal, misconceptions and errors: for example, one girl lines up a column of base-ten numerals on the left before adding, and consequently gets unexpected wrong answers, even though her use of an addition algorithm is otherwise quite correct.

[6] See, for example, De Vault and Kriewall, op cit; also J. H. Flavell, *The Developmental Psychology of Jean Piaget*, Van Nostrand, 1963; H. Ginsburg and S. Opper, *Piaget's Theory of Intellectual Development*, Prentice-Hall, 1969; Hans G. Furth, *Piaget and Knowledge*, Prentice-Hall, 1969; Roger Brown, *Social Psychology*, The Free Press, New York, 1965; Piaget's "check-ups", designed for the Nuffield Project are discussed in *Check-ups*, available from the Nuffield Mathematics Project, 12 Upper Belgrave St., London, S.W. 1, England; for the effect of Piaget's ideas on school mathematics, cf. Biggs and MacLean, *Freedom to Learn: An Active Learning Approach to Mathematics*, Addison-Wesley (Canada), 1969; also E. M. Williams and Hilary Shuard, *Primary Mathematics Today*, Longman, 1970.

the "pre-operational stage" (identified with children 2 years old to 7 years old), the "concrete operational" stage (children 7 to about 11 years old), and the so-called "formal operational" stage (children over 11 years old). There probably is some deep sort of truth to this correlation with age, but it cannot be construed as implying a rigid classification system for mathematical ideas and activities. For one thing (as I shall attempt to show presently), one can easily find adults who are at the various "immature" stages with regard to specific tasks, perceptions, or questions; secondly, one can take "formal operational" procedures and, by interpreting them in various ways, alter their epistemological classification in the Piagetian scheme. (For example, the common word game where one starts with, say,

SOUP

and changes one letter at a time – under the rule that, at every stage, one must have an accepted English word – until one ends with

NUTS,

can be used as a paradigmatic introduction to mathematical proofs in, say, arithmetic or algebra, with the result that formal proof-making, under this approach, must be re-classified as a more elementary kind of activity, and is demonstrably feasible with children at least as young as 8 or 9 years old). Thirdly, I believe the Piagetian stages are based upon an inadequate foundational theory of what is really happening in typical interviews (and I shall presently suggest an alternative theory).

A relatively rigid identification of "stages" with chronological ages, and with specific mathematical ideas and activities, is not the only error in the conventional wisdom. A second inadequacy appears in the ambiguity of the Piagetian concept of "conservation". If a child believes that the number of blocks is not changed when the blocks are rearranged, he is said to "conserve" – that is, what Piagetians call "conservation" is recognizable as a form of what mathematicians might call an assumption of *invariance*. Unfortunately, the conventional wisdom fails to specify carefully the transformations underlying the assumed invariance. Thus, a child who believes (as Miriam did), that there are more blocks in a long row of 8 red blocks than there are in a short row of 8 yellow blocks (cf. Figure 4) is said "not to conserve" – she does not assume invariance under operations that change the length of the rows. But Miriam did believe that you could move the two *rows* closer together in order to see if they would "match" or "fit together" (in a one-red-cube-corresponding-to-one-yellow-cube fashion). So Miriam *did* assume invariance of number under some forms of geometric rearrangement (i.e., to some extent she *did* "conserve").

There is also an implicit assumption, in the conventional wisdom, that "to conserve is mature" (with an implication that it is "good"), whereas non-conservation is "immature" (with an implicit suggestion that it is "foolish" or "silly"). Now, in a strict chronological sense, it is a fact that somewhat younger children "don't conserve", and somewhat older children do. In that sense, conservation is chronologically subsequent to non-conservation. But is it *wise* to assume invariance if one has no decision procedure for testing the assumption? Miriam seems to have been at an important transitional state – she had available to her two decision procedures ("counting" and "matching")[7] that let her confirm the invariance assumption, and she was about ready to make use of them, to allow herself to be convinced by their testimony.

But what about the Piaget experiments on volume? In these, children assume an invariance which they *cannot* confirm – and, indeed, which is contradicted by some of the available evidence. Thus, children will say that "all of the water" has been poured from one glass into a second, even though one can easily notice that the first glass is still wet. Is it now "mature" (i.e., implicitly "good") to assume that the volume of water in the second container is equal to the volume of water in the first? (Or is this better understood as a useful *first approximation*?)

I think there is a more serious objection to the conventional interpretations of Piaget. What seems to me to be central to these tasks is a conflict between *perception* and some *formal abstract system* which the child has "stored in his mind."

To emphasize this point I want to stress that one can try to use relatively primitive perception to determine "how big" something is. We commonly look at a house, or at a human being, and form an opinion, usually based mainly on perception, of the size of the house or of the person. Of course, perception can lead us astray. Some stage actors, actually small, can project an image of being much larger, and some large actors can make themselves

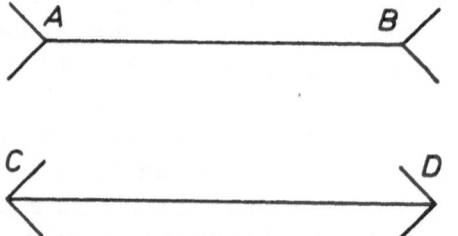

Fig. 11   Which Distance is Greater, $d(A, B)$ or $d(C, D)$?

---

[7] "Matching", as we have just remarked, is really a special case of invariance, under a restricted class of transformations.

look small. (Bruner has some data on size perceptions of coins.) A particularly notorious example would be optical illusions, such as shown in Figure 11
where the distance between A and B is actually equal to the distance between
C and D, but does not appear so.

But we also have available to us various formal systems which provide
systematic decision procedures – e.g., the use of a ruler to measure both
distances, in the preceding example.

Given alternative ways of deciding, there clearly exists a potential for
contradictory conclusions. These are what many Piagetian "check-ups" and
interviews identify: is there a conflict, and if so, which choice does a child
make?

We need also to distinguish conflicts between perception vs. a formal
system, from conflicts between two formal systems. We also need to distinguish the case of formal systems which a child has available and *could* use
(but does not), from the case where a child does not have any appropriate
formal system available at all. Thus, many children who claim that "all of
the water has been poured from a first container into a second" may merely
have failed to notice, or to take account of, the small amount of fluid that
still remains in the first container. If this is pointed out to them, they may be
perfectly capable of making use of this additional datum.

Indeed, the water pouring experiments may involve the comparison of
a simple "first-approximation" model of reality to a much more complex
"second approximation" model.

With this interpretation, we can drop such vague action words as "the
child does *conserve*", and replace them with a more careful description of
the formal cognitive systems which the child has available, and which of these
he commonly uses. It is admittedly weird to see a child apparently reject
clear-cut evidence, which is what young children do seem to do in many
Piagetian tasks. This becomes comprehensible when one reinterprets it in
terms of the repertoire of available formal systems, and possible conflicts
among these systems, or between a formal system and a relatively primitive
perception.

Indeed, the phenomena even begin to seem very natural when we use
ourselves as subjects. Rose Grossman, a mathematics educator in Yonkers,
N.Y., observed that for ordinary drinking glasses, the perimeter of the rim
is greater than the height, yet most people's perception leads them to the
opposite inequality. You can use this as a Piagetian task. Get a glass; try
to decide for yourself which is larger, the height or the perimeter of the rim.
Measure them (masking tape is convenient for this), and see if your formal
decision procedure produced the same conclusion as your relatively direct
perception. How far can you extend this? To cognac glasses? To various

liqueur glasses? Is there a point where you *do* get opposite inequalities from the two decision procedures? (How about various bottles?)

Thus, we can interpret Chris's drawings of the water level in the tilted or inverted jar by saying that he used a formal system – namely, geometric invariance under rotation – and he did *not* use an alternative formal system that would account for the effect of gravity. (One could have carried the interview further to find out if Chris could make use of this second formal system.)

One can further speculate on where these formal systems come from. Geometric invariance under rotation is surely *not* learned by six-year-old children as a result of formal academic instruction. But how could they cope with their everyday physical environment without developing some of the so-called "perceptual constancies"? In the case of the rotation of ordinary rigid objects, this presumably leads to the development of the formal system of geometric invariance under rotation. I believe the formal systems that are used can be described with much more clarity than is common in Piagetian discussions, and that as a result of this greater clarity it is possible to make quite plausible identification of the sources from which the formal systems have been developed[8].

I even believe that I can explain the formal systems being used by Miriam when she claimed one straw was longer than the other (cf. Figure 2), and can give a credible explanation of the experiences in Miriam's life that gave rise to these formal systems. However, following the admirable example of Fermat, I leave this as an exercise for the reader.

Syracuse University, Syracuse, N.Y., U.S.A.

---

[8] Note that I am using the phrase "formal system" to designate a decision procedure or cognitive structure which the child has in his mind, and which can often be described with considerable precision by observing the actions of the child in certain decisive situations. Were this written on paper, instead of stored in a child's head, I believe we would call it a mathematical system. It differs from the better examples of mathematical systems in the respect that it may involve contradictions, with resultant indecision or inconsistency in the child's behavior. It differs from Piaget's use of the word "formal" in the sense that a cognitive structure (in our present sense) is a set of mental constructs, a specific data-processing "sub-routine", as it were. It is not a developmental stage; rather, developmental stages in a child's life can be described in terms of which cognitive structures are available at various ages, and which are in fact used in various settings. (Thus "cognitive structure" or "formal system", as used here, is probably *part of*, but not identical with, "schemes" or "schemata" in Piaget's useage.)

ZOLTAN P. DIÉNÈS

AN EXAMPLE OF THE PASSAGE FROM THE CONCRETE TO THE MANIPULATION OF FORMAL SYSTEMS

INTRODUCTION

Everybody knows that mathematics is an abstract subject. It follows that most people who have studied the problem of learning such an abstract subject, would agree that some passage from the concrete to the abstract must be mapped. By concrete, we mean usually, our immediate contact with the real world. We come into contact with objects and events and we re-act to them. This is the concrete level at which all organisms behave until they are able to organize their re-actions to events into re-actions to sets of events. This is the first stage towards abstraction – when the organism begins to classify. It is of interest to try and investigate the details of the abstraction process, not only from concrete experiences to classification, but to the learning of extremely complex abstract systems such as exist in mathematics and which more and more people, including children, are called upon to learn. The difficulties, in the way of such studies, are great, one of the main difficulties being that mathematicians on the whole are not interested in learning, and psychologists on the whole do not know enough mathematics to be able to formulate the problem in a way in which a possible solution might be sought. At the Sherbrooke Psychomathematics Center, for the past few years, we have been studying the abstraction process as it proceeds from the concrete to the final stage of wielding a mathematical formal system. We have rather few laboratory results of this aspect of our work as yet but we have enough classroom evidence that we can postulate certain regularities that seem to occur and certain pre-requisites that appear to have to be satisfied before certain stages of learning can successfully be undertaken. We have, so far, sub-divided the process of learning mathematics into 6 consecutive stages. These are approximately the following:

(1) Initial inter-action.

(2) Discovery of regularities in situations and consequent play with sets of rules or constraints.

(3) The comparison of several games possessing the same structure, that is with the same sets of rules; this is the stage of the search for isomorphisms.

(4) The representation of isomorphic situations in one, all-embracing, usually graphical form.

(5) The study of the representation by the description of its properties. In

this way not only one rule-structure or one game is being studied, but all games with identical rule structures. This might be called the stage of symbolizing, since in order to describe something, symbols will be necessary.

(6) Out of all the descriptions, certain parts are taken as initial descriptions and rules derived whereby other parts of the description can be derived from the initial description. The rules must be so constructed that any descriptions derived do, in fact, describe properties of the representation of our structure. The initial descriptions are known as axioms, the rules of the game are usually called the rules of procedures, and the eventual descriptions which we can reach from the initial axioms are known as theorems. The method used to arrive at a theorem from the initial description, using the rules of procedure, is known as a proof. This final stage could be called the stage of *formalization*.

To sum up, the stages are the following:

(1) Inter-action.

(2) Rule construction and manipulation.

(3) Isomorphisms.

(4) Representation.

(5) Symbolization.

(6) Formalization.

In this paper, I shall try to give a geometrical example of how we can lead a child from initial inter-action to the manipulation of a formal system. I have chosen a very simple example, namely the equilateral triangle.

STAGE I

A triangle is the simplest two-dimensional polygon and the equilateral triangle is the most symmetric version of that simple figure. So, it seems a fitting introduction to many aspects of geometry. In effect, studying the triangle is studying, in a sense, cycles of three. If we put a pin through the centroid of a piece of cardboard cut off in the form of an equilateral triangle and we trace its perimeter on the sheet on which we have placed it, we can turn it around until the cardboard triangle again is seen to be inside the perimeter we have drawn. When we have done this for the third time, the cardboard triangle will assume its original position. So, a cycle of three is one of the important properties that we shall have to encourage children to abstract from the study of an equilateral triangle. There are many cycles of three in real life. For instance, most people have three meals a day: breakfast, lunch and supper. And they go to bed and the next day they again have breakfast, lunch and supper, and so on. The properties of this cycle are very similar to the properties of the cycle described by an equilateral triangle when it is turned around in a plane about its centroid. Three children can hold hands and dance around a certain object which they place in the middle of the circle which they have made. A "move" can be a movement which allows each child to reach the position occupied by the child on his or her left. Again this has to be done three times before each child comes back to his or her original position. It is possible, of course, to give children a large number of equilateral triangles to play with made of wood, plastic or cardboard preferably in different colours. It would also be possible to cut some equilateral triangle doors out of these cardboard pieces and simply ask the children to make anything they like with these large and small triangles, of different colours. They will make all sorts of patterns and realize that these triangles can fit together in certain ways but not in other ways. They will be able to make isosceles trapezoids with them which they will call boats or hats or they will make equilateral hexagons with them which they will probably call circles, to the horror of some mathematics teachers. Children could also be given some circular discs of different colours to make patterns with. They will realize, of course, that they cannot cover the whole floor area with circles, whereas they can with equilateral triangles. There will always be some spaces left between the circles. These spaces could be drawn and possibly cut out of cardboard which could also be played with in the same way as equilateral triangles have been played with. All such activities can be classed under inter-action with the environment, where of course, the teacher has already gone to some considerable pains to structure the environment suitably so that experiences leading to "equilateral triangle properties" should come into the life of the children. All such experiences would be Stage 1 Experiences, that is *inter-action with the environment*.

STAGE 2

As a result of these playful activities, certain properties of equilateral triangles will have been intuitively learned. Most children would be able to tell you that the representative objects used have three corners, and that there are three ways in which you can put them down to occupy the same space, unless you are allowed to turn them over, in which case there are six. They will have probably wanted to draw pictures on these triangles, possibly in the three corners. Some children might have discovered that if you join the midpoints of the sides of a triangle, then you make a kind of inner triangle and leave three similar outer triangles near the corners. They would have probably drawn some pictures in these and used them for various games which they will have invented if they have the very good fortune of being supervised by a teacher who allows children some creative outlets of this kind.

So, it will now be possible to suggest to children that they follow certain rules, that is they can invent some games or the teacher can invent some games which will lead them to the appreciation of the structure of the equilateral triangle. One good ball game can be played in the school yard by six children. You need a seventh child to give the orders. We can put three children on opposite sides of an imaginary line which could be acting as a mirror.

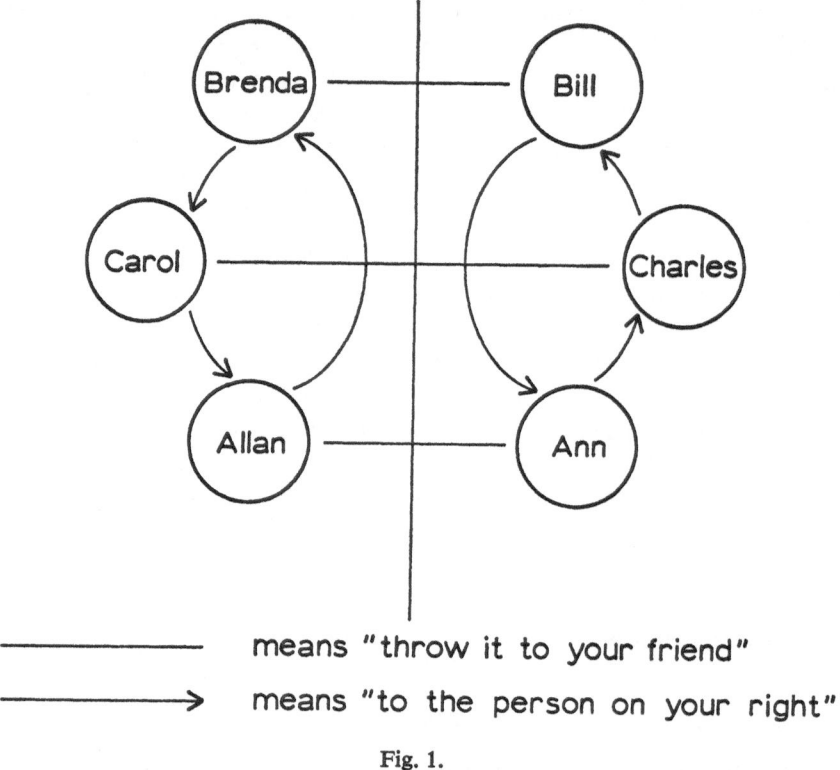

—————————— means "throw it to your friend"

————————→ means "to the person on your right"

Fig. 1.

For instance, Allan could be the image of Ann, and Ann, of course, the image of Allan. Carol could be the image of Charles, and Charles of Carol, and Brenda and Bill could be images. Brenda, Allan and Carol, on one side of the line would form a triangle and the reflection of this triangle would be Bill, Ann and Charles. Now, the children whose names begin with the same letter, could be designated as friends. Ann and Allan would be friends, Bill and Brenda would be friends and Carol and Charles would likewise be friends. Allan, Brenda and Carol, on one side of the line, would be facing toward the centroid of the triangle at the vertices of which they are standing; similarly for Ann, Bill and Charles. So, Carol would have Allan on her right,

Allan would have Brenda on his right, and Brenda would have Carol on her right. Similarly, Charles would have Bill on his right, Bill would have Ann on his right and Ann would have Charles on her right. Let us remember that Allan and Ann are friends, Brenda and Bill are friends and Carol and Charles are friends. Now, a seventh child who can be called Dick, could be giving the orders which would go something like this: throw the ball to your right. In this case, you do not throw it across the line but you throw it to the person standing immediately on your right. Throw the ball to your left would be another throw or throw it up and catch it would be another. Throw it to your friend would be another possible throw. The more complicated throws would be: throw it to your friend's right, then to your friend's left. For example, if Brenda had the ball, and Dick said: throw it to your friend's right, Brenda would throw it not to Bill, but to Ann, because Bill is Brenda's friend and Ann is on Bill's right.

If now, Dick said again: throw it to your friend's right, then Ann would throw it back to Brenda because Allan is Ann's friend an Brenda is Allan's right. So, the friend's right of the friend's right, is yourself, like the friend of the friend is also yourself. It will be easy to verify that the friend's left of the friend's left is also yourself. These types of throws would correspond to the reflections along the axes of symmetry of the equilateral triangle and throwing the ball to your right or to your left would correspond to one third of a turn either one way or the other way of the triangle. There is, of course, no need to keep to the rule of having children whose names begin with the same letter as long as the children know who is whose friend. And, anyhow, this can be determined by their positioning. So, to make the game more motivating, one can decide that the object of the game is to be out as little as possible. Now, Dick starts by being out so he has to try and get in, and getting in is achieved by getting somebody to make a mistake, so if somebody throws the ball to the wrong person, that counts as a mistake and so this person goes out and issues the command and tries to get in again. The last person to make a mistake wins the game.

We can play another game, on a lattice. We can have, for instance, three kinds of trees: fir trees, orange trees and coconut trees or any other kinds of trees that might be growing in the neighborhood where the game is being played. So, you set out a wood or an orchard in a triangular fashion, that is you make a triangular lattice and you decide that in order to make the wood or the orchard more interesting, we try and plant trees next to each other that are different (Figure 2). "Next" means "joined by a lattice line". In this way, for example, in the first row, we could have a fir tree, orange tree, coconut tree; again fir tree, orange tree, coconut tree and so on. Now, the second row, cannot start with a fir tree nor with an orange tree, so it has to start

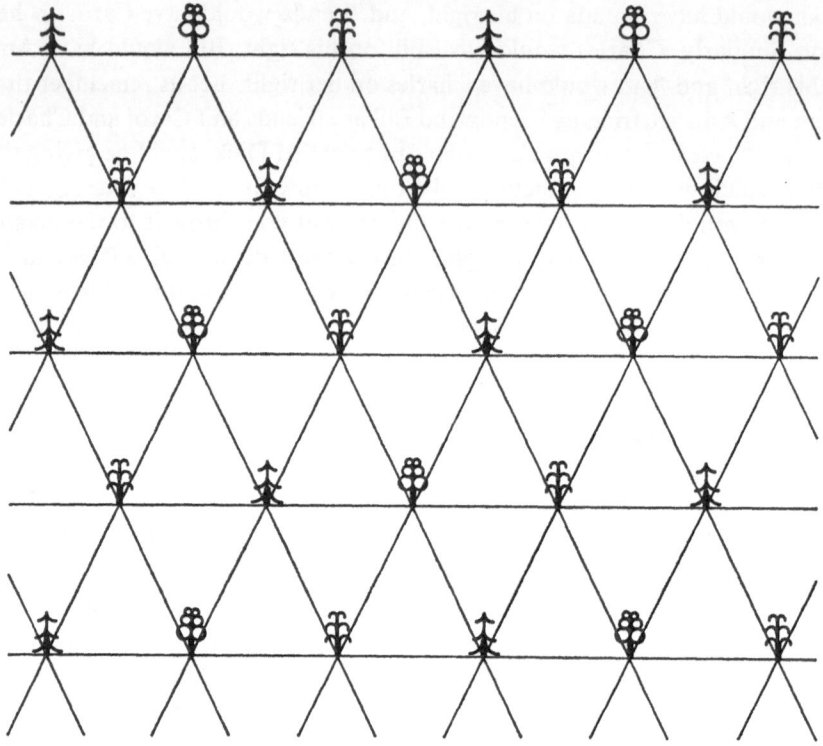

Fig. 2.

with a coconut tree. And so it would go coconut, fir, orange, coconut, fir, orange and so on. In this way, the orchard or wood is built up. We can then go for a walk in this orchard, now, if we are walking from an orange tree to a coconut tree, we can say we are in the orange-coconut position. If we are walking from a coconut tree to an orange tree, we are in the coconut-orange position and so on. So, if we are walking straight on, we would be passing from fir to orange, to orange to coconut, to coconut to fir and be back to fir to orange but, of course, further along the line of trees. All the positions which have the same description, would be identified as equivalent to each other. So, it would be discovered that there are six different descriptions and so the positions fall into six different equivalence classes. There would also be six different ways of getting from one position to another, including the move of staying where you are. So, for example, there would be a move which we might call ADVANCE which would take you to the next position ahead of you. There would be another move which would be REVERSE which would be the inverse of the move ADVANCE. There could be another one which is called TURN. Let us say that one would forbid the move which would turn a very sharp turn either to the right or to the left, that is from fir-orange you could move to orange-coconut by taking a sharp turn to the right or to the left or you could move from fir-orange to orange-coconut by advancing, so there is no need to have the sharp turn. So, when we say TURN, we mean turn either to the right or to the left, but turning gently only through a sixty degree angle.

Children will soon discover other moves such as BACK SKID, or FRONT SKID. A BACK SKID is imitating what a car would do if it was standing in the fir-orange position and skidded around to fir-coconut position, because the back of the car is still nearest to the same kind of tree as before, but the front is facing a different kind of tree. A front skid would take you, for instance, from fir-orange to coconut-orange, where you are still facing the same tree as before but behind you is another kind of tree.

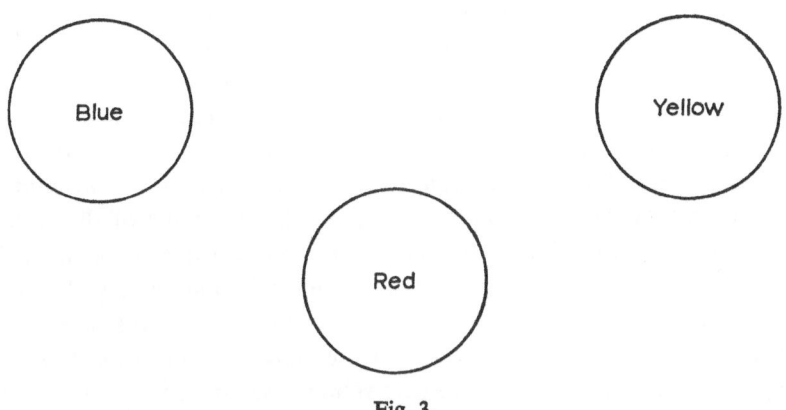

Fig. 3.

A third way to play an equilateral-triangle-game would be to designate three positions on the floor. They could be referred to by colours. Let us call one position the red one, next to it would be the blue one and next to it the yellow one, so that red to blue to yellow to blue, goes around in the clockwise sense. Then, we can put a child in each of these positions and tell them to hold hands. We can establish a number of different moves in this game. For instance, we might say that the one on red has to go to blue, the one on the blue bas to go to yellow and the one on yellow has to go to red. This could be the clockwise move. Naturally all you would have to do is to simply dance around from red to blue, blue to yellow, yellow to red. The counter-clockwise move would be obviously red to yellow, yellow to blue, blue to red. One of the rules of this game, is never to let go of your hand, that is, the children must always hold hands with each other; otherwise they lose a point or something. The more interesting moves come in when we try and change the position of two of the children without changing the position of the third. Let us say that the one that is standing on red has to stay still and the ones standing on blue and yellow have to change places but without letting go of each other. What happens, of course, is that red and yellow can make an arch and blue can go through the arch and yellow will, in the meantime, have passed on to where blue was and blue will take up yellow's position. Red, of course, at this stage will have his arms crossed and so will have to turn about himself and face outwards. If the move is properly carried out, all the three children will be facing outwards instead of inwards.

So, corresponding moves can be invented for the person on blue staying still and red and yellow changing places, and the person standing on yellow staying still and the red and the blue changing places. Here again, we have the moves corresponding to the rotations and reflections of the equilateral triangle. Another version of playing this game is to take a piece of cloth in the form of an equilateral triangle which could be one colour on one side and the other colour on the other side and each child can hold one corner of the sheet. When we do the move where red stays still and blue and yellow change places, then the sheet turns over from one colour to the other. To specify the states of the game in each case, we could pick one of the children, say Allan, and decide that wherever he is will be the name of the state in which the game is. Of course, Allan may be on red in two different ways. He may be facing outwards or he may be facing inwards, and he may also be on blue, outwards or inwards, or he may be on yellow, outwards or inwards. So, there are six states in the game: red outwards; red inwards; blue outwards; blue inwards; yellow outwards; yellow inwards. This way of establishing the moves uses *fixed axes in space*.

There is, of course, another way of establishing the names of the states. We could, instead of picking one child and the three positions, have picked the three children and one position. Let us say that the children's names are Michael, Nellie and Olive. Let us pick red as the appropriate place where the state is determined. So, if Michal is on red and is facing outwards, that will give the state: Michael – out. Or, if Michael is on red facing inwards, that state will be called Micahel – in. So, if we are always referring to the position marked as red on the floor, there will be the following six states: Michael – in, Michael – out, Nellie – in, Nellie – out, Olive – in, Olive – out. Then, of course, there will be the moves where Michael stays put, where Nellie stays put, and where Olive stays put and then there will be a move where Michael goes where Nellie was, Nellie goes where Olive was and Olive goes where Michael was; as well as the opposite of this move. This way of establishing the moves uses *mobile axes*. The Michael-to-Nellie-to-Olive-to-Michael move would sometimes be a clockwise and sometimes a counterclockwise move, depending on whether for example the children happened to be placed as

or as

It must be remembered that the words "clockwise" or "counterclockwise" assume
  (i)   the fixed position of the onlooker,
  (ii) the system of reference fixed in space.

If we wish to use mobile axes, then for example in the "hand-holding game", it would be sufficient to give the instructions as "go to the next place on your right" or "go to the next place on your left" as *left* and *right* have the idea of mobile axes built into them.

STAGE 3

Now, to come to the third stage, that is to the stage of comparing these games to each other, we must of course, play them in conjunction. What we can do is, for instance, to have the six children's ball game being played and, at the same time, the three children holding hands in the form of a triangle game being played. It will have to be established which move in the ball game corresponds to which move in the hand-holding game.

The clockwise move could be made to correspond to throwing the ball to the right. The counter-clockwise move to throwing the ball to the left. The red move could be made to correspond to throwing the ball to your friend and the friend's right and the friend's left correspond to the other two moves where blue stays still while red and yellow change places and yellow stays still while red and blue change places.

Instead of marking the places of the floor – red, blue and yellow – we could take the triangular sheet and possibly paint the corners red, blue and yellow or to make the correspondence with the fir tree, orange tree, coconut tree game more obvious, we could paint a fir tree on one corner, but on both sides of the sheet; an orange tree at the other corner but on both sides; and a coconut tree on the third corner but again on both sides and we could pick two positions, say red and blue, and we would say that the sheet is in the position: coconut-fir when the coconut is at red and the fir is at blue and so on. So, in this way, there would be six states in which the sheet can be. These would correspond to the six states in which the person can be walking between different trees in the wood. It would be quite easy to determine the correspondence between the game in which the sheet is moved around and the game in which you are walking from tree to tree.

As a result of these games, children will have been having experience of the abstract structure of the isometries of the equilateral triangle. They will be beginning to think less of the particular properties of each game and gradually more and more of the common properties that they all have. Every one of these games has six positions. They have two third-order moves and three second-order moves and one do nothing move and these moves are related to each other in very similar ways. It is possible to establish a dictionary between the six words in one language, namely in one game, and the corresponding six words in the other language. And, if we say, that the succession of two words in one language can be replaced by a third word in that language, then if we translate this replacement statement by the dictionary into another language, then what we are stating about the succession of certain two words in this other language being replaceable by a certain third word is this other language, should in fact, be true. In other words, the dictionary should translate true statements about the replacements in one language into true statements about the replacements in the other language. And, likewise of course, false statements should be translated into false statements. The children will have fun establishing dictionaries and even more fun in establishing dictionaries that do not work. In other words, ones that do not translate true replacement statements into true replacement statements, but sometimes they translate them into false ones and vice versa. These will be discarded as not very suitable dictionaries because it makes the game go wrong.

STAGE 4

So, we have now described the Stage 3 of the procedure. Now, the children have achieved a certain measure of abstraction and they will need a peg on which to hang it. This peg is the representation. We will need six spaces or points which we can draw on a sheet of paper or on the floor, each of which represents one state in the game. It is possible to draw this in two rows of three. If we call these A, B, C in the first row and X, Y, Z in the second row, then we find, for instance, that it is the second-order move such as throwing the ball to your friend or red staying still and blue and yellow changing places or the turning move or the backskidding move in the orchard game, can be represented by moving across from one row to the other row, that is from A to X or X to A, from B to Y or from Y to B and C to Z or Z to C. So, this can be represented by a certain type of arrow, let us say for instance, a thin arrow. So, $A{\longrightarrow}X$, $Y{\longrightarrow}B$ and so on. Now, we will need a representation for a third-order move. This will be advancing in the orchard game or red going to blue, blue going to yellow, yellow going to red in the hand-holding game or throwing the ball to your right or to your left in the ball-throwing game. It will be found that if we plot these moves on our map, we shall find ourselves with an arrow going from A to B, one from B to C and then one from C to B. These arrows could be thick ones to distinguish them from the thin ones. Or, we could use different colour arrows, or we could distinguish them from one another in any way we please (Figure 4).

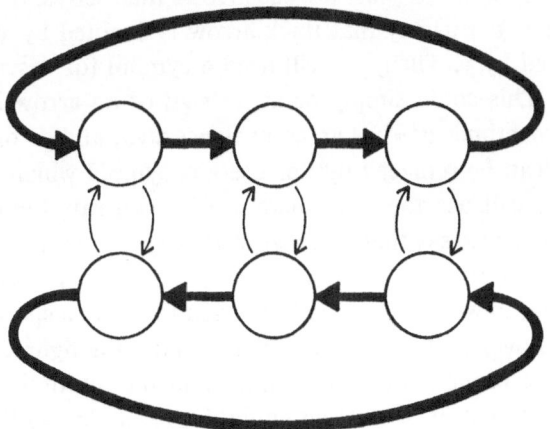

Fig. 4.

It will be interesting to find that as we plot the map further, we will not find that the arrow goes from X to Y, but that it goes from Y to X, and not from Y to Z but from Z to Y and then from X to Z instead of Z to X. So, the second cycle, that is the cycle of the X's, Y's and Z's, seems to go around in the opposite sense from the cycle of the A's, B's and C's. It will be necessary for children to be able to fill in the empty spaces of this diagram, by the corresponding states of any of the games which they have learned possess this particular structure, having rubbed out the A's, B's, C's and the X's, Y's and Z's.

If they are not able to do this, it means that they are not ready for this abstract form of representation and they should play more isomorphic games or maybe they should play other games from quite another part of mathematics. In other words, it is no good trying to force the child into a higher order of abstraction before he is ready for it. If however he is easily able to fill in the spaces and the moves of an abstract representation by the concrete examples of the games from which he has abstracted this representation, then he is ready to go on.

So, Stage 4 consists in passing from the individual games and the correspondence established between these games, to a common representation which will represent all the games of the same structure at the same time. This is the peg on which to hang his new abstraction.

STAGE 5

The next stage in the proceedings is to investigate the properties of this representation. We would need a sign for the third-order move, that is for the thick arrow, and a sign for the second-order move, that is for the thin arrow. We could, of course, just use the arrows themselves, or we could call them x's and y's. Let us say that thick arrow is denoted by x and the thin arrow is denoted by y. Then, we will need a symbol for describing: "is replaceable by". This could simply be an $=$ sign or an arrow in both directions $\longleftrightarrow$ sometimes used as an equivalence sign, and so on. Or, simply, we could say "can be replaced by" or use any symbol which appears to us convenient. We will use the $=$ sign here. So we can say, for example, that $xxyxx=y$. We can also say that $xxx=yy$ or that $xyx=y$ and so on.

The series of moves xxx is replaceable by the series yy because in both cases we get back to where we started. It is good to have a special symbol for such series of moves and such a symbol is usually the figure 1, called "the neutral". We are reminded of the number 1 in the multiplication game in which it plays the neutral role. So we could, for example, say that $xxx=1$ or that $yy=1$ or that $xyxy=1$. There are many such descriptions, in fact, there would be an infinite number of such descriptions because we can make the paths which we follow as long as we please and there will always be different ways in which to reach our destination from our original starting point. So, we are now able to describe our representation, that is our abstraction in terms of a *new language*. We could now ask various questions about these replacements. Clearly, we have introduced an *equivalence relation* and we might ask just how many equivalence classes our equivalence relation generates. It is very obvious that there are six, corresponding to the six moves which we can make in each of the games. Each of these moves can be expressed by means of x's and y's. The "do nothing" move, for instance, can be expressed as yy; the clockwise move, for example by x; the counter-clockwise move by xx; the throw it to your friend move or the red stays still and blue and yellow change places moves can be represented by y and then throw it to your friend's right, for example, might be xy to your friend's left would probably yx.

STAGE 6

So, here we have a way of expressing the six different moves, Now, it is possible to show, although we have no space here to show it, that any series of x's and y's can be replaced by one of the other of these six short series of moves by a certain limited number of admitted replacement rules. Now, this is where we are getting on to Stage 6 because we are trying to restrict the way in which we describe our system. Since there are an infinite number of properties to describe, we need to restrict ourselves to a certain finite number of them and find a method of generating all the other descriptions. So, this is the beginning of Stage 6: the formalization of the system.

One of the initial pieces of description could be, for example, $xxx=1$, another one could be $xyx=y$, and yet another one could be $yy=1$.

> *Systems of Axioms*
> (1) $xxx=1$
> (2) $xyx=y$
> (3) $yy=1$.

This means that we are *allowing* these replacements. We need to put another rule in the proof game, namely that the symbol 1 can be introduced anywhere along the line or suppressed anywhere where it might occur. For instance, if we have a sequence $xxyxx$ we can turn that into a sequence $xxyx1x$ or $x1xy1xx1$ and so on. These 1's are of course DO NOTHING's and it is reasonable that they can be introduced or omitted at will. So then, with this additional rule, we claim that starting from any sequence, we should be able to reach one of the following short sequences: *yy, x, xx, y, xy, yx* after a finite number of introductions and suppressions of 1 and replacements of 1 by *yy* or *yy* by 1; 1 by *xxx*; *xxx* by 1; and *xyx* by *y* and *y* by *xyx*.

We might try to see how *xxyxx* can be reduced to *y* by the use of these rules. Clearly, the middle *xyx* can be replaced by a *y* so that leads us to *xyx*, which of course, itself can be reduced to *y* itself. So, in two moves, we have reached *y* from *xxyxx*. So, we now have a theorem in which we say that *xxyxx* belongs to the same equivalence class as *y* that is, is replaceable by it. We can show this simply by looking at the three simple properties or initial descriptions or axioms and we do not have to even look at the representation or diagram.

THE BINARY FUNCTION

There is just one more flaw in the formality of the system. We have introduced no formal way of saying that we are dealing with a binary operation. We also have no way of referring to any *generic* series of moves in the game.

For the latter we could easily use "frames" such as □, △, ⎕, etc. such as are now commonly used for "place holders".

For showing that we are operating on two "moves" by the binary function by stringing the moves together into one move, we could just use a letter *F* (for function). So, instead of writing *xx*, or *xy*, we would write *Fxx* or *Fxy*. Naturally we could apply the binary function *F* successively, as for example in

$$FFxxFxy$$

(the binary of the binary of *x, x,* and the binary of *x, y*) or

$$FFxyz$$

(the binary of (the binary of *x, y*) and *z*).
It will be seen that parentheses are unnecessary.

The associative principle and the principle of the neutral element can then be expressed as follows:

General Axioms

$$\mathsf{F\,F\,\square\,\triangle\,⎕ = F\,\square\,F\,\triangle\,⎕} \qquad \text{(a)}$$

$$\mathsf{F\,1\,\square = \square} \qquad\qquad \text{(n)}$$

We also have the *particular axioms*:

$$F F x x x = 1 \qquad\qquad (1)$$
$$F F x y x = y \qquad\qquad (2)$$
$$F y y \quad = 1.$$

Let us now see how we "prove a theorem" in this formal system.

THEOREM.

$$F F y x y = F x x$$

Proof:

$F F y x y$					
$F 1 F F y x y$	by (n)				
$F F F x x x \, F F y x y$	by (1)				
$F F x x F x F F y x y$	by (a)	$\square = F x x,$	$\triangle = x$		$\square = F F y x y$
$F F x x F F x F y x y$	by (a)	$\square = x,$	$\triangle = F y x,$	$\square = y$	
$F F x x \, F F F x y x y$	by (a)	$\square = x,$	$\triangle = y,$	$\square = x$	
$F F x x F y y$	by (2)				
$F F x x 1$	by (3)				
$F F x x F F x x x$	by (1)				
$F F F x x F x x x$	by (a),	$\square = F x x,$	$\triangle = F x x,$	$\square = x$	
$F F F F x x x x x$	by (a),	$\square = F x x,$	$\triangle = x,$	$\square = x$	
$F F 1 x x$	by (1)				
$F 1 F x x$	by (a)	$\square = 1,$	$\triangle = x,$	$\square = x$	
$F x x$	by (n)				

If we also assume the transitive property of the equivalence relation as forming part of a system, then we can say that we have shown that $F F y x y$ and $F x x$ belong to the same equivalence class.

To make the formal game easier to "play", it is advisable to use pieces that are removable. Any small objects will do to represent the elements of the system. The spelling rules for writing strings of operators or "words" must be learned. The ground rules for these are:

(i) $x, y, 1$ are words

(ii) any two words may be written one after the other, as long as F is written in front of the joined string.

Then for example the associative principle becomes the following "manipulative rule": *In a string of two F's and three words the second F may be interchanged with the first word, and vice versa.*

The neutral law can be stated as: *In a string F, 1 followed by a word. F1 can be omitted, and vice versa.*

The game is in one sense extremely abstract, embodying properties of a large number of games, but also extremely concrete, as it can be played as a

Stage 2 game, leading in to another learning cycle. After learning many such games, children will begin to classify them, and we are on our way to playing games with abstract systems leading to such notions as completeness, freedom from contradiction, independence and the like.

Université de Sherbrooke, Sherbrooke, Quebec, Canada.

ARTHUR ENGEL

GEOMETRICAL ACTIVITIES FOR THE
UPPER ELEMENTARY SCHOOL

INTRODUCTION

In Germany, high school starts with grade 5. In grades 5 and 6 about one hour per week is devoted to intuitive geometry. The aim is to develop the *geometric intuition* of the child and to familiarize him with many important *geometrical concepts*. The course appears somewhat unsystematic, since it consists of a great number of geometrical activities which are supposed to develop various *intellectual* and *manual* skills. In grade 7 there is a gradual transition to a more systematic development of geometry.

In this paper I shall present a selection of activities which I have used in grades 5 to 7 for the past 16 years. Well known topics will be mentioned only in passing; less familiar ones will be considered in greater detail. At this level I prefer topics which are not treated later, but which are still interesting, important and challenging. In fact, they are more interesting and instructive than many specialized curiosities which are proved in higher grades. These examples will show that, even at an early age, one can reach rather deep results in a short time and starting from scratch. It is in geometry that children are for the first time confronted with nontrivial mathematics.

COMBINATORIAL ACTIVITIES

For young children, activities with a combinatorial touch are best suited. They are easy to comprehend, and they often have a recreational twist which appeals to the natural curiosity of the child. Here are some examples:

(a) A square can be cut into 4, 6, 7, 8, 9, 10, 11,... squares. Show how! Figure 1 shows the solution. It needs no explanation.

Problem 1. Solve the same problem for the equilateral triangle.

(b) Two rectangles are drawn in the plane. Into how many parts can they subdivide the plane? Think of all the possibilities!

Figure 2 shows subdivisions into 2, 3, 4, 5, 6, 7, 8, 9, 10 parts.

Problem 2. Study in a similar way subdivisions of the plane by

(a) a rectangle and a circle
(b) a rectangle and a triangle
(c) a triangle and a circle
(d) two triangles.

Problem 3. Study subdivisions of the plane by 1, 2, 3, 4, ... circles. Do you see a pattern? Can you explain the pattern? What happens if one circle is added?

Problem 4. Cut a cake by 1, 2, 3, 4, ... straight cuts. Do you see a pattern? Explain the pattern. What happens if an additional line is drawn?

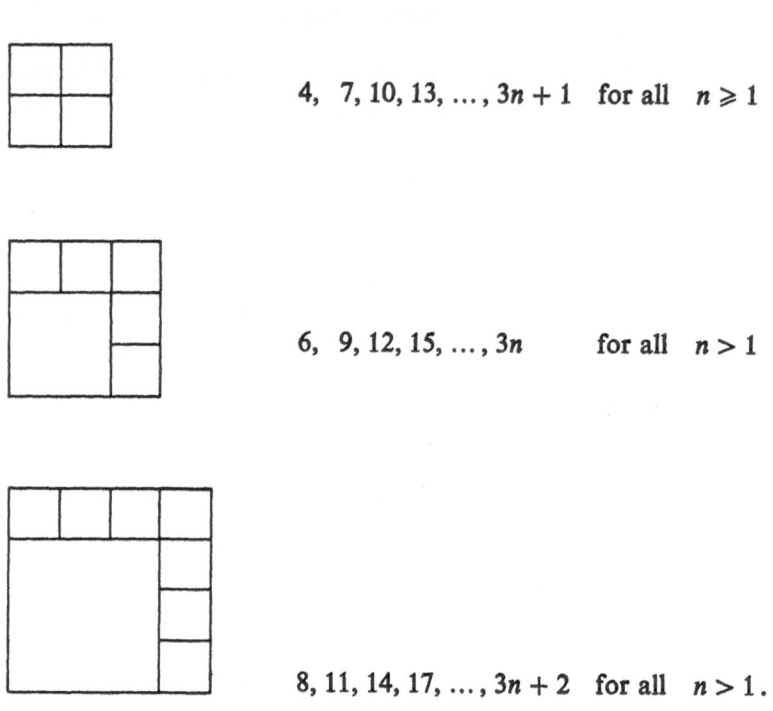

$4, \ 7, 10, 13, \ldots, 3n + 1 \quad$ for all $\quad n \geqslant 1$

$6, \ 9, 12, 15, \ldots, 3n \qquad$ for all $\quad n > 1$

$8, 11, 14, 17, \ldots, 3n + 2 \quad$ for all $\quad n > 1$.

Fig. 1.

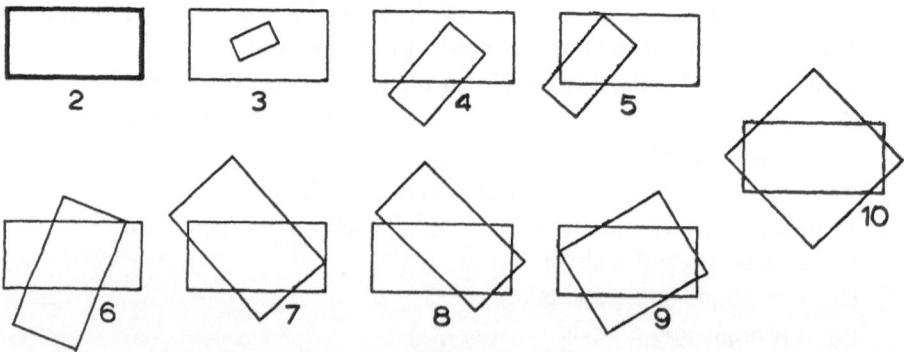

Fig. 2.

Problem 5. (This is a challenging problem in space visualization.) Into how many parts can a cube and a sphere subdivide space? If a challenging problem is posed to children without any help, they fumble around aimlessly and loose interest. They must be taken on a guided tour.

You are in a big cubical room; there is a small sphere in the room, its center coinciding with the center of the cube. We have a subdivision of space into 3 parts. Now the sphere starts expanding. When it touches the 6 walls there are still 3 parts only. If the sphere increases slightly, the number of parts jumps to 9. Now the sphere cuts each face in a circle. This circle expands and it becomes the incircle of the face. Now one part disappears and 8 parts appear. That is, we have 16 parts, the maximum possible number.

Problem 6. It is beyond the power of visualization of a child to see the subdivisions of space by two cubes. What is the maximum number of parts?

Covering problems are the source of many instructive activities. Here are some examples. Later we will meet more.

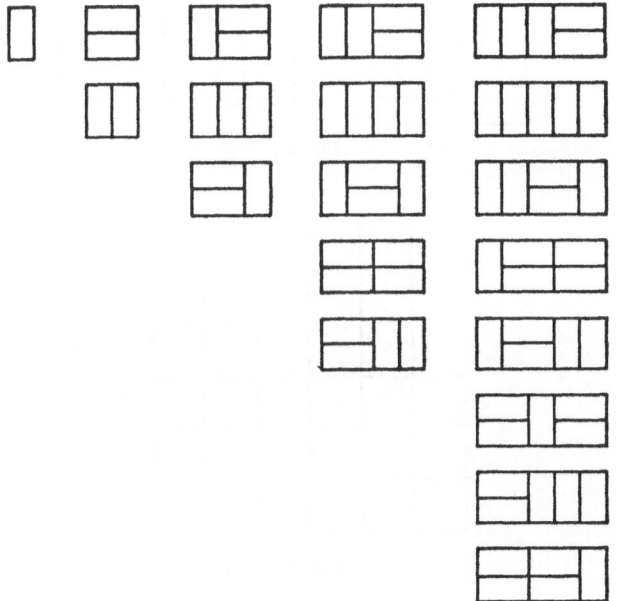

Fig. 3.

(a) Children are given a supply of 2×1 rectangles. They are challenged to pave a $2 \times n$ road and find the number p_n of distinct patterns. By trial and error they find p_1, p_2, p_3, p_4, p_5.

By now a pattern emerges. Suppose you have a 2×6 road. You can start in two ways:

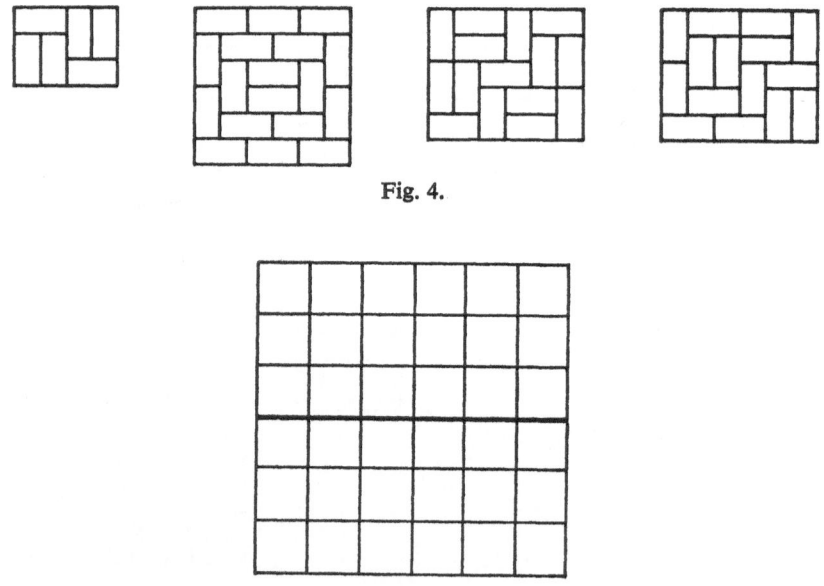

either ☐ or ⊟

In the first case, a 2×5 road remains which can be paved in p_5 ways. In the second case, a 2×4 road remains which can be paved in p_4 ways. Hence

$$p_6 = p_5 + p_4.$$

This activity does not fill one hour. When it is completed, one must have other activities in store which use the 2×1 rectangles.

Here are two possibilities: The rectangles are now bricks, and the children are challenged to build earthquake-proof walls. Figure 4 shows two unsuccessfull and two successful attempts. The first wall can slide vertically and the second can slide horizontally.

Fig. 4.

Fig. 5.

Children have more fun with two-person covering games. Two persons place alternately 2×1 rectangles on the board (Figure 5). Each rectangle must cover two squares on the same side of the heavy line in the middle. The player who makes the last move wins. There are two obvious winning strategies for the player who makes the second move. Whenever his opponent puts down a rectangle, he places a rectangle symmetrically either with respect to the heavy line or with respect to the center of the board. Children quickly recognize the strategy used by the teacher, especially if they are familiar with symmetry.

Problem 7. You have an unlimited supply of black and white 1×1 square tiles. You are to pave a $1 \times n$ road with the restriction that no two black tiles are next to each other. Let p_n be the number of distinct patterns. Find $p_1, p_2, p_3, p_4, \ldots$.

Problem 8. Figure 6 shows a tile consisting of 5 rookwise-connected unit

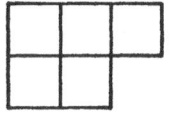

Fig. 6.

squares.* These tiles can be used to tile enlarged scale models of themselves, as is shown by Figure 7. Suppose you have an unlimited supply of congruent tiles consisting of n rookwise-connected unit squares. Prove that if they tile

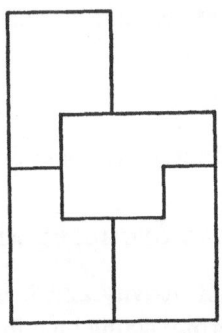

Fig. 7.

a rectangle then they also tile enlarged scale models of themselves. Suppose you can tile an $a \times b$ rectangle. This rectangle can be used to tile an $ab \times ab$ square, and the square can be used n times to cover an enlarged copy of the original tile.

Problem 9. The plane can be covered by congruent triangles, quadrangles, and centrally symmetric hexagons.

Solution: Two congruent triangles pave a parallelogram, which paves a

* Any two squares are said to be rookwise-connected if, in the game of chess, a rook can move from one to the other.

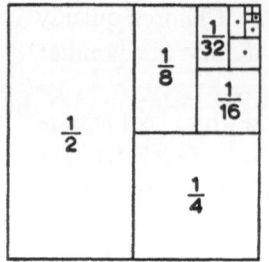

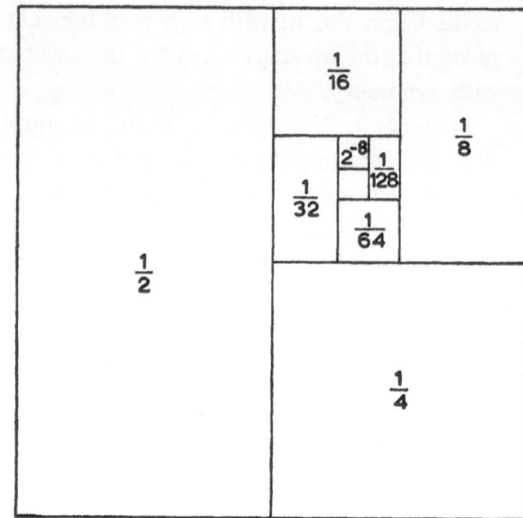

Fig. 8.

strip, which in turn paves the plane. Two congruent quadrangles pave a centrally symmetric hexagon, which paves a staircase strip, which in turn paves the plane.

Infinite tiling problems are fascinating and instructive. Take the tiling problem corresponding to

$$\tfrac{1}{2} + \tfrac{1}{4} + \tfrac{1}{8} + \tfrac{1}{16} + \tfrac{1}{32} + \cdots$$

Figure 8 shows two of infinitely many possibilities.

COMBINATIONAL GEOMETRY AND SYMMETRY

Classification is an important activity and it is accessible at an early age. The discrete case of a finite group acting on a finite set is an unlimited source of fascinating and important activities; it is also one of the best ways to develop space intuition and the idea of equivalence. The following examples will show its rich mathematical content. I will not go into applications.

(a) You have an unlimited supply of congruent sticks in two colors, black and red. How many distinguishable tetrahedra can you build?

There are 12 distinguishable tetrahedra. Figure 9 shows six. To get the remaining 6 switch colors.

Problem 10. In how many distinguishable ways can you color the vertices of a tetrahedron with (a) two (b) three colors?

Problem 11. In how many distinguishable ways can you color the faces of a cube if two colors, black and white, are available?

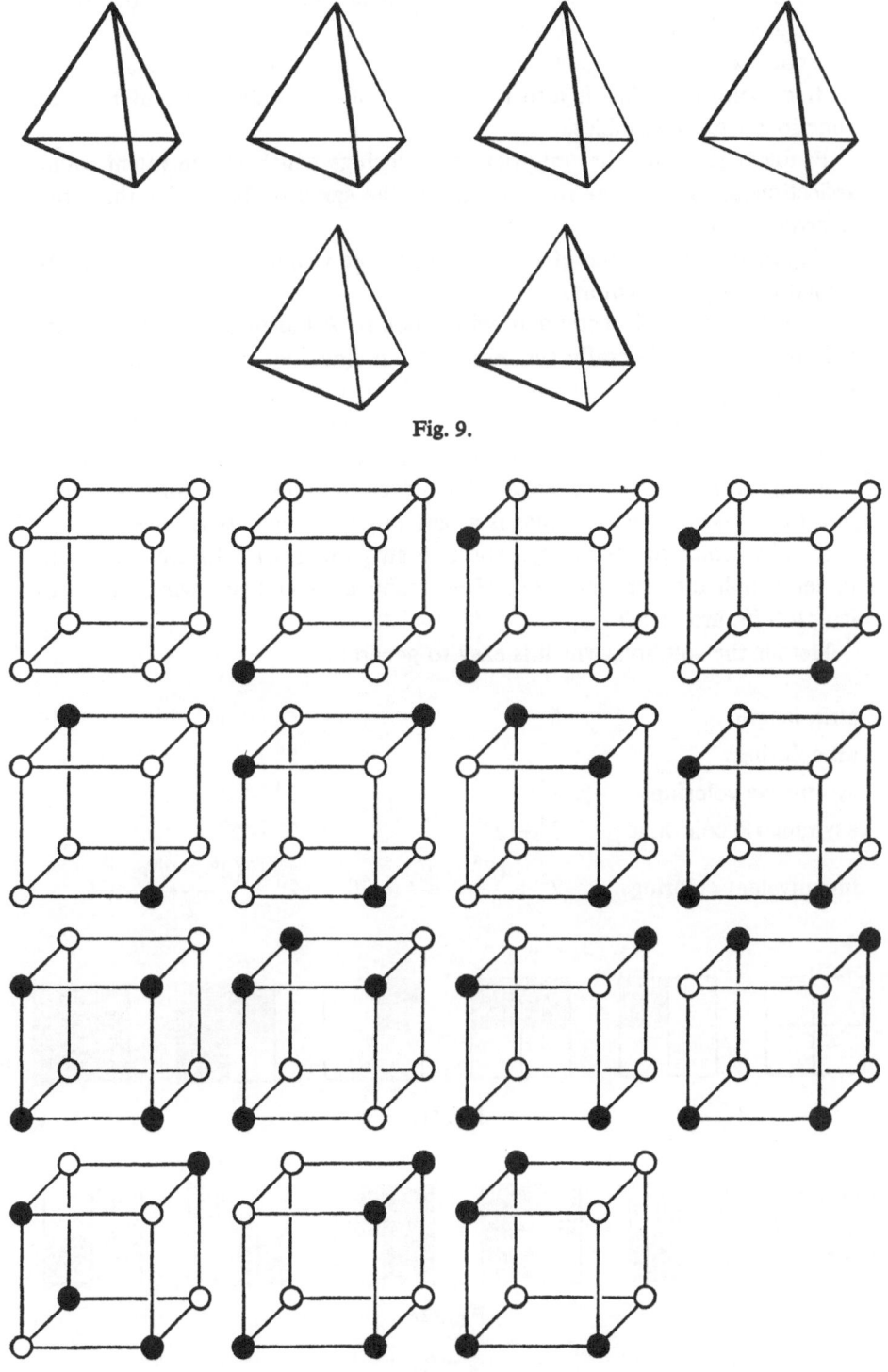

Fig. 9.

Fig. 10.

(b) There are two colors available, black and white. In how many distinguishable ways can you color the 8 vertices of a cube?

Under rotation there are 23, under rotation and reflection there are 22 distinct colorings. This is also the number of essentially different Boolean functions with 3 variables.

Problem 12. Find the only pair of colorings which is equivalent under reflection but not under rotation. Solve the same problems for the tetrahedron in (a).

Figure 10 shows some of the colorings. The remaining cases can be obtained by switching colors.

(c) Color the 2×2 chessboard with 2 colors. All distinguishable colorings (Figure 11) are bilaterally symmetric. So it does not matter which group operates on the board, the cyclic group C_4 or the dihedral group D_4.

(d) With 3 colors the situation is different. The number of distinct colorings is 24 under C_4 and 21 under D_4. There are 3 pairs of asymmetric colorings. It is a waste of time to ask children to find all 24 colorings. But to find the 6 asymmetric colorings is a perfectly good activity (Figure 12).

(e) *Two-Color Strips (Ties)*. The two strips in Figure 13 are equivalent under a half-turn or reflection. How many distinct black-white strips of length 5 (6) are there?

Here is the solution, which is easy to generalize:

strip length	5	6
all colorings	2^5	2^6
symmetric colorings	2^3	2^3
asymmetric colorings	$2^5 - 2^3$	$2^6 - 2^3$
inequivalent colorings	$2^3 + \dfrac{(2^5 - 2^3)}{2} = 20$	$2^3 + \dfrac{(2^6 - 2^3)}{2} = 36$

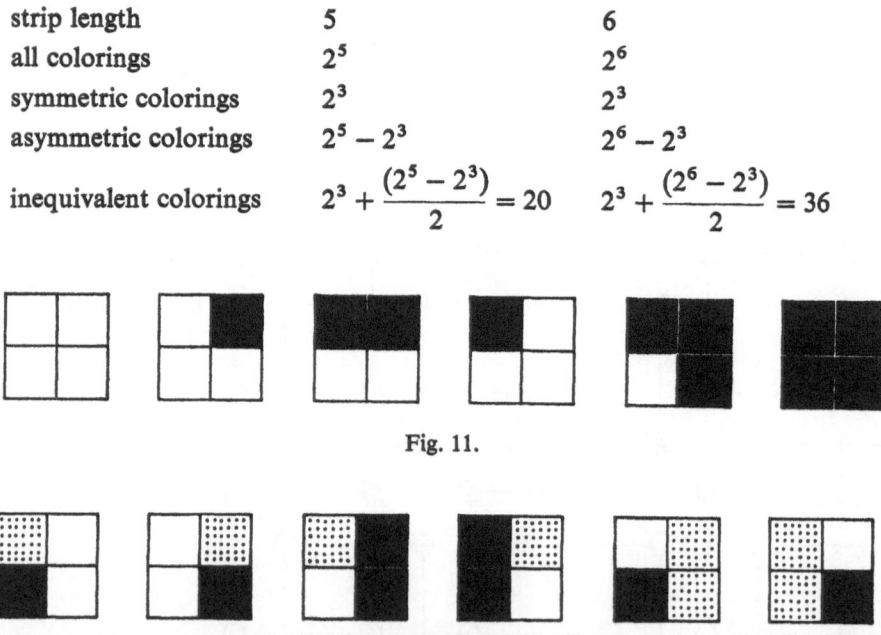

Fig. 11.

Fig. 12.

Problem 13. Show generally: The number of distinct 2-color strips of length n is $\frac{1}{2}(2^n + 2^{n/2})$ for n even, and $\frac{1}{2}(2^n + 2^{(n+1)/2})$ for n odd.

(f) Bend a strip of prime length p into a ring. In how many distinct ways can the ring be colored by two colors? (No flipping over of the ring!)

Fig. 13.

All colorings: 2^p.

Colorings symmetric under rotation: 2.

Colorings asymmetric under rotation: $2^p - 2$.

Nonequivalent colorings under rotation: $C_p = 2 + (2^p - 2)/p$.

Problem 14. Show that with a colors there are $C_p = a + (a^p - a)/p$ non-equivalent colorings under rotation. Deduce Fermat's theorem from this result.

For non-primes the formula is more complicated.

Obviously, we are dealing here with the famous *necklace* problem: You have an unlimited supply of pearls in two (a) colors. How many distinct necklaces of n pearls can you make, if a necklace may not be flipped over?

(g) *Distinguishable friezes of 0's and 1's.* Consider all periodic doubly infinite sequences of 0's and 1's. Let S_n be the set of distinct (under shift) sequences of period n. Find S_1, S_2, S_3, S_4, S_5.

$$S_1 = \{\ldots 0 \ldots, \ldots 1 \ldots\}$$
$$S_2 = \{\ldots 00 \ldots, \ldots 11 \ldots, \ldots 10 \ldots\}$$
$$S_3 = \{\ldots 000 \ldots, \ldots 111 \ldots, \ldots 010 \ldots, \ldots 101 \ldots\}$$
$$S_4 = \{\ldots 0000 \ldots, \ldots 1111 \ldots, \ldots 1010 \ldots, \ldots 1100 \ldots, \ldots 1110 \ldots,$$
$$\ldots 0001 \ldots\}$$
$$S_5 = \{\ldots 00000 \ldots, \ldots 11111 \ldots, \ldots 10000 \ldots, \ldots 11000 \ldots,$$
$$\ldots 11100 \ldots, \ldots 11110 \ldots, \ldots 10101 \ldots, \ldots 01010 \ldots\}$$

What has this problem to do with necklaces under rotation? The general formula is

$$\# S_n = \frac{1}{n} \sum_{d/n} \varphi(d)\, 2^{n/d}$$

(h) Children enjoy solving problems like this: Find all distinct necklaces with 7 white and 3 black pearls. A necklace may be rotated and flipped over. Identify the asymmetric necklaces.

This is an outstanding activity. The child gets a sheet of paper with pictures

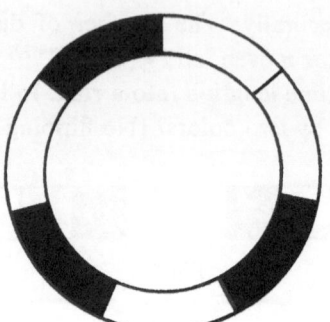

Fig. 14.

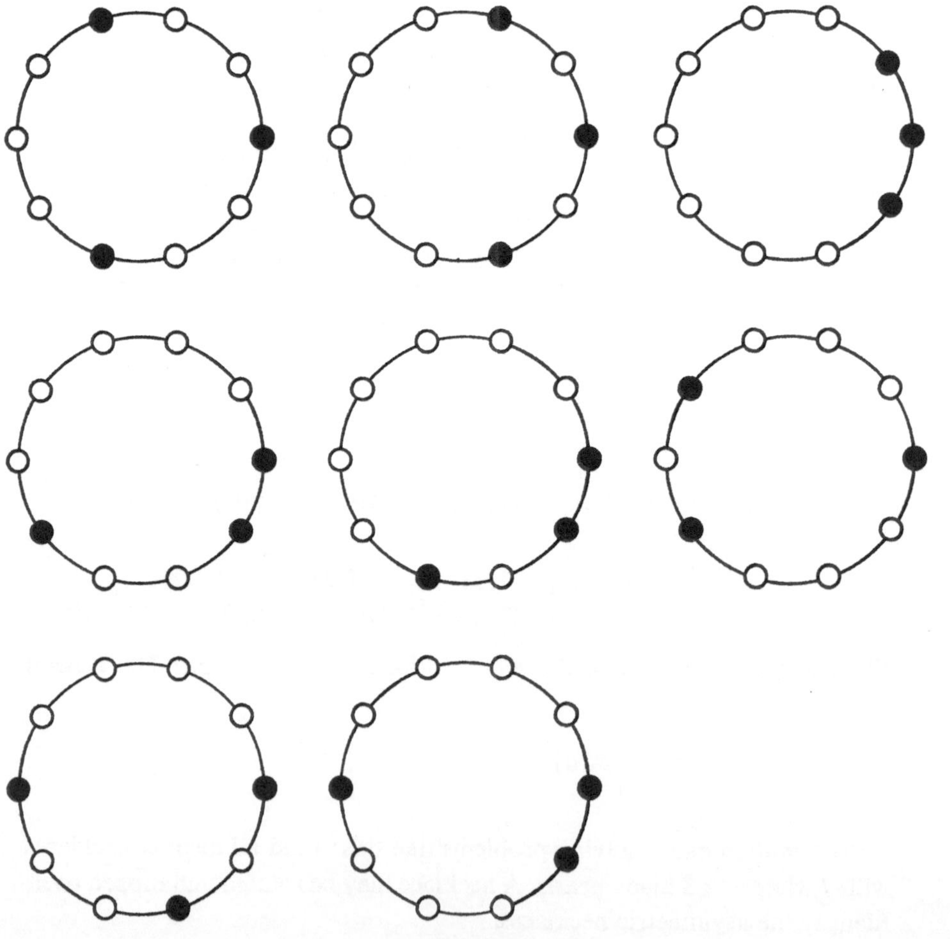

Fig. 15.

of necklaces with 10 white pearls. He colors 3 of the pearls with a black pencil. It requires a considerable effort on his part to see if two colorings are equivalent or not.

Figure 15 shows the 8 necklaces distinct under the dihedral group D_{10}.

Fig. 16.

Since 4 of the necklaces are asymmetric, there are 12 distinct necklaces under the cyclic group C_{10}.

(i) Here is another activity with a beautiful result: Take n equally spaced points on a circle and draw all distinguishable n-gons. Children find by trial and error the solution for $n=1, 2, 3, 4, 5, 6$. For $n=6$ the number of distinct polygons should be told or else they will overlook some cases. Figure 16 shows the result. For $n=6$ there are two asymmetric 6-gons.

(j) You have a cube. In how many ways can you make a die? On a die the points on opposite faces add up to 7. Place a cube in front of you. Put one dot on the top face and 6 dots on the bottom face. Next put two dots on the front face and 5 dots on the back face. Now the cube is rigid and the face for the three dots can be chosen in two ways: left or right face. Hence there are two distinct labelings of a die shown in Figure 17. Look at some dice to see if both labelings are used.

Suppose you drop the restriction that the sum of the points on opposite

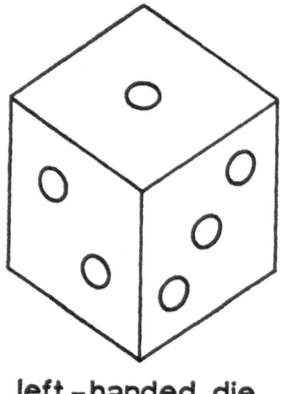
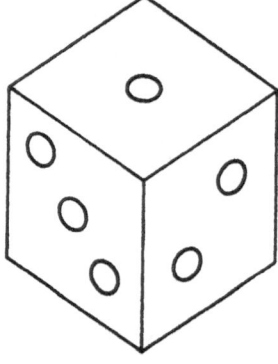

left-handed die   right-handed die

Fig. 17.

faces be 7. In how many distinguishable ways can you label (color) the 6 faces of the cube?

Rotate the cube so that 1 is in front. Face 2 is either adjacent or opposite.

If adjacent, rotate face 2 to the top. The cube is now rigid. Then all $4!=24$ different ways of labeling the remaining faces are distinguishable.

If face 2 is opposite face 1, rotate face 3 to the top. The cube is now rigid, and there are $3!=6$ ways to label the remaining 3 faces.

Hence, the cube can be labeled in $4!+3!=30$ distinct ways.

Problem 15. Two faces of a cube are to be colored black, two white, and two red. What is the number of distinct colorings?

GEOMETRICALLY DEFINED POINT SETS

Here is another class of activities. A particle moves in the plane or in space. But its movements are restricted by various strings or rigid rods. Find the set of points the particle can reach.

(a) Figure 18 shows a fence *BACDA*. A goat *G* is tied to the fence at *A* by a string of length 3. Shade and describe the area it can graze.

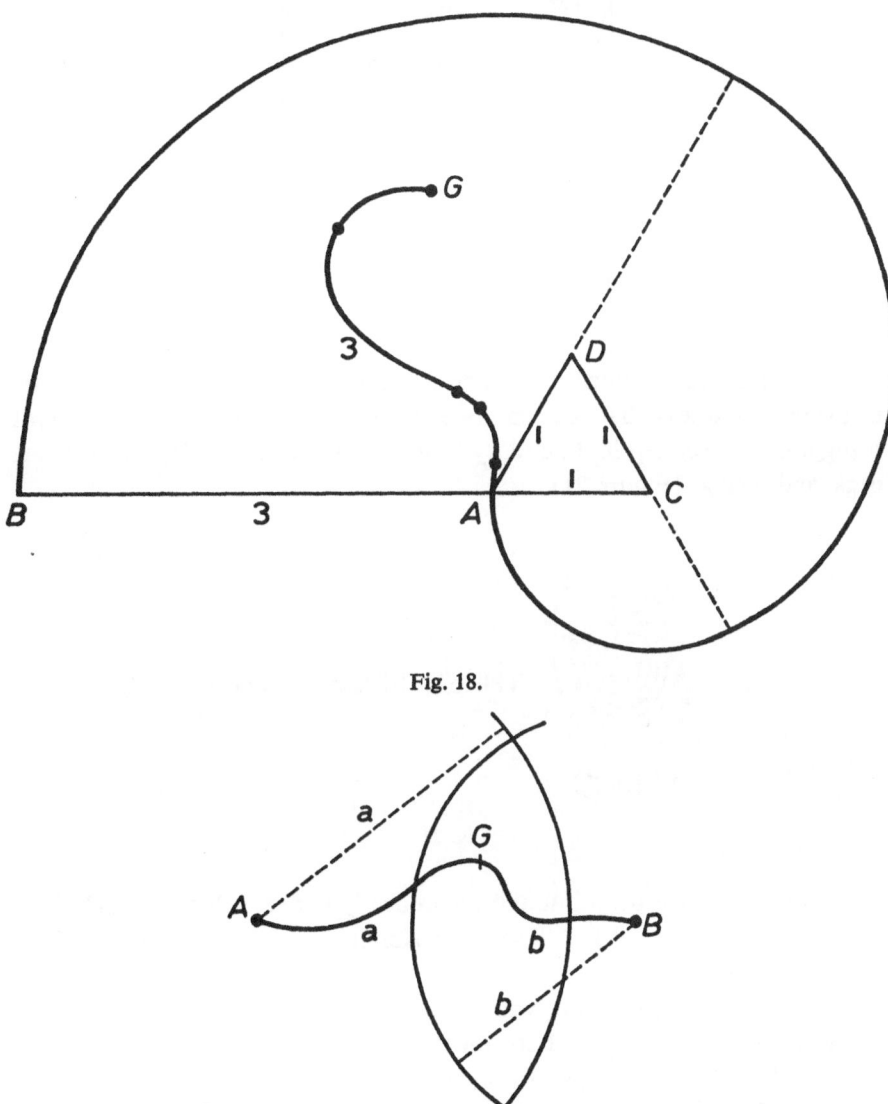

Fig. 18.

Fig. 19.

(b) A goat is tied to points A and B by ropes of lengths a and b. Shade its grazing area (Figure 19).

(c) A goat is tied to the vertices of a unit square by strings of unit length. Find the grazing area. Suppose you want it to reach $A \cup B$. Which string must be cut? How about $A \cup B \cup C$? (Figure 20).

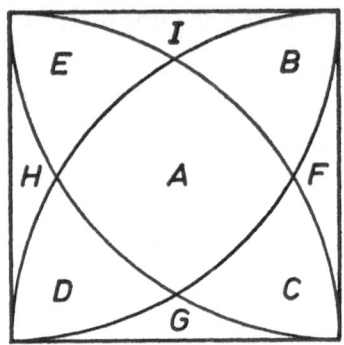

Fig. 20.

(d) A man has a semicircular lawn in front of his house. He wants to tie a cow in such a way that she can graze the entire inside of the lawn without being able to overstep its boundary. Can he do it with a string, three short pegs, and a ring? (Figure 21).

STRING   THREE PEGS   RING

Fig. 21.

Figure 22 shows a beautiful and theoretically perfect solution which does not work in practice. Why? There is a simple, practical solution which is almost perfect. Find it!

(e) A dog is tied to a fence with a leash of length c. He can crawl under the fence and run around it both on the outside and on the inside. Which points of the plane are unsafe for a cat?

Surprisingly, there is a great difference between the outside and the inside. Suppose c is increased until the inside safety zone shrinks to a point. Which

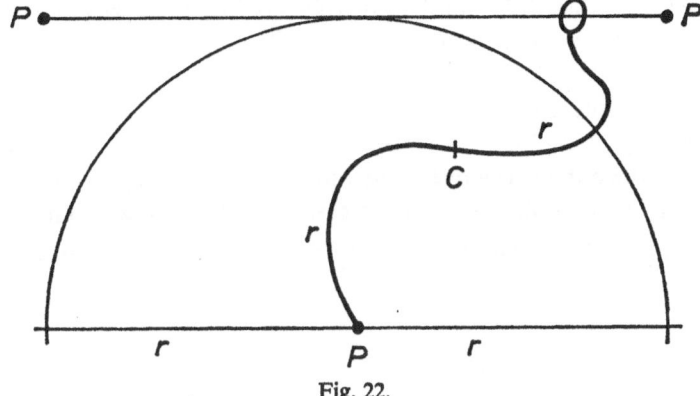

Fig. 22.

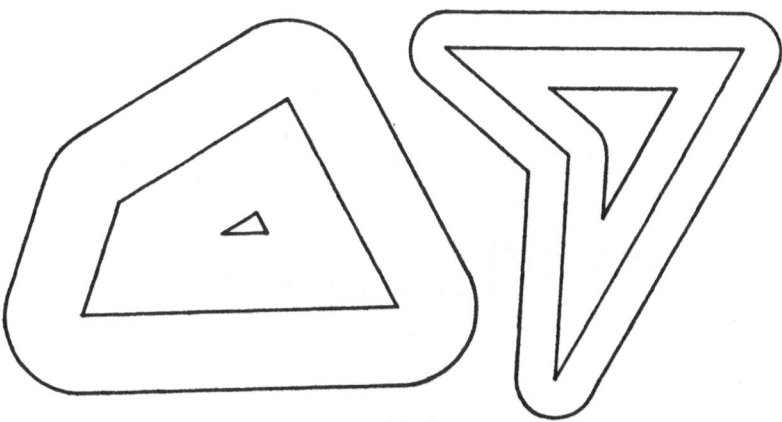

Fig. 23.

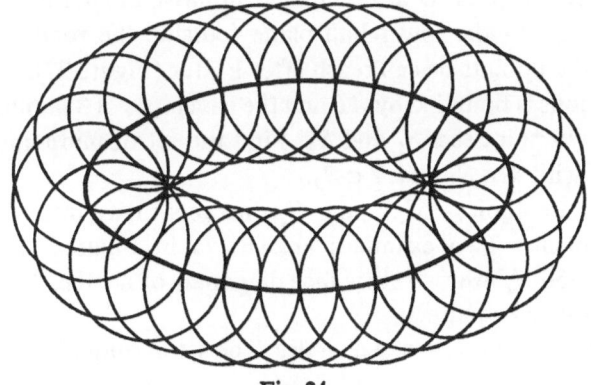

Fig. 24.

point is it? How do you construct this point? Figure 23 shows the unsafe points for a convex and a nonconvex quadrilateral.

(f) A dog is tied to an elliptical fence by a leash of length c (Figure 24). Find the set of unsafe points for a cat.

(g) A strong flying insect is leashed to the surface of a cube. The end of the leash on the cube can move freely along the surface by magnetic rollers. Describe the set of points in space it can reach (Figure 25). This is a very challenging problem in space visualization.

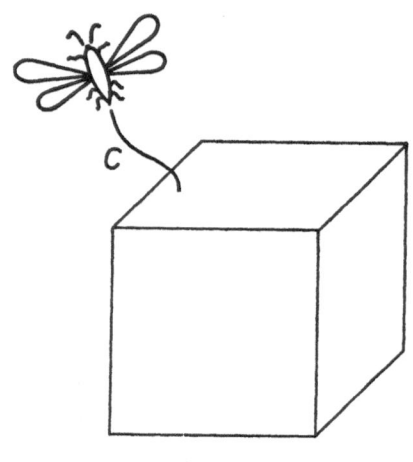

Fig. 25.

(h) Given two fixed points A and B in the plane. Find the set of all points C such that the triangle ABC has acute angles.

C must lie in the black area in Figure 26.

(i) A rider leaves point O at noon. He can move along the roads AB and OC 20 km/h and in the lower half-plane (outside the roads) at 10 km/h. Find the set of all points he can reach after 1 hour (Figure 27). The region of possible locations is bounded by AB and the envelopes of 3 families of circles.

(j) Given two point sets S_1 and S_2. Find the set of midpoints of all line segments XY with $X \in S_1$ and $Y \in S_2$.

By appropriate choices of S_1 and S_2 one gets many interesting and instructive problems. Two examples are shown in Figures 28 and 29. In Figure 28, S_1 and S_2 are two skew face diagonals of a cube. The set of midpoints is the shaded square.

In Figure 29, S_1 and S_2 are two circles. The set of midpoints is the shaded ring.

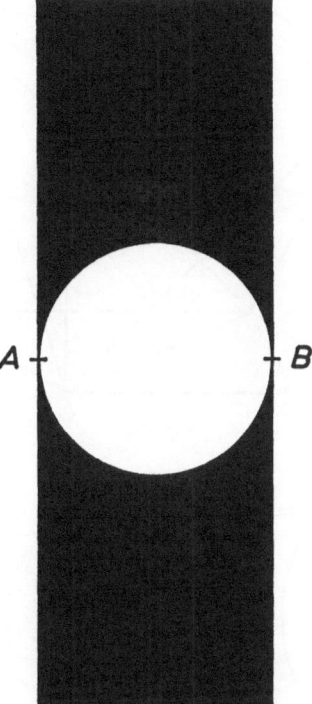

Fig. 26.

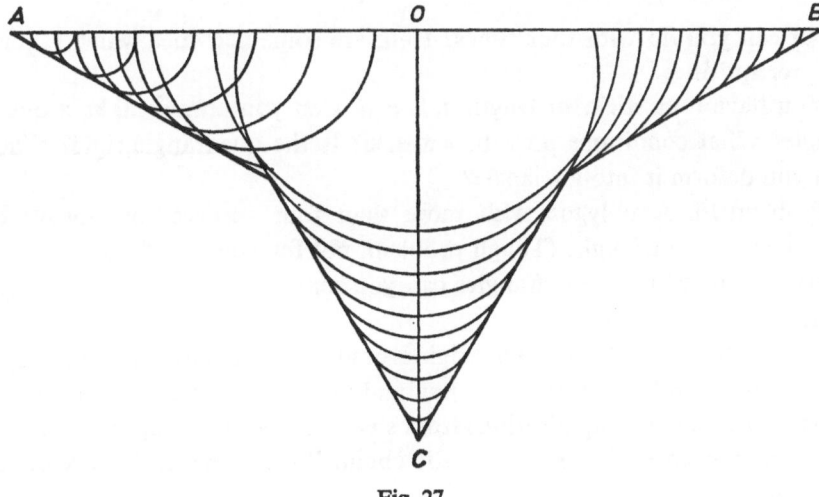

Fig. 27.

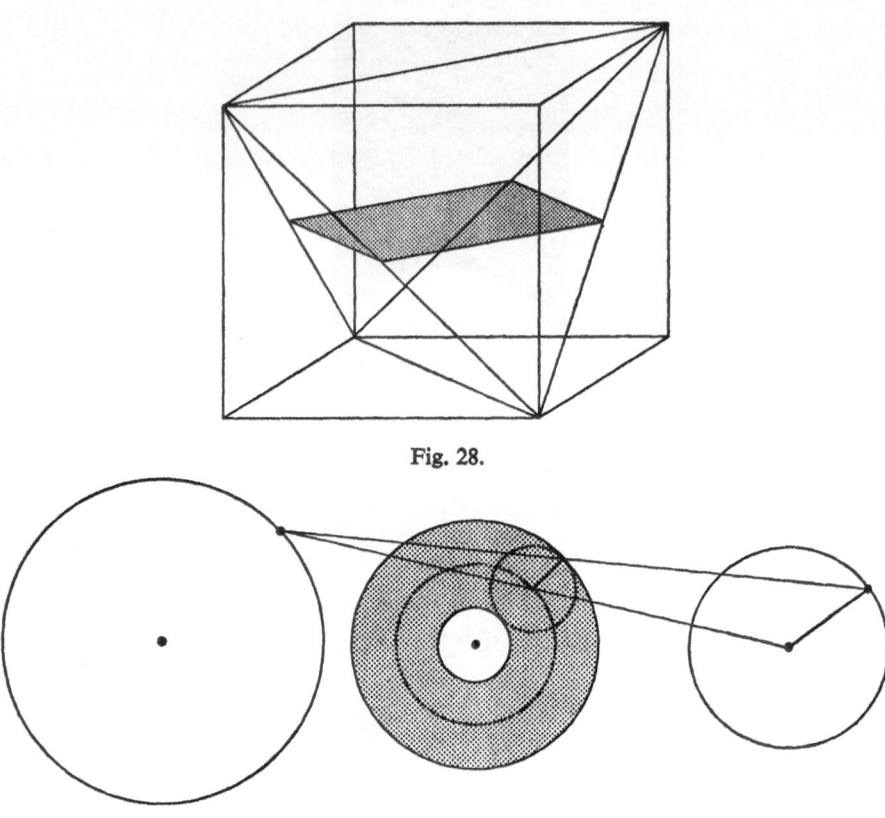

Fig. 28.

Fig. 29.

FOUR STICKS

What can you do with four sticks? Here are some activities which require no prerequisities:

You have four sticks of length a, b, c, d. Can you always make a quadrangle? What conditions must be satisfied? Is the quadrangle rigid? When can you deform it into a triangle?

Problem 16. A polygon with more than four vertices can always be deformed into a triangle. (Tough problem, not for young children).

How many distinct quadrangles can you make (a) with turning over (b) without turning over?

The fact that a quadrangle is not rigid is of utmost technological importance. The *four-bar linkage* is the simplest and most basic mechanism. It has literally thousands of applications from steam shovels to ploughs. If none of the lines is fixed we have a kinematic chain. We are free to fix any of the four links.

In Figure 30, the longest link *a* is fixed. The result is a *crank-and-rocker mechanism*. When the shortest link *d* makes a complete revolution the rocker *b* only oscillates between two extreme positions indicated in the figure. How do you find these extreme positions?

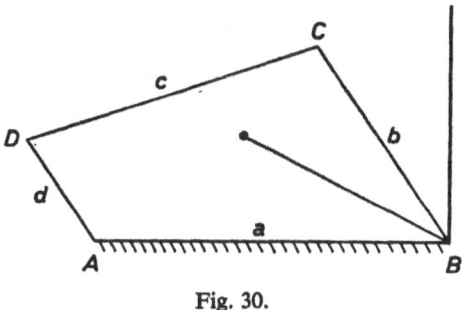

Fig. 30.

When can the shortest link *d* make a complete revolution? Show that the inequality

$$a + d \leqslant b + c$$

(shortest link + longest link ⩽ the other two links) must be satisfied.

Fixing link *c*, instead of *a*, gives a similar mechanism. By fixing link *b* opposite the shortest one you get a *double-rocker mechanism*.

Figure 31 shows two positions of the mechanism. Link *d* makes a complete

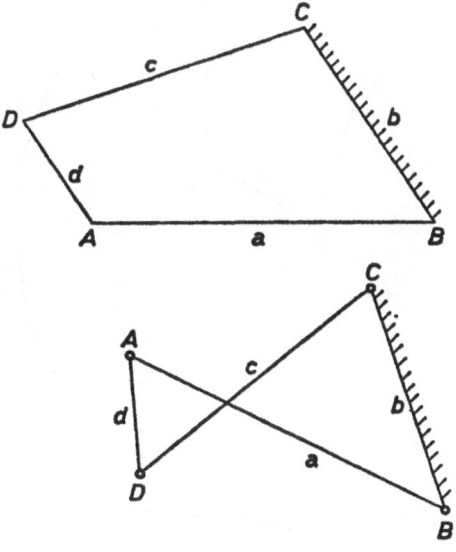

Fig. 31.

revolution during the operation of the mechanism, whereas *a* and *c* are the oscillating rockers.

Finally, by fixing the shortest link *d* we get *a double-crank linkage* (Figure 32). Links *a* and *c* make complete revolutions. What about *b*?

OPTIMIZATION

You have three sticks of lengths *a*, *b*, *c*. Can you always make a triangle? What condition must be satisfied? Here the *triangular inequality* appears for the first time, and it is used immediately to solve nontrivial problems.

(a) Six consumption centers *A*, *B*, *C*, *D*, *E*, *F* are located at the vertices of a regular hexagon. They are to be supplied from one supply depot *P*. Where should you place *P* so as to make the sum of the distances from *P* to all consumers a minimum? (Figure 33). A straightforward application of the triangular inequality gives

$$(PA + PD) + (PB + PE) + (PC + PF) > 2r + 2r + 2r = 6r$$

for each $P \neq 0$. Hence *O* is the optimal point for *P*.

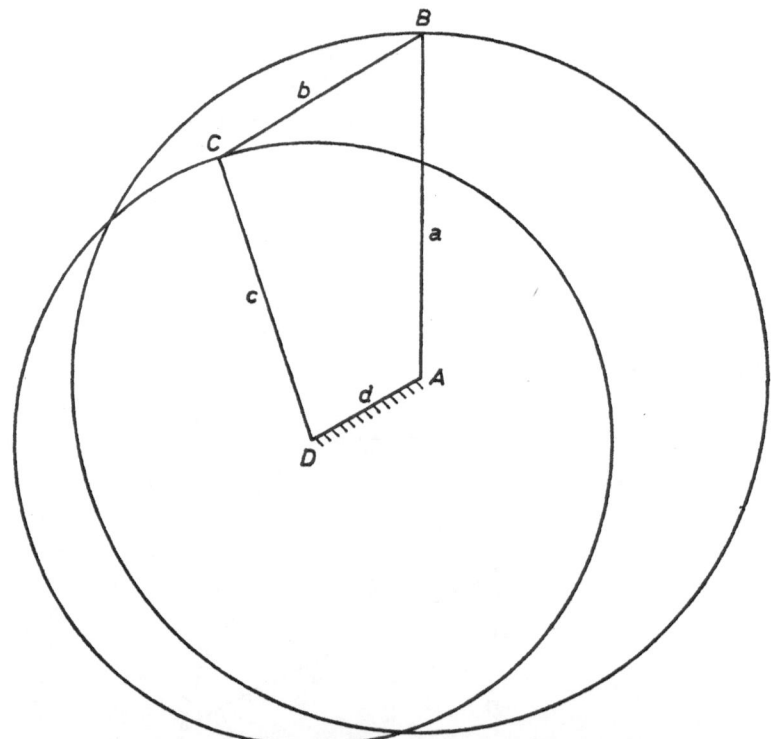

Fig. 32.

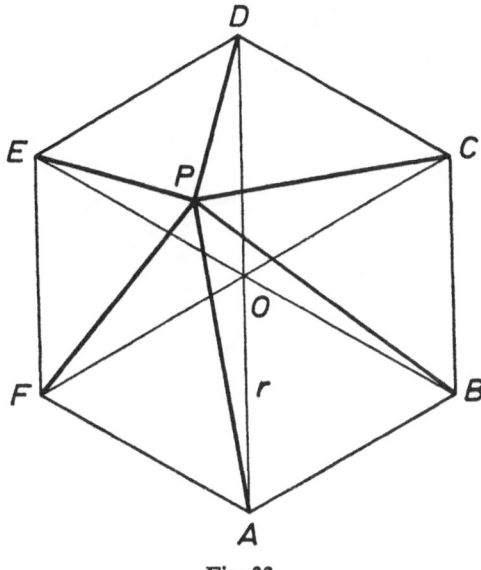

Fig. 33.

Problem 17. (Tough). Solve the same problem for a regular pentagon.

(b) Four consumption centers are the vertices of a convex quadrangle *ABCD*. Find the optimal location for a supply depot. Two applications of the triangular inequality show immediately that the point *O* of intersection of the diagonals is the optimal location (Figure 34).

But suppose *A*, *B*, *C*, *D* are not the vertices of a convex polygon. Suppose *D* lies inside the triangle *ABC*. First show, by two applications of the triangular inequality to Figure 35, that

$$a + b > c + d.$$

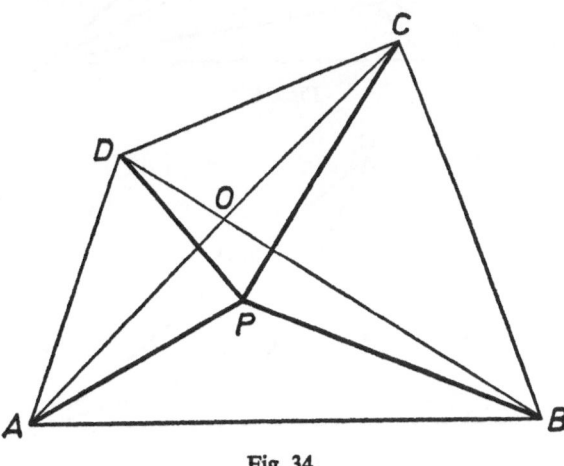

Fig. 34.

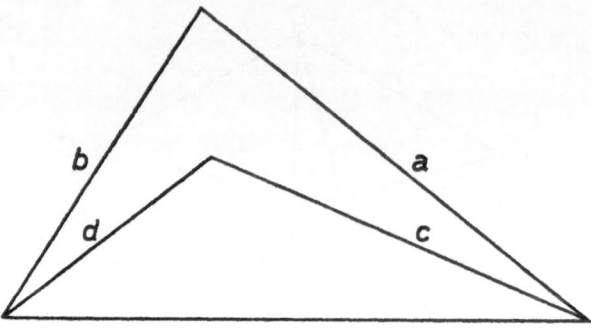

Fig. 35.

Now look at Figure 36. By the previous result, $BP + CP > BD + CD$, and by the triangular inequality

$$AP + DP > AD.$$

By adding the inequalities we get

$$PA + PB + PC + PD > DA + DB + DC$$

Hence, D is the optimal location.

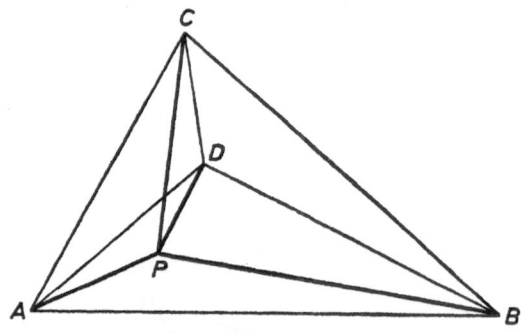

Fig. 36.

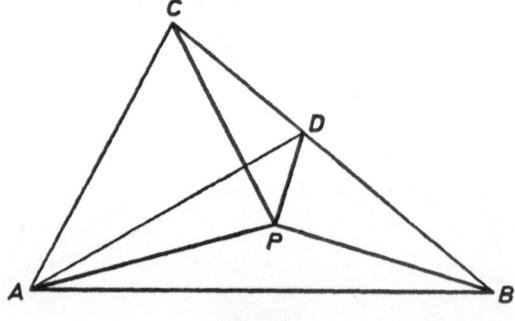

Fig. 37.

Suppose D lies on a side of the triangle ABC (Figure 37). We have immediately

$$\frac{\begin{aligned}PA + PD &> DA\\ PB + PC &> DB + DC\end{aligned}}{PA + PB + PC + PD > DA + DB + DC}.$$

This means D is again the optimal location.

(c) But what if the points A, B, C, D all lie on a straight line? This leads to a highly interesting problem which can be solved completely for any number of points: n friends live at $x_1 < x_2 < x_3 < \ldots < x_n$ on the same street. Find a meeting place so that the total distance travelled is minimal.

For $n=2$ any point $x \in [x_1, x_2]$ will give minimum distance $x_2 - x_1$. Let $n=3$. For x_1 and x_3 any point in $[x_1, x_3]$ will do. Of these points x_2 is optimal for x_2 itself. Hence x_2 is the optimal point.

Let $n=4$. For x_1, x_4 any point in $[x_1, x_4]$ will do. For x_2, x_3 any point in $[x_2, x_3]$ will do. Hence, any point x in $[x_1, x_4] \cap [x_2, x_3] = [x_2, x_3]$ is an optimal meeting place (Figure 38).

Generally: for n even, any point in the innermost interval $[x_{n/2}, x_{n/2+1}]$ is optimal. For n odd, the innermost point $x_{(n+1)/2}$ is the optimal point.

Let us generalize further: At x_1, x_2, x_3, x_4, x_5 live 20, 50, 70, 80, 100 people, respectively (Figure 39). Find the optimal meeting place.

(a)
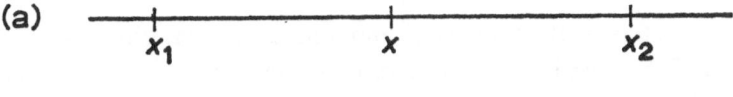

(b)
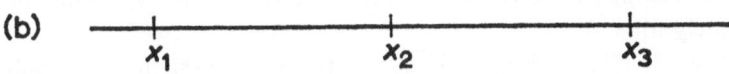

(c)
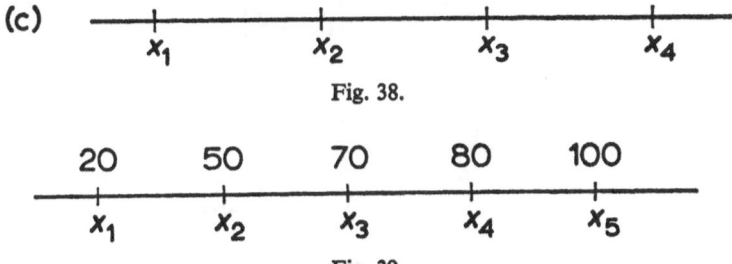

Fig. 38.

Fig. 39.

We start with a trial point on the line and decide by majority voting whether this point should be moved to the right or to the left. The optimal point is x_4, since a 180:140 vote is against moving it to the left and a 220:100 vote is against moving it to the right.

Let us generalize further to two dimensions. Figure 40 shows 7 con-

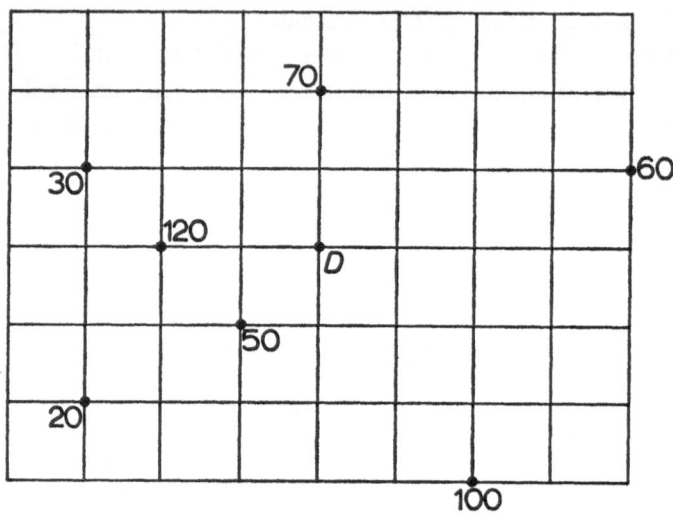

Fig. 40.

sumption centers with daily consumptions indicated. Find the optimal location D for a supply depot. Again, by majority voting, the best location D is found.

The last problem shows that in the taxicab metric many optimization problems become trivial. In the Euclidean metric the corresponding problem is extremely difficult.

Problem 18. (Easy). Figure 41 shows the street system of a small town

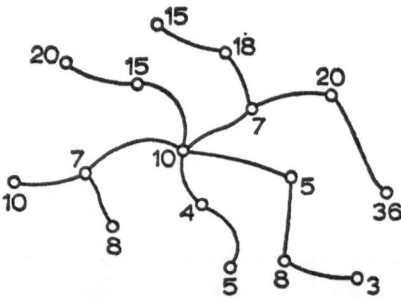

Fig. 41.

together with the consumption centers and their daily demand. Find the optimal location for a supply depot.

(d) Now suppose the cost of travel is the square of the distance travelled. Find the optimal location of a meeting place. For two points one sees by geometric insight that the mid-point is best (Figure 42).

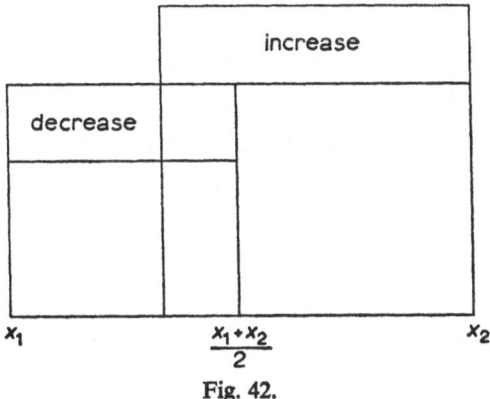

Fig. 42.

In general it is the centroid. A proof is not possible at this stage. But one can check by numerical cases. For instance, take $x_1=2$, $x_2=3$, $x_3=7$.

Location of meeting place	2	3	4	5	6
Cost of travel		26	17	14	17	26

At the mean $\bar{x}=(2+3+7)/3=4$ the cost is a minimum.

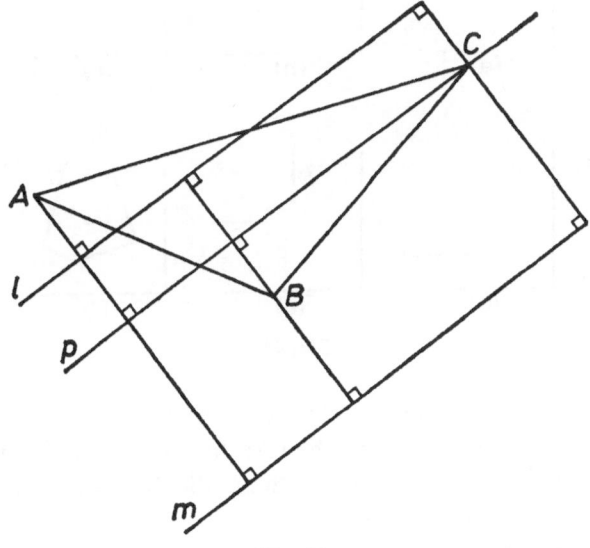

Fig. 43.

(e) *A, B, C* are three oil wells. A pipeline *p* is to be laid with prescribed direction parallel to line *m* (Figure 43). The sum of the distances from *A, B, C* to *p* should be as small as possible.

The optimal location *p* of the pipeline can be found again by a majority vote of the oil well proprietors.

Problem 18. For what directions of line *m* does the best pipeline pass through *A, B, C*?

Problem 19. Suppose now that the direction of *p* is not fixed. What is the best location for *p*?

MINIMAX AND MAXIMIN

These activities are inspired by game theory and linear programming.

(a) Consider the points of the plane with positive coordinates. Choose a closed subset *S* of these points and tell the children: You may choose any point $(x, y) \in S$. The smaller of the two numbers *x, y* is your gain. Which points of *S* are optimal for you? That is, the student is challenged to find a point of *S* with the smaller coordinate as large as possible:

$$\max_{(x, y) \in S} \min(x, y).$$

The game is played with different choices of *S* (Figures 44a–f) until the

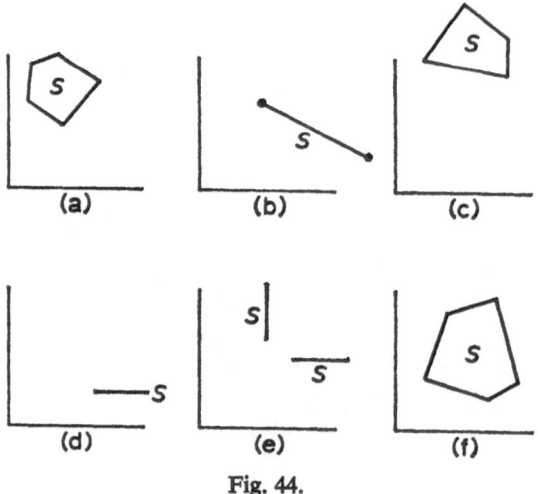

Fig. 44.

children come up with a simple algorithm for finding the optimal points: Take a right angle with vertex on the first diagonal $y = x$, like in Figure 45. Translate it in the direction of the origin parallel to $y = x$ until it hits *S*. The points hit first are the optimal points.

(b) Again, the child chooses a point $(x, y) \in S$. But this time the larger

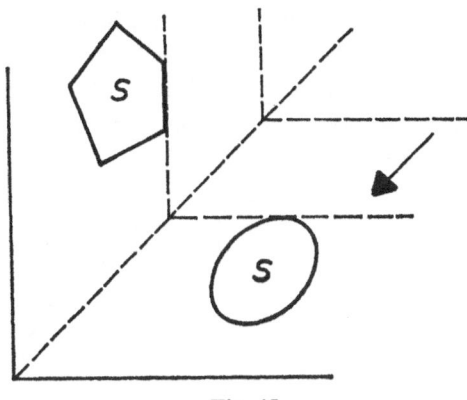

Fig. 45.

of the two number x, y is his loss. So he tries to make max (x, y) as small as possible. He is looking for $\min_{(x,y) \in S} \max(x, y)$.

Problem 20. Give a simple algorithm for finding the optimal points.

(c) Children choose a point $(x, y) \in S$. Their gain (loss) is the difference $|x-y|$. Find the optimal points.

(d) Children choose a point $(x, y) \in S$. Their gain (loss) is the sum $x+y$. Find the optimal points.

(e) Two children play. The first player chooses a point $P \in S$, his opponent knowing the choice. The second player chooses a point $Q \in S$. The first player forfeits to the second the distance PQ. What are the best choices for the first player? How does the second player find his optimal choices?

(f) Suppose S is a circular disc with radius 1. Each of two children chooses a point of S. But this time each child is ignorant of the other's choice. The first player pays to the second the distance of the two chosen points.

The first player can avoid losing more than 1 per game by always choosing the center of the disc. And he will lose more, on the average, if he chooses any other point P. The second player needs only choose a diameter HT at random. Then he tosses a coin. If head comes up his choice is H and if tail comes up his choice is T. His expected gain is $\frac{1}{2}(HP+PT)$. But $HP+PT \geq \geq HT=2$.

Hence, the expected gain is ≥ 1. The equality sign is valid only if P happens to lie on HT. The probability for this is 1 if P coincides with the center of the disc and is 0 otherwise (Figure 46).

IMPOSSIBILITY PROOFS

There are many crackpots who try all their life to do something which can be proved to be impossible. Angle trisectors will never die out. For this reason a child should meet convincing impossibility proofs early in his life. Best suited for this level are parity proofs, and coloring proofs from combinatorial geometry. These proofs require no prerequisites; they are short, elegant, and totally convincing.

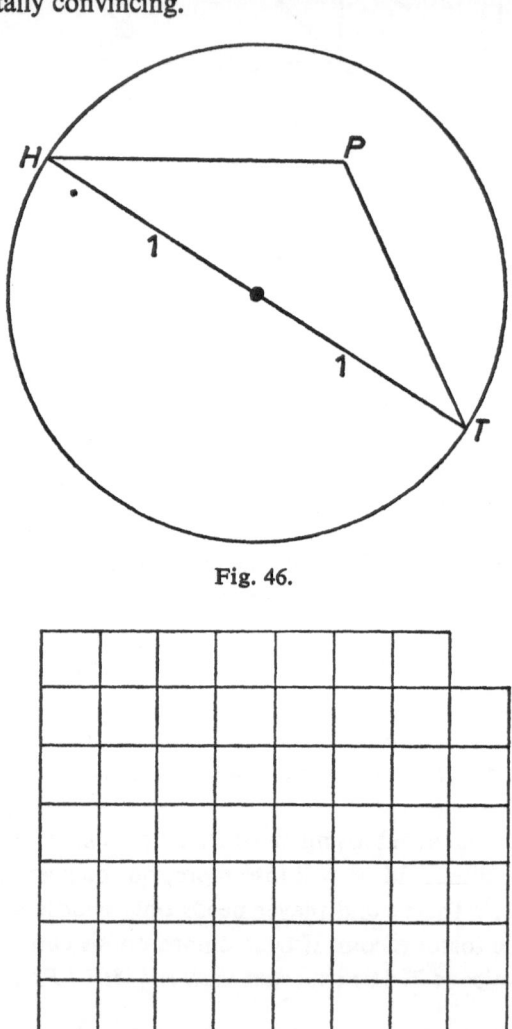

Fig. 46.

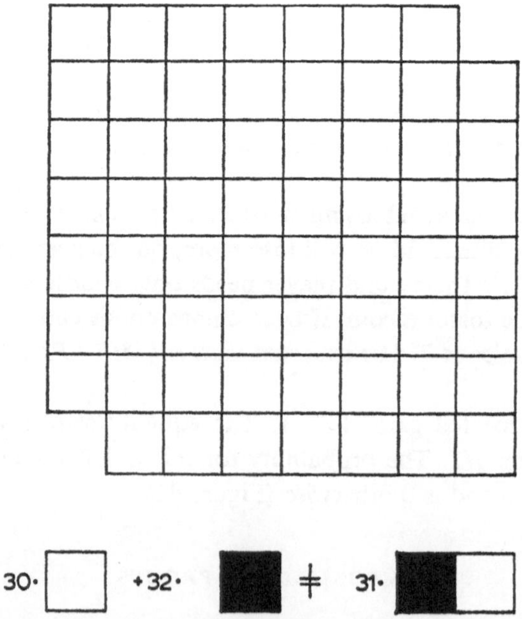

Fig. 47.

Since this is a well known subject, I only mention three examples of increasing difficulty. It is easy to devise any number of similar examples.

(a) The classical example is the following: Can you cover the 62 squares of the mutilated chessboard in Figure 47 by 31 dominoes, each covering 2 adjacent squares of the board? This is impossible and the concise proof is summarized in one line below the board.

(b) It is easy to cover an 8×8 chessboard (Figure 49) with 16 T-tetrominoes (Figure 48b). But can you cover the chessboard with one square tetromino (Figure 48a) and 15 T-tetrominoes?

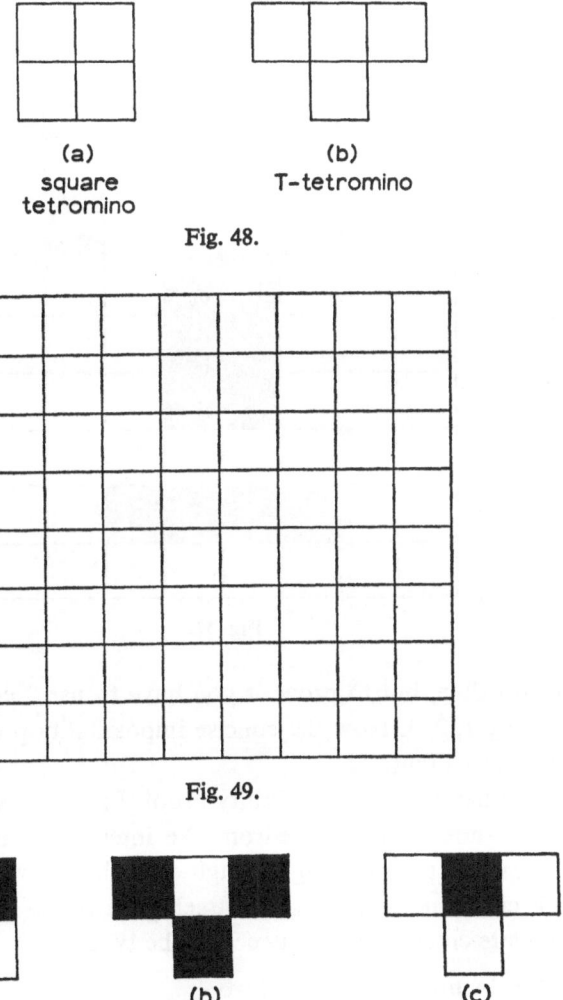

(a)
square
tetromino

(b)
T-tetromino

Fig. 48.

Fig. 49.

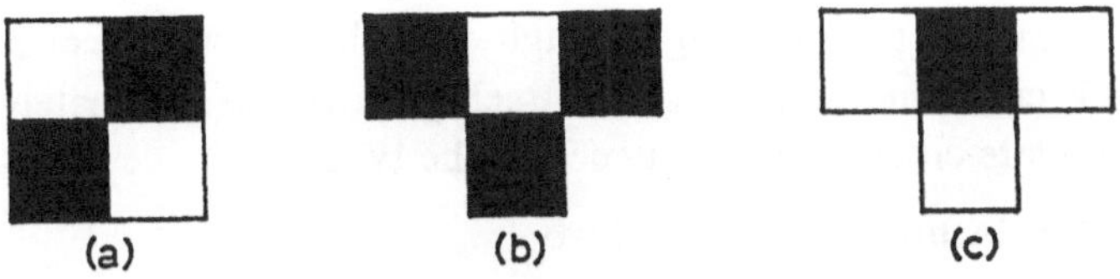

(a)   (b)   (c)

Fig. 50.

This is not possible. The standard chessboard has 32 black squares. The square tetromino covers two of these (Figure 50a). The 30 remaining black squares cannot be covered by 15 T-tetrominoes. For each tetromino covers either 3 black squares or 1 black square (Figures 50b, c). But a sum of 15 odd numbers (3's or 1's) is an odd number, and cannot be equal to the even number 30.

(c) Look at the funny face in Figure 52. Can you cover its 72 white squares with 24 tetrominoes of the type shown in Figure 51?

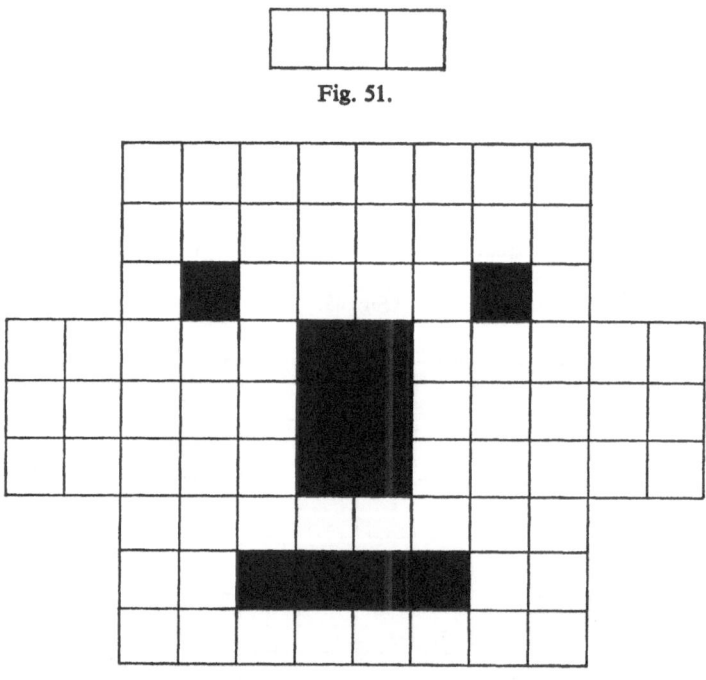

Fig. 51.

Fig. 52.

It is not possible, but to prove it you have to use 3 colors. Let us label the 3 colors by 1, 2, 3. Now the concise impossibility proof is summarized in one line below Figure 53.

Here is another classic impossibility proof. Figure 54 shows the Schlegel diagram of a rhombic dodecahedron. We interpret it as a road map for 14 cities. Is there a path passing through each city exactly once? No! There are 6 black cities and 8 white cities. Each path passes alternately through black and white cities. But a sequence of the type

bwbwbw...   or   wbwbwb...

cannot contain 6 b's and 8 w's.

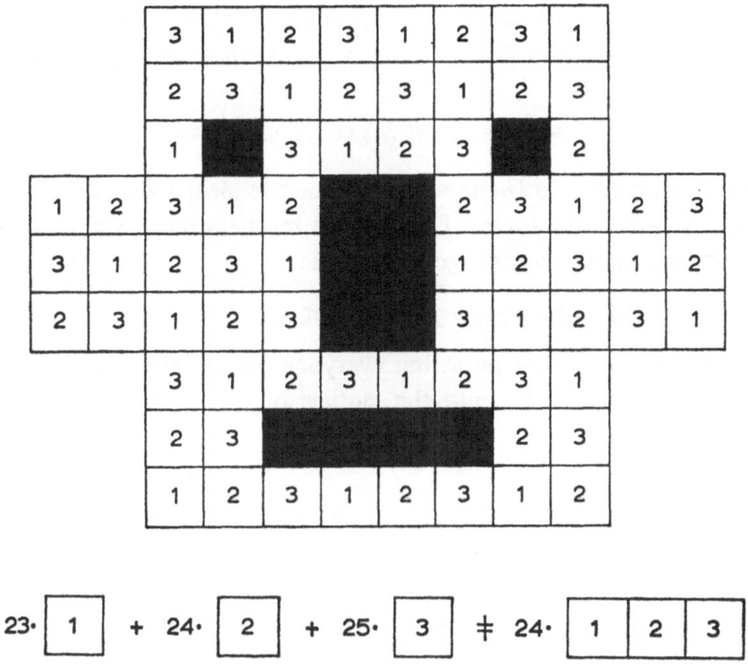

		3	1	2	3	1	2	3	1		
		2	3	1	2	3	1	2	3		
		1	■	3	1	2	3	■	2		
1	2	3	1	2	■	■	2	3	1	2	3
3	1	2	3	1	■	■	1	2	3	1	2
2	3	1	2	3	■	■	3	1	2	3	1
		3	1	2	3	1	2	3	1		
		2	3	■	■	■	■	2	3		
		1	2	3	1	2	3	1	2		

$$23 \cdot \boxed{1} \; + \; 24 \cdot \boxed{2} \; + \; 25 \cdot \boxed{3} \; \neq \; 24 \cdot \boxed{1\ \ 2\ \ 3}$$

Fig. 53.

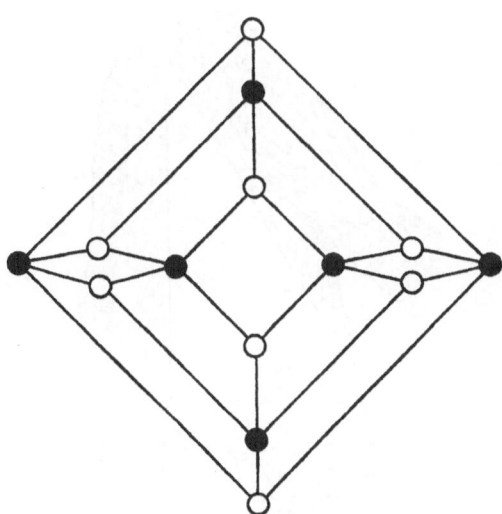

Fig. 54.

Problem 21. Our last impossibility proof is of an entirely different kind. Figure 55 shows a space polygon. Show that such a space polygon cannot exist.

PROBLEMS IN A STORY SETTING

A teacher should dramatize important geometric ideas by some story which the children will never forget. Experienced teachers make extensive use of this effective device. Here is a good example:

The Tragic Mistake of the Poor Tailor of Sikinia

In Sikinia people are very poor, but everyone owns a ferocious dog. These dogs tear triangular holes into the clothes of passers-by. One expensive

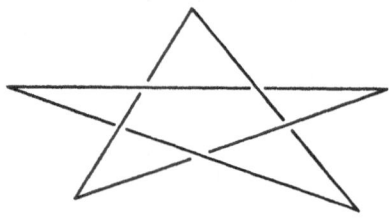

Fig. 55.

breed of dogs tears holes in the shape of perfect squares. The victims don't throw their clothes into trash cans. They go to a tailor to mend them. There

Fig. 56. Mink coat.

was a poor tailor who was making a living patching up holes. When the hole was a square he cut a square piece of cloth. To test it he would fold it along each diagonal. Is this a good test? Is there a folding test for a square?

The secret dream of our poor tailor was to become rich by mending mink coats. One day he had his big chance. A lady came with a mink coat which had a huge triangular hole on the back (Figure 56). Our poor tailor had never mended furs before, but only regular cloth. And he made a tragic mistake. On mink, hair grows on one side only. The other side is clean shaven. You cannot turn it over like cloth which looks the same on both sides. But our poor tailor had to learn this the hard way. He cut a patch to fit the hole, but it fit only on the wrong side (Figure 57). What to do now? How can we help our poor tailor?

Fig. 57. A part to mend the coat. It fits in the wrong way.

Symmetric pieces can be turned over. So we must cut up the patch into symmetric pieces. But you should do it in as few cuts as possible. Two cuts are always sufficient, as is shown by Figure 58.

One cut is sufficient for the two shapes in Figures 59 and 60. Can you find the cut?

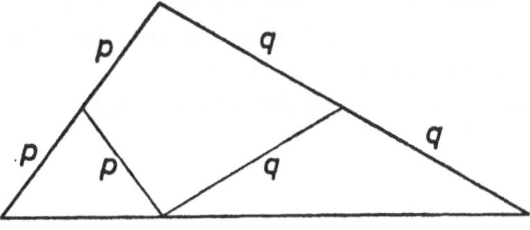

Fig. 58.

Sharing a Pie

Cain

There was once a couple, Adam and Eve. They had a son, Cain. To develop Cain's intellect Adam liked to pose problems. One day he showed Cain a triangular pie Π, and he said: You may choose any point O in the plane of Π and reflect through O to Π'. The intersection $\Pi \cap \Pi'$ is yours.

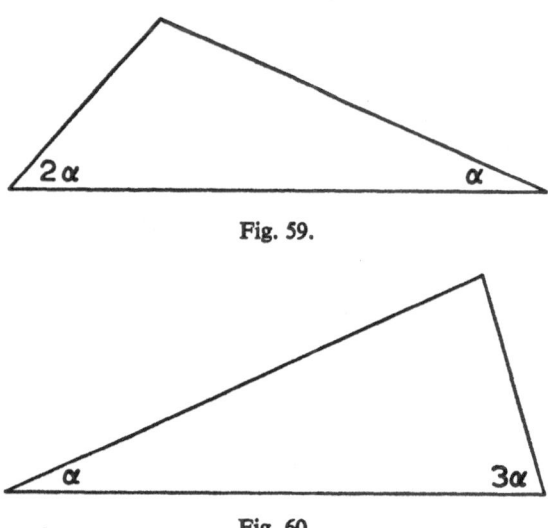

Fig. 59.

Fig. 60.

This is a rich geometrical situation: the children are confronted with a battery of about 30 geometrical problems arising from it. Here are some:

(a) Cain got nothing. Where did he choose O?

(b) Cain got a piece in the shape of a parallelogram. Where did he choose O?

(c) Cain got a hexagon. Where did he choose O? What shape is the hexagon?

(d) Can he get a triangular piece of pie?

(e) What else can he get?

(f) Where should he choose O to get as much as possible?

(g) For what shapes of the pie can Cain get all of it?

(h) The pie was a polygon and he managed to get all of it. Deduce as many properties of the polygon as you can.

These problems, with the exception of (f), require no prerequisites for their solution.

Solutions

(a) He chose O outside Π (Figure 61a).

(b) He chose O inside Π, but outside or on the triangle of midpoints (Figures 61b, d).

(c) He chose O inside the triangle of midpoints (Figure 61c).

(d) No! He cannot get a triangle. $\Pi \cap \Pi'$ has O as its center of symmetry. But a triangle cannot have a center of symmetry. Why not?

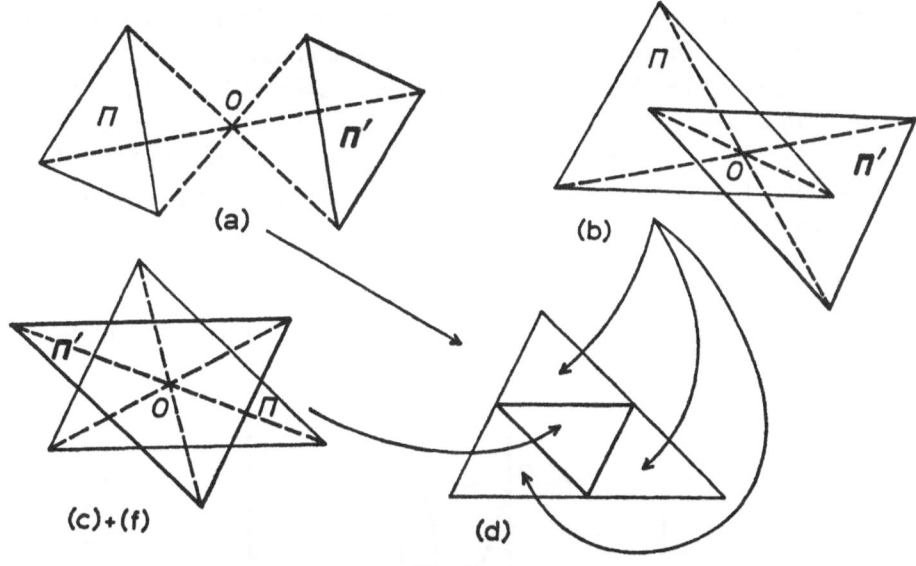

Fig. 61.

(e) $\Pi \cap \Pi'$ can be a hexagon, a parallelogram, a line segment, or a point.

(f) To get as much as possible ($\frac{2}{3}$ of the whole pie), O must coincide with the centroid. This cannot be proved at this stage. It is a question for empirical exploration.

(g) Obviously Π must have a center of symmetry. Cain should choose this center as his point O.

(h) The polygon has a center of symmetry. Hence, it has an even number of sides. Opposite sides are equal and parallel.

Cain and Abel

After some time Abel arrived, a very clever boy. Adam liked him very much and gave him preferential treatment. This behavior is dangerous and it will eventually end in a tragedy.

One day Adam showed the two brothers a pie. And he said to Abel: You may cut off a piece as big as you can. The remainder is for Cain. But your cut must be straight and it must go through a point O which is chosen by Cain.

Around this geometrical situation a whole battery of geometrical problems arises. Here are some with solutions:

(a) Name a class of fair pies. Pies with a center of symmetry are fair if Cain is smart enough to place O in the center.

(b) Has every fair pie a center of symmetry? Yes! But the proof is not so simple and can be postponed for higher grades.

(c) Does there exist a shape which is favorable to Cain? This is a trivial question for us, but children often have to think about it.

(d) Suppose Abel does not want to take advantage of his brother's blunders. Can he always be fair for every shape of the pie, and for every choice of point O? Yes, the existence of a fair cut through every point of the plane is obvious, but its construction is not obvious. A continuity argument is involved here.

(e) The pie is *rectangular* and Cain failed to choose the center. Which cut is most favorable to Abel?

Reflect the pie Π at O to Π' (Figure 62). Then it becomes completely

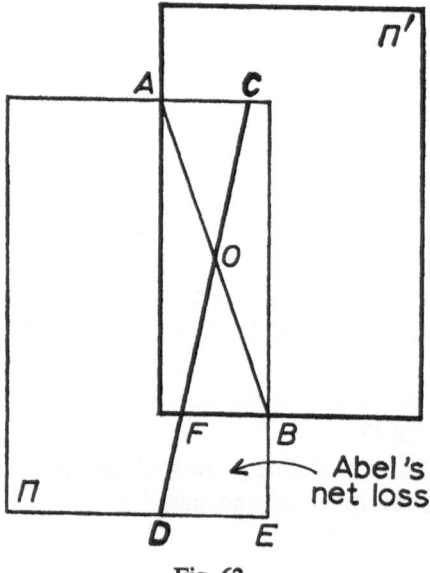

Fig. 62.

obvious that the cut AB, which is bisected at O, is best for Abel. Indeed, take another cut, say CD. This cut results in a net loss to Abel of the piece $DEBF$.

(f) The pie is *circular* and Cain failed to choose the center. Find the optimal cut for Abel.

Reflect Π at O to Π'. Again the cut AB, which is bisected at O, is optimal for Abel. Another cut *CD* leads to a net loss for Abel equal to the shaded area. (Figure 63).

(g) The method used in (e) and (f) works for any *convex pie*. The best cut has its center at *O*, and it is easy to construct by reflecting Π at *O*.

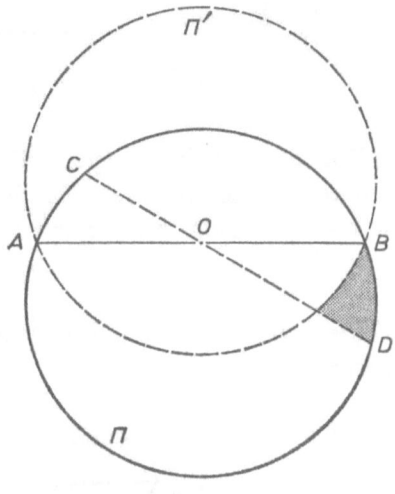

Fig. 63.

(h) For *nonconvex pies* complications arise. In Figure 64 there are 3 cuts bisected at *O*. Which one is the best for Abel?

Turn the cut around *O* and make a balance of gains and losses.

(i) One *square pie* is fair. Cain chooses the center. Two congruent square pies, as in Figure 65, are also fair, since Cain can choose the point *O* which

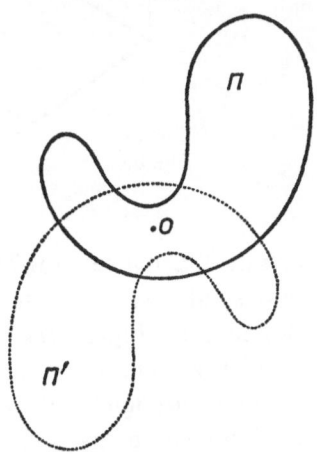

Fig. 64.

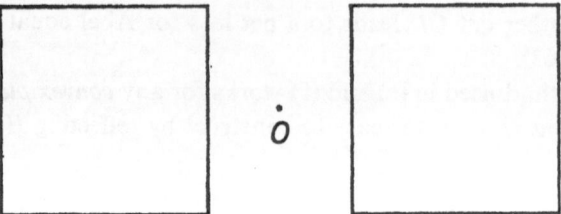

Fig. 65.

guarantees him one half of the pies. But suppose Abel may translate one of the pies before Cain chooses the point D. Can he increase his share? What if he may rotate a pie? Investigate the case where the second pie in Figure 65 shrinks to a smaller size. Study the case of two unequal circular pies.

(k) *Pies in the sky*. Three different pies are floating in the sky. They have the shape of boxes (Figure 66). Abel wants to bisect all three simultaneously

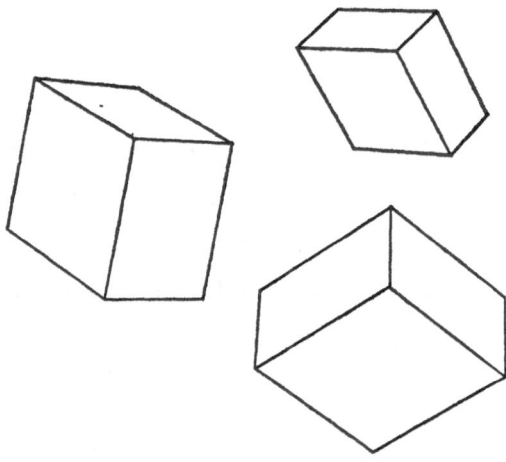

Fig. 66.

by a plane cut. Can he always do it, and if yes, how? For what locations of the pies are there infinitely many solutions? If the problem causes trouble, one should solve its plane analog first (see Figure 67). The solution is extremely simple and elegant. Each box has a center of symmetry. Each plane through its center bisects a box. The plane through the three centers bisects all boxes simultaneously. There are infinitely many solutions if the centers are collinear.

Abel runs for his life

In Figure 68 Abel is at *A* and Cain is 100 yards due North at *C*. Cain chases Abel. Both are running at the same speed. Abel is running East. Cain's path is more complicated. Every 15 yards he locates Abel's current position and runs in the direction of that position. In the long run he will

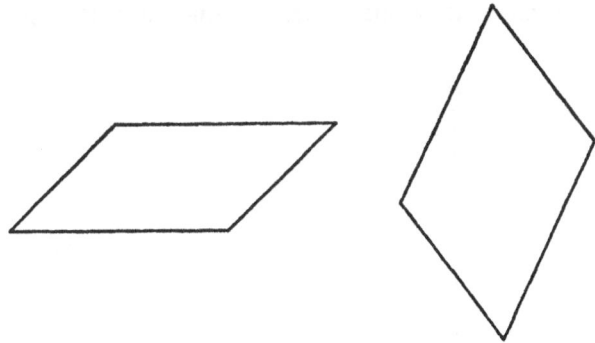

Fig. 67.

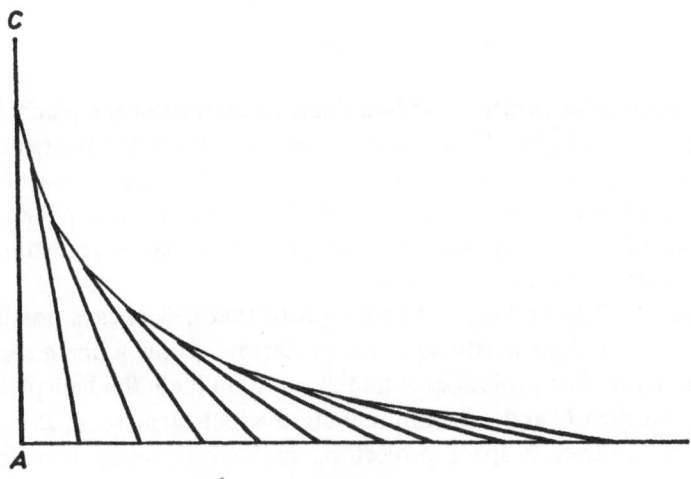

Fig. 68.

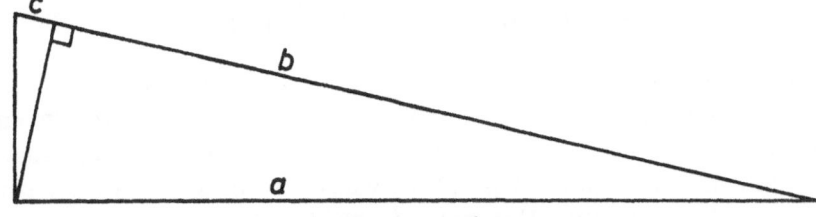

Fig. 69. $b < a < b + c$.

cut the original distance by one half. By using a different strategy Cain could reduce the distance to less than c for every $c>0$. Look at Figure 69 and prove this.

RECTANGULAR PROJECTORS

In some activities students become familiar with important concepts, in others they find empirically an important result. But in some activities only geometric intuition is developed. Here is one such example: Figure 70

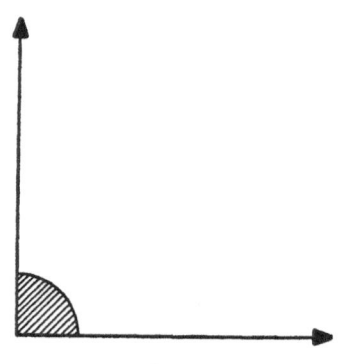

Fig. 70.

shows a rectangular projector. It illuminates a quarter of the plane. Four of these projectors are placed in the plane. Can you always illuminate the whole plane? Or can you find some positions for which this is not possible?

Children assume different positions and try to cover the plane by projectors placed at these positions. They usually succeed, sometimes after many fruitless attempts.

At the end of the activity comes the proof that it is always possible. Let A, B, C, D be the four positions of the projectors. Draw a line g separating A, B from C, D. The projectors A and B can illuminate the half-plane of C, D. The projectors C and D can illuminate the half-plane of A, B.

Problem 22. Take 8 space projectors, each illuminating one octant of space. Can you always illuminate the whole space?

ASSUMPTIONS AND CONSEQUENCES

An important class of problems consists in making assumptions and drawing consequences. A teacher should have hundreds in store. In one hour one can use up 5 to 10 of these problems. It takes a teacher many years until he has a store of problems which can be solved with a bare minimum of prerequisites. Here are three examples:

(a) The opposite edges of a tetrahedron are equal. Show that all faces are congruent. (See Figure 71.)

(b) All faces of a hexahedron are congruent parallelograms. Show that the faces are rhombi, i.e., the hexahedron is a rhombohedron. (Figure 72). Is

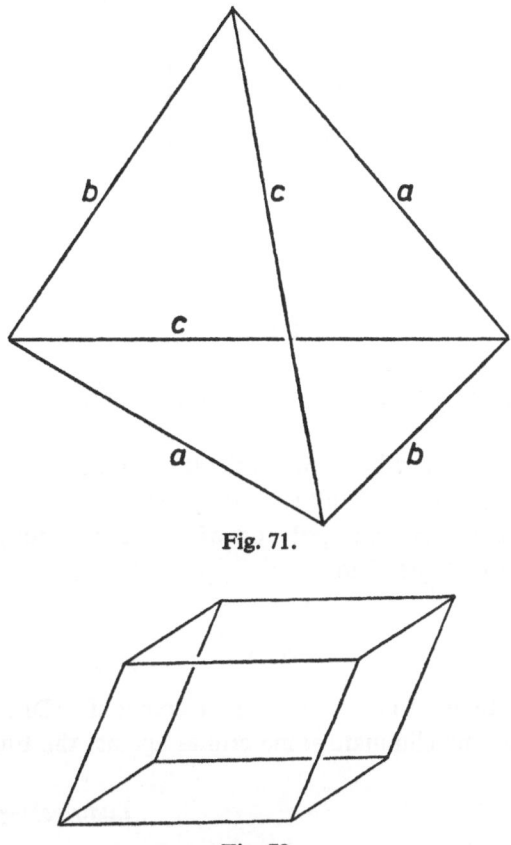

Fig. 71.

Fig. 72.

the rhombohedron symmetric? Yes, it has three planes of symmetry, each passing through a pair of opposite edges.

(c) Definition of a regular space polygon: It is a closed sequence of congruent segments which do not lie in the same plane. The angles between two successive segments are equal. Show that regular space polygons with 4, 6, 7, 8, 9, 10, 11, 12, ... vertices exist.

The case of an even number of edges is especially easy. Here are two solutions:

(1) Take a strip of congruent squares and fold it at right angles to form a staircase (Figure 73).

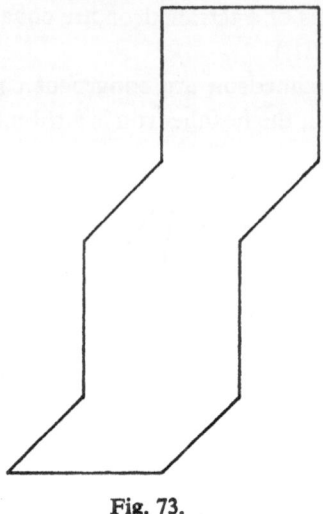

Fig. 73.

(2) Take a regular plane polygon and raise every second vertex by the same amount.

Problem 23. It is somewhat more difficult to find regular space polygons with 7, 9, 11, 13, ... vertices. Find such polygons!

Problem 24. A regular space polygon with 5 vertices does not exist. This is a famous and tough problem.

NOTE

I would like to thank Father Larry N. Lorenzoni, SDB, CSMP Senior Editor for helping me eliminate some crimes against the English language.

Ludwigsburg College, GFR

T. J. FLETCHER

THE TEACHING OF GEOMETRY
PRESENT PROBLEMS AND FUTURE AIMS

SUMMARY. Some general observations are made on present problems in the teaching of geometry at pre-college level, and some aims for the future are suggested. These remarks are inevitably made against a background of British experience. Aids for the teaching of geometry are discussed; and two lines of development, perspective drawing and nomograms, are described as a school introduction to geometrical ideas which are important at college.

1. GENERAL CONSIDERATIONS

Let us commence by considering two of the most widely quoted remarks attributed to Greek geometers. Menaechmus is said to have told Alexander that there is no royal road to geometry [5]. In other versions of the story it is Euclid who makes the remark to Ptolemy. The last century has seen many advances in geometry and the last decade has seen spectacular changes in the teaching of algebra at school level, but it seems to me that geometry remains as difficult a subject as ever and that a revolution in the teaching of geometry on the same scale as the revolution in the teaching of algebra is hardly to be anticipated. Furthermore, bearing in mind experiments which are currently in progress, on the present evidence there is no obviously desirable direction for any further revolution to take. Progress is more likely to come from the effective mobilisation of existing resources than from any new, previously untried, revolutionary remedy.

The second famous statement is the reply which Euclid is said to have given [15] to a pupil who asked after learning the first theorem "What shall I get by learning these things?" The well-known reply was, "Give him threepence, since he must make gain out of what he learns". I take the view that the pupil's question was a reasonable one, and a question which the teacher must be prepared to face at any time, and that better answers than Euclid's are possible.

The cry "Euclid must go!" has gained a certain notoriety in recent years. Our reaction to this in England was merely mild surprise since as far as we were concerned Euclid had already been gone for a long time. In fact 1971 sees the centenary of the Mathematical Association, which was originally founded under the name of the Association for the Improvement of Geometrical Teaching – improvement meaning the replacement of Euclid's text by something more suitable for school pupils. It is important to note this objective, because in England the problem was not to free Euclid from logical blemish, the problem was to replace it by a teaching strategy that was more acceptable. The Euclidean tradition has been under fire for a long time and it is unnecessary to go over the various reasons or to describe in detail the stages of its decline in elementary teaching.

A factor of perhaps even greater importance in determining the direction which geometrical teaching has taken has been the apparent decline of interest in geometry itself among professional mathematicians. The schoolmaster sees that many areas of geometry to which he devoted much time in his youth no longer form part of the university course. A generation ago the schoolmaster felt much closer to the geometry which was done in universities and he could feel that advance in geometry meant finding more theorems about the higher geometry of the triangle or of cubic curves. But now it seems to the schoolmaster that the university mathematician believes that the mine is worked out, at least as far as the geometry which the schoolmaster can understand is concerned. The universities' entrance requirements provide further confirmation that they are no longer interested in the geometry of former years.

In the past a number of areas of mathematics have been developed and have become useful working tools, but subsequent mathematical developments have passed them by and they survive only as fossils as far as creative mathematicians are concerned. Spherical trigonometry and descriptive geometry are cases in point. Many seem to feel that other branches of geometry which they studied when young are moving the same way.

It can be claimed that the contemporary mathematician's reduced interest in geometry is apparent rather than real, that many flourishing branches of mathematics employ geometrical language and imagery, and that a properly disciplined geometrical imagination is as big an asset to a mathematician as ever. This may be so – but little has been done to show schools what this implies in terms of classroom practice.

The problem of what geometry to teach to the future mathematician is certainly difficult, but it must not obscure the problem of what geometry to teach to everybody else. The school pupil who is destined to become a mathematician is taught alongside the future surveyor, the crystallographer, the cartographer, the architect, the atomic engineer as well as alongside some who may be as gifted as he is mathematically but whose future occupations will make little or no use of this gift.

By using the resources for individualised learning which are now being developed, schools of the future may be able to provide a more personal curriculum than schools did in the past; but desirable as this may be, the school course for future mathematicians has to be seen in relation to the courses which are provided for everybody else, and it must be an aim of education to see that the specialist student not only ends up knowing something, but that he also knows how to communicate his knowledge to his fellows in terms which make sense to them.

Geometry in practice makes use of a wide variety of techniques, no one of which is powerful enough to make the rest superfluous. Every branch of geometry has a selection of problems which are most easily solved by using its own methods; equally well if one is restricted to given methods then problems can be found which are difficult to do that way.

Some current programmes of reform attach great importance to giving an overall, coherent view of mathematics. There can be no doubts about the desirability of this; but the student also needs the ability to solve problems and he needs an appreciation of the way mathematics fits into the broader pattern of our culture.

There is far more worthwhile elementary geometry than any one student is able to cover. The student needs a wide variety of geometrical experience, but exactly what experience is of secondary importance. He also needs a sense of direction, which can be supplied by certain unifying themes. Many college texts assure the reader in the preface that the only pre-requisite is "a certain level of mathematical maturity" – but they do not usually say how the reader is to acquire this if he has not got it already. Maturity can only be acquired with time and with the opportunity to think about a variety of relevant things in your own way.

These considerations imply a geometry course which consists of a core plus options. The core which is developing in the new programmes in British schools consists mainly of transformation geometry and vector geometry, and this is a tendency which seems widespread in other countries as well. Transformation geometry usually begins by developing intuitive ideas of symmetry. These lead on to the isometries of the plane and eventually to the ideas of the relevant transformation groups. This is an attempt to implement the Klein Erlangen programme in a school context. It may be observed in passing that no British programme envisages that this should be done axiomatically. Indeed the new courses are so far from being axiomatic that anyone would be put to it to say precisely what the axiomatic structure was. This can, and does, mean that at the end of the course there is really no agreement as to what a "proof" of a "theorem" is, although there should be a very good understanding that in certain circumstances given certain facts then other facts follow. I am not worried by this, for I have said that geometry is a difficult subject and that what matters at school level is the accumulation of a broad range of experience on which to build later. At the present stage of knowledge to attempt an early axiomatisation which is well motivated is to seek for a royal road which is no more to be had than it was at the time of Euclid.

The advantages of the transformation outlook and the arguments for regarding it as part of the geometrical core are clear. The importance of ideas of transformations, mappings, "jections" of various kinds does not need arguing, and geometry is an area of study in which these ideas can be seen at work. I would plead however that the student should see them at work in sufficient depth, producing non-trivial results which excite his imagination and command his respect. For older school pupils this might be achieved by a course such as Jeger [4] or by a few lessons of a quite old fashioned kind on inversive geometry.

There can also be little disagreement about including work on vectors in a contemporary school course, even if to do no more than provide a background to linear algebra. There is however an interesting difficulty. The amount of geometrical knowledge which is needed as a background to the immensely powerful ideas of linear algebra is very small indeed, although experience indicates that many students enter college lacking it. It is important therefore that school teachers should be very clear about what is involved, and ensure that the big central ideas are stressed repeatedly as they arise in various contexts. For this it is essential that much of the pupils' vector work should be (very simple) work in three dimensions. There is little point in troubling to find vector proofs of results which are more easily proved in other ways.

The course devised by the Scottish Mathematics Group [13] provides an interesting unification of the vector and transformation approaches by starting with (i.e., intuitively assuming as axioms) coverings of the plane by parallelograms etc. In this connection it is essential to refer to the school text by Papy [11] which is concerned with the simultaneous development of the real number system and the affine plane. His mathematical strategy merits most careful study, although the pedagogical approach is very different in Papy and in British texts – of which the Scottish series may serve as an example.

I have already suggested that the strictly logical pursuit of geometry is extremely difficult for beginners, and I do not think it feasible to approach school geometry as if it were merely an application of logical principles which have been previously understood. I think even that the weaker point of view, the traditional one, that geometry can be justified in the school curriculum as a vehicle for the teaching of logic is suspect. It is suspect for reasons which are very well known – the beginner cannot disentangle the logic and the physical science, and the logical basis on which geometry has been taught up till now has gaps which it is dishonest to conceal. Geometry is only a vehicle for the teaching of logic to sophisticated pupils, and the appropriate level of sophistication has so far only been attained towards the end of the secondary school course, it has never been available at the beginning. This gives little cause for regret, because we now see that there are far more possibilities in the use of algebra as a vehicle for the teaching of logic than were previously supposed; and algebra can provide the main line of logical development in the school course.

Apart from this, logic can be taught explicitly, as a study in its own right; although the appropriate methods still require much investigation. There does not seem much hope of teaching logic merely as a technique which can be subsequently applied. This has failed in the past with adults and it assumes that learning is far more transferable than it is with most students. Furthermore, teaching techniques followed by applications is a line of development which is now being rejected in most other subjects of the school curriculum. This applies in the teaching of languages, science, art, religious education and almost everything else, including in some cases craft skills. The sequence of techniques-followed-by-applications is being rejected for a teaching approach which is more subtle and certainly more difficult to carry out – a contextual approach which gives more recognition to the character and needs of human beings. This involves devising learning situations by which students generalise from experience. Abstractions are too important to be told to the student, he must come to see them himself. In other words the pupil develops understanding not so much by following a logical exposition as by making for himself a sequence of conceptual reorientations. The problem of teaching is to set up learning situations from which the pupil acquires the experience which compels the reorientation.

In later sections of this paper I exemplify this approach by suggesting two lines of development which might be considered, perhaps as an option, in a school course, and which lead to the ideas which are now known to be crucial in the foundations of geometry, and to be crucial in the relationship of geometry to abstract algebra.

2. AIDS AND EQUIPMENT

An up-to-date programme requires a range of appropriately designed teaching aids. In particular a geometry course which envisages making a number of optional studies available on an individualised basis will certainly need reading material, and reading material of different kinds suited to the problems in hand. Thus one must consider not only textbooks of a traditional kind but also a range of books going from highly didactic programmed material to freer reference books which cover historical background or further applications of mathematics, and place the work of the course in a wider setting. Thus in association with a geometry course there might be books and pamphlets at varying levels of difficulty on cartography, mechanical linkages, finite geometry, technical drawing and model making tasks for the student as well as aspects of pure geometry which the main course does not cover.

Good historical material should also be available, where appropriate with reproductions of original sources, and, of course, pictures as well as text. I do not advocate teaching the history of mathematics as an integral part of the course as a general principle, and I would say that in the teaching of modern algebra history is largely irrelevant. But by contrast it seems to me that history is very nearly essential if the actual content of geometry is to be properly understood – although I admit that this is a contentious statement! What is important in the mathematical education of every man is to show the place of mathematics in our culture [8, 9], and in this story geometry must have a very prominent role.

Film, especially animated diagrams, can also be used. The moving diagram has a contribution to make to the teaching of geometry which as yet has hardly been realised. A film which duplicates a book is a waste of time. Filmed lectures have a limited value if they are made by mathematcial celebrities or if they are particularly well done, but what requires especial thought is the use of film to convey peculiarly geometrical ways of thinking.

It is clear that in future if much traditional geometry is to be taught at all for the most part it will have to be taught in a summary form. The need to summarise areas of available knowledge is frequently recognised in post-graduate mathematical education, but it is seldom attempted lower down. Admittedly it is only justified where the student can bring certain experience to bear, but this frequently is the case in the later years of the school, and the animated film might very well be a useful means of summarising geometrical knowledge.

Computer animation of acceptable quality is now available and so technical animation is not necessarily the time-consuming task which it used to be. There are some advantages in silent animation, one being that it is possible to provide silent viewing facilities to individual students at low cost; and if the student can vary the speed at which the diagram is projected and reverse it as he wishes then this is an immense gain.

It is important to emphasise that moving film is required to bring out aspects of geometry which cannot be brought out by other means, and not merely to save the student the trouble of turning pages. The greatest loss in the recent decline of geometry has been (to my mind) the aesthetic loss. Geometry can be an extremely attractive part of mathematics. It was the first love of many who later came to teach mathematics in the past, and if the student is to succumb to the charms of geometry then a certain time is needed for dalliance. Present day courses must convey the aesthetic appeal of mathematics by every possible means. The big claim that I would make for the geometrical animated film is that it can show the aesthetic fascination of geometry to a number who might otherwise fail to see it.

Stereoscopic illustration would also be required in a full range of individualised learning packages. This can supplement, but not replace, solid models. In some cases it is appropriate for the student to make solid models for himself. It is important that stereoscopic technical illustrations should be of the highest possible technical quality, otherwise some viewers can not see them at all. Perhaps the best of all is the material prepared using polaroid processes. Some surprisingly good coloured stereoscopic illustration is now made for children's books using lenticular sheets, but I do not know how good this process is for geometrical diagrams. Anaglyph is comparatively cheap, and easily stored, but once again only the very best quality is good enough. Davies' book [3] is an example of how persuasively three-dimensional good anaglyph illustrations can be. Stereo-animation is technically possible, but very little has been made so far.

3. ARCHITECT'S GEOMETRY

This section and the next sketch out two independent lines of investigation which eventually converge on the same central ideas. The lines of investigation are perspective drawing and nomograms, and the central ideas are Desargues' theorem, the axioms of projective geometry and the profound inter-relations between geometry and the number system [12, 14, 16], the study of which began in Greek geometry but only reached its present form at the turn of the century with the work of Hilbert [6] and others. We may remind ourselves that Desargues was an architect and engineer.

At any particular time in a mathematics course the work which is being done must make sense to the student at the time, and it must also have a mathematical future – that is to say the teacher must have a clear idea of where it leads. The complex of ideas outlined in the previous paragraph can satisfy both criteria.

I suggest that the Klein Erlangen programme and the axiomatisation of projective geometry, with an understanding of the significance of Desargues' theorem (with geometry over a skew field) and Pappus's theorem (with commutative multiplication) are two big geometrical developments of the past century which should have an honoured place in the education of every mathematician. Reference has already been made to the place of the Erlangen programme at school level, and the work now to be outlined is to be seen as the early stages of two possible approaches to the second topic. It is not the intention that the final stages of the story should be reached at school; the argument is that each line of development is educationally worthwhile, and that it puts the student in a sequence of positions where he is achieving immediate aims and also has the opportunity to make the next theoretical jump forward for himself if he is able to. This can provide a good foundation for later, more theoretical developments. The main points in a college course on projective geometry are not appreciated if the student is encountering everything for the first time. He needs to be well aware of the problems before he is offered elegant solutions.

In both of the lines of development to follow *construction* plays a leading role, and this is deliberate, because construction problems frequently motivate the beginner in geometry more readily than theorems. The purpose of the exercise at least is clear, and the sequence of ideas arises naturally from the sequence of physical operations which have to be performed. When the proof of a theorem is required the sequence of ideas is merely a logical sequence, and this is a more abstruse thing with which the student has understandable difficulties.

The history of perspective drawing has many mathematical lessons to teach, so a course might contain a historical outline along the lines of Kline, and it might also include warnings that celebrated artists made mistakes in the past [7]. The student can get going very quickly if he is set tasks of drawing his own pictures of railway lines, houses, garden fences or indeed almost anything which takes his fancy. There is no need to explain everything before he is allowed to do it as mistakes are inexpensive and easily corrected. The study of photographs and classical paintings has something to offer – how do various things appear? Can you draw similar things for yourself?

When the time has come to codify a few elementary rules the introduction of a book for artists rather than something which is clearly a school text on mathematics might be a more subtle soft sell. The book by White [17] would do excellently, or a book such as Norling [10]. A valuable adjunct would be a book of half completed drawings to which the student would be invited to add more. Using guesswork based on his previous experience he can add windows, doors and chimneys to the outline of a house, put on a roof, add a wing, draw a swimming pool in the garden, or extend a fence, keeping the posts properly spaced. We might have a drawing with the family standing around at different places in the garden. Given the height of one estimate the heights of the others. Explain your method!

From this activity the main problems, the technical terms and the theory can crystalise out. The essential ideas at the beginning are the ground plane (GP), the picture plane (PP) and the horizon. The main problem is the projection of information on GP to PP. The simple rules for drawing parallels can be formulated, and as we come to see how a "direction" is related to a "point at infinity" our geometrical vocabulary is enlarged in a natural manner. The guiding principle of the teaching at this stage is for the student first to do something and then to consider how he did it; to ask what principles he was applying, and to ask how explicit recognition of the principles gives power to do more. Later on the practical problems get harder, and we might consider how to draw shadows cast by the sun, or the patch of light on the garden when the light shines through the windows at night.

As the theory is made explicit Desargues' theorem appears very early, because it describes the relation between any triangle on GP (or indeed in

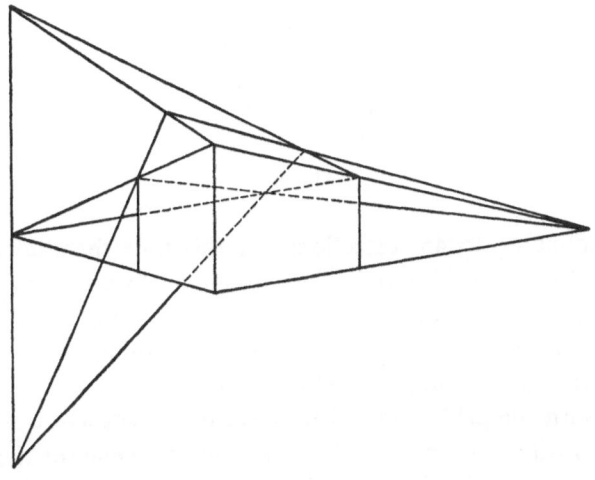

Fig. 1.

space) and its image in PP. Desargues' theorem for coplanar triangles is a more complicated matter, but it appears naturally over and over again in perspective drawings. For example, Figure 1 is a partially completed drawing of a house. The dotted lines indicate how the position of "the far top corner" may be located in the drawing. Desargues' configuration is immersed in this figure, and occurs repeatedly in similar situations. Also, without going prematurely into the theory, we can consider the use which the architect makes of the plane version of Desargues' theorem in drawing lines through points which are off his paper [1, 12, 19].

Almost all of the work described so far can be done with one or two vanishing points only. The houses drawn will be "rectangles" and the need to draw "squares" will be bypassed by maintaining a discrete silence. But eventually the student will recognise one of the big problems of Renaissance painting – to draw *squares* in perspective. Study of suitable paintings by old masters will show many cases where the secret was the drawing of a tiled floor on which the figures were stood like chessmen, and thereby disposed correctly in space.

A square grid can be drawn by Pelerin's construction (Figure 2). This involves the introduction of "distance points" (DP). These are the vanishing

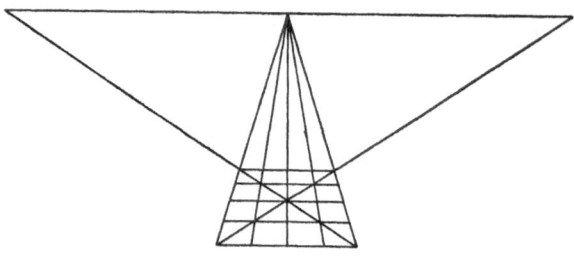

Fig. 2.

points corresponding to 45° directions. By this time the student needs to appreciate that the vanishing point corresponding to a particular direction is located on PP by "pointing" from the observer's position. The location of the DP's is a particular case of this. All these things and more are explained very clearly and simply in White [18].

Equipped with the grid provided by Pelerin's construction we are in an extremely interesting situation, and it is well worth while for the student to spend some time investigating just what can be done with it and just what can not. Perpendiculars can be constructed (since we have established the orthogonal involution on the line at infinity), and many more possibilities are opened out. The student can now attempt to draw reflections in mirrors and other more intricate tasks. Many suitable examples occur in Yates [19]. The situation here is particularly instructive, because Yates's examples are based on the assumption that the geometer has a straight edge and one given square. This square in the context of his book is tacitly assumed to be a "proper" square, i.e., we are talking about squares in GP. But the constructions transfer to PP, and the perspective image of the square can be used just as well as the original. Considerations of this kind can make it clear to the student how to the geometer a "right angle" is in a certain sense a convention and is not something inextricably tied up with physical measurement. When sketching a building an architect may call certain things "right angles" when they are certainly not right angles as far as measurements on the drawing go. But the things which he calls right angles fit coherently into a geometrical whole, and the same language can describe GP and PP.

The techniques assembled so far are still not sufficient to bisect angles (in general) or to rotate lengths freely in GP. For this and related purposes the architect introduces measuring points (MP). These might be considered as an option in a school geometry course. The architect's rules for MP's are simple enough, at least as far as the MP's for the GP are concerned (these are located on the horizon line), but in a mathematics course it is the reasons which matter. These can be explained in a relatively simple way [1, 18], but good illustrations and perhaps models are essential.

Equipped with the idea of measuring points the student might undertake a little military intelligence work from suitably chosen photographs. Given one measurement on the GP he can calculate any other. Similar exercises could be based realistically on architects' drawings.

The theoretical point of interest is that the introduction of measuring points is tantamount to the introduction of compasses. Indeed the architect has to use compasses to construct MPs, since the MP associated with a particular VP is the point on the horizon the same distance from VP as VP is from the eye position. (Actually there are two such points, but in most practical situations only one is needed.) It is interesting to relate this to the Steiner constructions using straight edge and one fixed circle with its center. Given one circle only we can construct all others in the same plane. Once again Yates's examples are applicable, and it is interesting to consider various locations for the fixed circle. It might naturally be the circle on GP surrounding the basic square in Pelerin's construction, or a circle on GP with center where the observer is standing. The standard Steiner straight-edge constructions as described in Yates apply in GP, but they all transfer to PP using there of course the perspective image of the true circle which is in GP.

Yet another approach is to draw the one circle as the circle in PP which stands on the two DP's. This circle includes also the rabattment of the eye position (i.e., the point on PP, on the center line, as far below the horizon as the observer's eye is from the horizon horizontally) and the use of this circle as an "absolute" circle raises further teaching possibilities. To what extent can it provide a heuristice introduction to ideas of the absolute conic?

This is already further than one would reckon to go with all but the ablest of school pupils, but the practical problems are by no means all over yet. Draw a box in perspective. Open the lid, and draw various positions of the lid as it opens. This requires further extensions to the theory of measuring points. These are very much the concern of the architect, but their mathematical interest is limited, no more major principles emerging. The final refinements which the architect needs may be taken by mathematicians as an optional extra.

4. NOMOGRAMS

Nomograms are the alignment charts used by engineers and others to facilitate numerical calculations. Usually a straight edge is laid across two scales, and the intercept on a third scale gives the required mathematical function of the variables associated with the first two. The theory of nomograms is intimately bound up with the foundations of plane geometry, but it is difficult to find any practical book which discusses them from this point of view. There are two basic nomograms, for addition and subtraction, and they may be derived from the basic diagrams for the induction of the additive and multiplicative groups on the line which are to be found in standard books on projective geometry.

Here the suggestion is that nomograms may be investigated and used over a period of years, the basic properties being derived from naive metrical geometry. Ultimately the introduction of numerical co-ordinates into geometry from a sophisticated point of view [12, 14, 16] will be seen as the re-formulation of long familiar ideas. (It would be interesting to know the history of nomograms, but the familiar, readily available references are silent on this.)

Nomos was a Greek word with a variety of meanings. A classical scholar, with only vague ideas of what a nomogram was, once remarked to me that *nomos* could mean a law, or it could mean something you wander about in. A nomogram is clearly a diagram embodying a mathematical law, but if we think of it as something which the student can wander about in, and learn by so doing, we have perhaps a better point of view.

The first nomogram for the student to meet is the addition nomogram on three equally spaced parallel lines. If the u- and v-scales are on the outer lines and the w-scale on the middle line (with half the interval between its graduations) then $w = u + v$. The student can draw his own diagram, experiment with it, and eventually extend the scales beyond zero so as to include negative numbers. This can emphasise "inverse elements", "the rule of signs" or whatever formulation of the theoretical ideas is chosen for development at this stage.

As a practical side-line the additive nomogram can be adapted for various purposes at subsequent points in the course. By "over-writing" the scales a nomogram can be produced for right-angled triangles with $a^2 + b^2 = c^2$, in dynamics for $v^2 = u^2 + 2fs$ (with an assigned value of f), and in many other places. The behavior of the nomogram for different spacing of the scales can also be investigated.

The multiplicative nomogram can be built up as in Figure 3. Starting with the u- and w-scales each point on the u-scale may be joined to the corresponding point on the w-scale, and it will be found that they all cross the line which is to be the v-scale at the same point. Further experiment shows that the mapping $u \rightarrow w = 2u$ identifies another point on the v-scale, which we may label "2". Continued experiment identifies other points on the v-scale which to the consternation of the student is not evenly spaced. When this initial

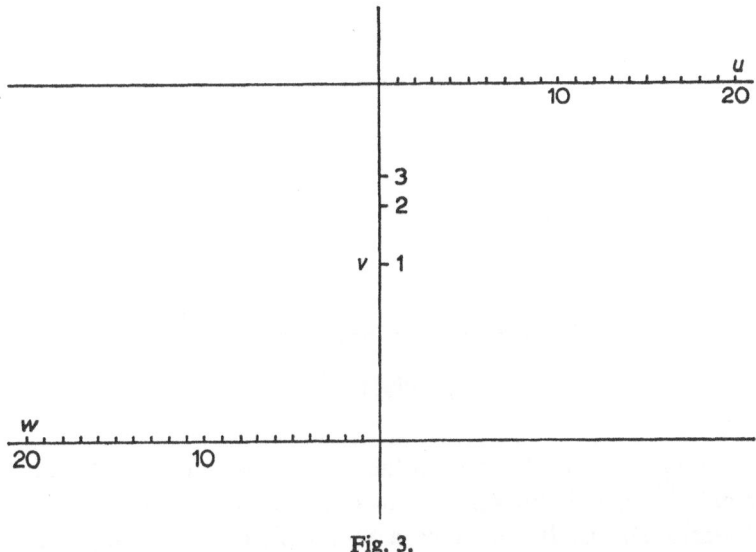

Fig. 3.

shock has been overcome as many points may be put on the scales as are felt necessary. If this investigation is taken at an appropriate time in the course the student may be led to extend the u-scale into the negative region on his own initiative, to extend the other scales as well, and to see how the multiplicative rules of sign are "forced" by the situation.

Later on, by over-writing the scales, the multiplicative nomogram can be adapted for other uses, e.g. to embody a formula such as $V = \pi r^2 h$, and when logarithms are known many further adaptations of the basic additive and multiplicative nomograms become possible. Ideas may be found in many practical books for engineers, some of which should be available in the classroom.

A nomogram embodying the optical formula $(1/u) + (1/v) = 1/f$ is worth constructing in its own right. Only later will this be seen as a special case of an additive nomogram. In the Cartesian plane $x = 0$ and $y = 0$ may be used for the u- and v-scales, and the line $x=y$, with suitable graduation, as the f-scale.

The suggestion is that all of the nomograms described can be approached experimentally, but of course we must set about proving things as and when the students are ready. Textbooks on nomograms usually prove their basic results simply enough, as far as they are concerned, by using similar triangles. It is useful for the teacher to know that the essential results are more easily proved as far as children are concerned by using area, because in traditional courses area is done some while before similarity.

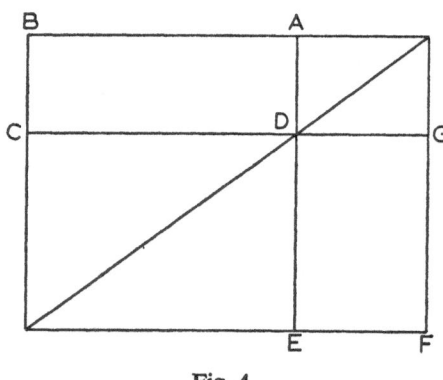

Fig. 4.

As a methodological observation the area proofs can all be based on the "gnomon" figure (Figure 4), in which the important property is area $ABCD=$ area $DEFG$. It would be a useful piece of pedagogic theory to know just how this figure, which was the basis of much of Euclid's treatment of ratio, fits into a modern context. In Figure 4 the figure is drawn as a rectangle, but the property is of course an affine one.

The devising of a respectable theory of area, which can be introduced early in the secondary school course, is a problem to which there may be no completely satisfactory solution. It should however be possible to arrange an approach which is honest at the time, and is made fully respectable later by the minimum adjustment. Somewhat similar remarks apply to angle, although the problems here are certainly harder.

Nomograms being familiar, in their simplest form, later on generalisations may be undertaken. This may be provoked by seeing more general nomograms in use, or by raising a question from perspective drawing – what does a nomogram look like in perspective? Experiment will then show that given a certain basic amount of information the rest of a nomogram is determined entirely by incidence constructions. Thus an additive nomogram is specified if any three concurrent lines are taken as u-, v- and w-scales, if a fourth (non-concurrent) line is used to assign the origin on each scale, and if the unit point is then taken on one. Likewise a multiplicative nomogram is specified if any three non-concurrent lines are taken as u-, v- and w-scales and a fourth line of general position is used to assign the unit points.

With these constructions the student may experiment, calibrating a number of points on the scales. The ideas have practical value as they sometimes facilitate the layout of a nomogram, making it possible to fit scales with the required ranges most conveniently into the space available.

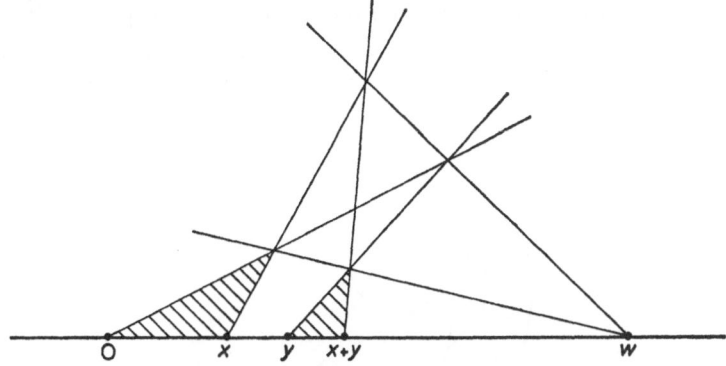

Fig. 5a.

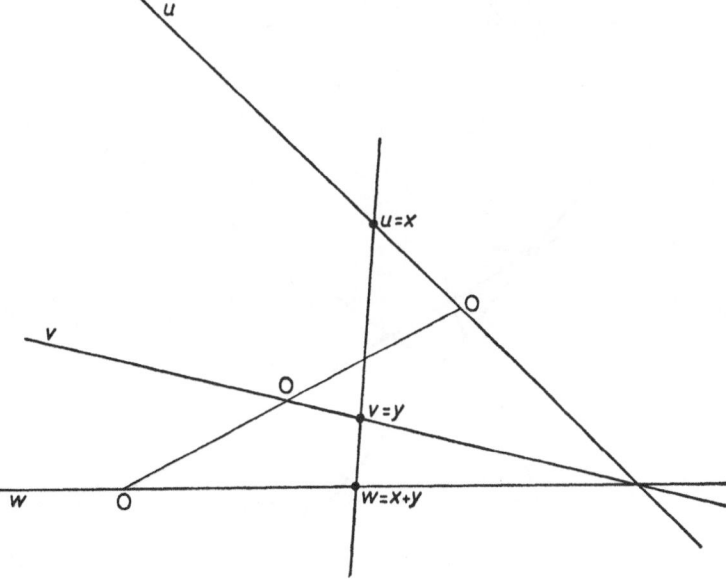

Fig. 5b.

As an exercise the student might relate the construction of nomograms still more closely to perspective drawing by considering the use of a perspective drawing of a garden fence as an additive nomogram, or investigate in how many ways additive and multiplicative nomograms can be picked out of the square pattern drawn by Pelerin's construction.

The theoretical points raised by this last round of experiment are considerable. We have the experimental evidence that nomograms work, but what are we to regard as a proof? We may pass on to see how the configurations of Pappus and Desargues are involved in the constructions. One aspect of this involvement is shown in Figure 5a and 5b. (These can be drawn very appropriately on overlays for use on an overhead projector.) Figure 5a shows the construction involved in putting an additive group on the line in the projective plane. This can be seen as a special case of Desargues' configuration. However, if different aspects of the figure are high-lighted we see

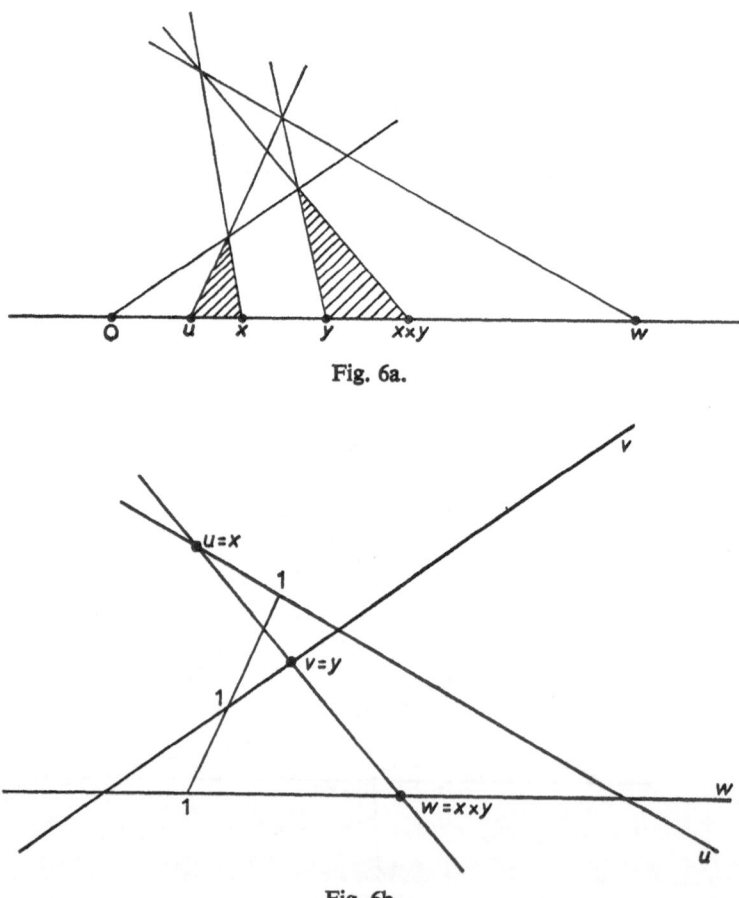

Fig. 6a.

Fig. 6b.

that we are dealing with the engineer's additive nomogram. Similar remarks apply to Figure 6a and 6b, for multiplication.

It is hardly possible to undertake the calibration tasks suggested three paragraphs back without becoming acutely aware of Pappus's theorem, which arises when checks are made on fresh points which have been marked on the scales. A special case is involved in the addition nomogram and the general case in the multiplication one.

Some students might be interested in nomogram scales on curves of the second degree, but this is a more specialised topic which adds little to the complex of theory which is the declared destination of the enquiry which we have outlined above. In the past much time was devoted to a detailed study of the conic sections; how much will be done in the future remains to be seen.

At the point now reached in the students' investigations there is a wealth of experience requiring effective organisation. When is this to be undertaken? In my view it should be undertaken at college, but it is certainly possible that knowledgeable teachers with capable students could undertake it earlier. I would doubt the wisdom of trying to do it too soon; but I am certain that if a college course in geometry is to be the illuminating, logical, aesthetic experience which it should be, then the student must come prepared with a fund of experience on which the lecturer can draw. Some ways of providing this experience I have sought to suggest.

Department of Education and Science
Darlington, County Durham, U.K.

NOTE ADDED IN PROOF. For the history of nomograms see M. D'Ocagne, *Traité de nomographie*, Paris, 1899.

[The author is employed by the Department of Education and Science (England and Wales). The views expressed are his own and are in no way attributable to the Department.]

BIBLIOGRAPHY

[1] C. K. Clutterbuck, *3-D Scale Drawing*, English University Press, 1966.
[2] H. S. M. Coxeter, *Introduction to Geometry*, Wiley, 1961.
[3] A. G. J. Davies, *Solid Geometry in 3-D for Technical Drawing*, Chatto & Windus, 1967.
[4] M. Jeger, *Transformation Geometry*, Allen and Unwin, 1966.
[5] T. L. Heath, *A Manual of Greek Mathematics*, Oxford, 1931.
[6] D. Hilbert, *The Foundations of Geometry*, Open Court, 1910.
[7] W. M. Ivins, *Art and Geometry*, Harvard University Press, 1946.
[8] M. Kline, *Mathematics in Western Culture*, Oxford, 1953.
[9] M. Kline, *Mathematics – A Cultural Approach*, Addison-Wesley, 1962.
[10] E. Norling, *Perspective Drawing*, Walter T. Foster, California.
[11] G. Papy, *Mathématique Moderne*, 2, Didier, 1965.

[12] T. G. Room, *A Background to Geometry*, Cambridge, 1967.
[13] Scottish Mathematics Group, *Modern Mathematics for Schools*, Blackie/Chambers, 1965.
[14] A. Seidenberg, *Projective Geometry*, Van Nostrand, 1962.
[15] H. W. Turnbull, *The Great Mathematicians*, reprinted in J. R. Newman, *The World of Mathematics*, Simon and Schuster, 1965.
[16] O. Veblen and J. W. Young, *Projective Geometry* (2 vols), Ginn, 1910, 1918.
[17] Gwen White, *A Book of Pictorial Perspective*.
[18] Gwen White, *Perspective*, Batsford, 1968.
[19] R. C. Yates, *Geometrical Tools*, Education Publishers, St. Louis, 1949.

HANS FREUDENTHAL

GEOMETRY BETWEEN THE DEVIL AND THE DEEP SEA

PHILOSOPHIES

For many years New Mathematics has meant new subject matter, and this is what it still means in many cases. Where people tried to teach it, it became meaning new teaching techniques. There are few people to which it means a conscious philosophy.

Of course, there is no educator without a teaching philosophy, or should I say, without two teaching philosophies, an explicit philosophy he professes, and an implicit one he acts out. The latter is the more interesting and the more important. I do not like explicit philosophies but I cannot understand educators unless I make their implicit philosophies explicit. This is why I start with philosophy. It is too bad, but I cannot help it.

I will try first to answer three philosophical questions. Since they are questions of a philosophy, that is of human attitudes, they admit a gamma of answers, and among these answers there are a few which I like better than the remainder.

My first question is: what is mathematics? I mean not how mathematics is defined but what is it to the human mind? Les us ask another question: what is language? There are two answers: language as a stock of words, prefixes, suffixes, flections, grammar and syntactical rules, and language as an activity of speaking, reading, writing, and expressing oneself. Linguists consider it as a great discovery of de Saussure's distinguish these two aspects of language, ready made and acted out language. Both aspects are operational in language teaching. As long as children are small they will be taught a language as an activity, whether it is their mother tongue or a foreign – this does not make too much difference. As they mature into rational beings, the teaching philosophy shifts from the activity to the ready made subject matter interpretation of language. You know that it does not work too well. Why not? Because it is a wrong philosophy? No, there are no wrong philosophies. Both the stock and the activity interpretation of language are correct but they apply to different people. The one is good for linguists who take a language as a subject matter and the other for the users of the language, who in fact, are the majority.

What is mathematics? Of course you know that mathematics is an activity because you are active mathematicians. It is an activity of solving problems, of looking for problems, but it is also an activity of organizing a subject matter. This can be a matter from reality which has to be organized according to mathematical patterns if problems from reality have to be solved. It can also be a mathematical matter, new or old results, of your own or of others, which have to be organized according to new ideas, to be better understood, in a broader context, or by an axiomatic approach.

A great part of mathematical activity today is organizing. We like to offer the results of our mathematical activity in a well organized form where no traces betray the activity by which they were created. This objectivation is a habit of mathematicians from the oldest times. It is a good habit, and it is a bad one. We freeze up the result of our activity into a rigid system, because this is objective, because it is rational, and because it is beautiful, and this we teach.

Few people know that mathematics is an activity. Little children are taught mathematics as an activity, but as they mature into rational beings, we are prone to teach them a well organized prefabricated deductive system of mathematics, because rational beings may be supposed to understand deductive systems. You know that it does not work very well.

My second question is: what is education? It is good luck that the first piece of education in history we know about, is a lesson of geometry, the socratic lesson Menon's slave was taught on doubling the square. Socrates taught the slave not the solution of the problem nor solving the problem, but finding the solution by trial and error. He did not teach a ready made solution but the way of reinventing the solution. Two millenia later Comenius said: "The best way to teach an activity is to show it." This is a socratic idea, though it involves more than a socratic lesson. While Socrates taught his lesson, the slave listened, whereas Comenius will show the student an activity to explain afterwards what it means and finally to have it imitated by the student.

Modern educators are likely to subscribe to a variation of Comenius' device. Not

"The best way to teach an activity, is to show it" but rather

"The best way to learn an activity, is to perform it."

It is a mere shift of stress though it is an important one.

The traditional answer to the question "what is education?" is "transfer of culture". I think it still prevails. Recently I discussed the task of the university (or rather the Task of the University) with a bunch of students, mostly from the humanities. Its task is, they said, to provide information, and if they challenged the university, then they did so because the university provided not enough information, or wrong information, or biassed information, or because it suppressed information. The university should offer its disciples an encyclopedia of all available information, and from this they would choose what they liked. Maybe this was just the look of the disciplines they studied (if they studied at all and did not refer to information about the study). Anyhow they seriously believed that elsewhere studies were conducted according to such medieval didactics; that the future physician was told by his professors how to cure illness, the engineer how to build bridges, the mathematician how to calculate. In fact this was once the practice of teaching. Today, I told them, scientific aptitudes are taught in hospital wards, laboratories, and practice rooms, and for the advanced student participation in research is a chapter of his curriculum. They agreed that cycling, swimming, skiing cannot be learned from a textbook, but I could not convince them that science was as well an activity that could be learned, not in lectures and from books, but by acting it out.

No astrologuer is needed to read the writings on the wall. The students I told of who could understand the learning process only as an information flow, are the same who reproach the university that it provides wrong information and who would use the information channels of the university to funnel what they consider as true information. If this were the point, they would even be right. The only answer they should be given is that learning is a different thing but receiving information.

From Comenius onwards learning has been becoming more and more an activity. This has been a consequence of the changes in science and society. For this reason I may dismiss the question whether people learn better by active building up the subject than by passive reception of a ready made matter. In fact much depends on the kind of subject matter what is better. It does not matter whether the authoritarian method of cultural transfer is bad or wrong or inappropriate in the learning process. It is much worse: this method is simply impossible, it does not agree with the character of modern society. The authority of the older cannot be maintained in the learning processes of emancipating youth. Our cultural assets are too dangerous to be offered the youth as ready made material. The instruction we provide should create the opportunity for youth to acquire the cultural heritage by their own activity. They should learn that the selfreliance they claim elsewhere extends to their own role in the learning process.

My third question is about mathematical education. Should mathematics be taught according to a deductive system? Most mathematicians are prone to answer in the affirmative. Maybe it is not always possible, but in principle it should. In any case they will try.

How can you teach logarithms before a theory of real number? How can you teach probability without an axiomatic definition of the fundamental concepts? How can you teach angles before analytic functions? How can you compute the area of cylinder, cone, and sphere before superficies of curved surfaces has been defined?

How can it be done? It can. It happened in history and didactically it is the correct way. Logarithms were invented when the ancient theory of real number was eclipsed by more naive approaches. Probabilities were computed for centuries without axiomatic principles. Angles are as old as mathematics itself. Fortunately Archimedes has never been bothered by H. A. Schwarz' paradox on the area of curved surfaces.

Actually students can learn to work with the slide rule without any axiomatic or any creative definition of real numbers and can discover the logarithmic property of the area of the graphic for $y = 1/x$ before they know integrals. They can work with probabilities as soon as they know fractions. They can understand and work with angles in mathematics and reality. They can find out areas and volumes by rational means with no general theory on areas and volumes.

Of course, mathematics involves deductivity. Working with the slide rule and the protractor is no mathematics, measuring areas and volumes is no mathematics. But accepting a deduction is no mathematics either, unless you adhere to the interpretation of mathematics as a ready made subject.

In which order, if not in a deductive one, should mathematics be taught? The answer is simple: in that one in which it can be learned, which means, the order in which it could be invented by the student. This is not at all a revolutionary idea. It is the socratic lesson. In a thought experiment the teacher has been reinventing the subject matter as though he himself were the student, and this is what he teaches. Modern didacticians would require that rather than teaching the reinvention, the teacher should have the student reinventing himself the subject matter. This is a modern reenforcement of the socratic idea.

Let us take the example of teaching Peano's axioms on natural numbers to a boy who knows quite a bit of school algebra. The deductive approach would be formulating the axiom system to derive a series of theorems from it. The thought experiment, on the other hand, guided by historical experiences, shows that nobody can invent Peano's axioms unless he has first formulated the priciple of complete induction. Further, nobody can formulate complete induction unless he knows what it is, and there is no knowledge about complete induction, unless it has been experienced as a proving activity. So the student should be brought into one or more situations where he can reinvent the trick of inductive reasoning. There are quite a lot of them, Pascal's trinagle, number of permutations, fundamental arithmetical laws, and so on. With virtually no help students learn to find and so use the principle of complete induction, though it is a hard thing to formulate it. From complete induction one could proceed to Peano's axioms if the thought experiment shows a way to do it. Of course, the student cannot hit on the idea of framing complete induction into an axiomatic system, unless he has already acquired a certain experience of axiomatic systems, that is of axiomatizing. If he has not, Peano's axioms is certainly not the best example to start axiomatics with.

The previous example shows a remarkable structure of the learning process which completely differs from the deductive structure of mathematics. The van Hieles called it the levels in the learning process. The student applies certain new rules unconsciously until at a certain moment he becomes conscious about them. From analyzing the structure of the subject matter he switches to analyzing his own mathematical activity. He grasps the laws and regularities of his activity, which enables him to apply them more consciously. Maybe he even succeeds in formulating their principles in general terms.

The van Hiele levels of the learning process are often characterized by a logical feature: the activity on one level is subjected to analysis on the next, the operational matter on one level becomes a subject matter on the next level.

Mathematics on the lowest level used to be allowed to start as an activity. Experiments, in particular by Dienes, have shown that children can perform rather involved mathematics on this level, that is to say they can develop an activity which on a higher level would be interpreted as mathematics. This bottom level is an indispensable precondition of mathematics. This should be stressed against people who rightly object that it is no mathematics at all. It is bottom-level, it is indispensable premathematics.

Adding numbers represented by counters is no mathematics. As soon as to add 5 to 8 the child physically or mentally divides the 5 into 2 and 3, it performs an act which on a higher level can be interpreted as mathematics. As soon as the child grasps this principle and tries to formulate it, it reaches the first level of mathematics.

WHAT IS GEOMETRY?

For long times mathematics has been synonymous with geometry. In fact, there existed other branches, too, algebra, trigonometry, calculus, which, however, were not much more than collections of haphazard, badly founded rules, whereas geometry was a perfect logical system, where everything rigorously followed from definitions and axioms. You know that things have changed; today mathematicians are prone to reject traditional geometry, because it is not a rigorously deductive system.

The deductive structure of traditional geometry has never been a convincing didactical success. People today believe that geometry failed because it was not deductive enough; to my opinion it failed because its deductivity could not be reinvented by the learner but only imposed.

There have been many efforts to reform teaching geometry. In my country Tatiana Ehrenfest-Afanassjewa, the wife and collaborator of the famous physicist Paul Ehrenfest, was much influential. Her "Übungensammlung" of 1931 is still a treasure of geometrical exercises. Tatiana Ehrenfest did not reject deductivity. On the contrary, some deductive system, influenced by Helmholtz' axiomatics, was the goal of her method. But her geometry started as a science of physical space, of the space in which the child lives and moves, as an organization of the learner's spatial experiences.

It sounds oldfashioned to claim that geometry should be related to physical space. In fact, Tatiana Ehrenfest knew mathematical axiomatics and cultivated it, but she was also a physicist who knew about the profound physical implications of space theories. Many didacticians today would reject physical space as a field to be organized by mathematics. We are mathematicians, or are we not? We are builders of mathematical structures, and if physicists and other people can use them, they may take them, but no deductive system can be built on the quicksand of reality, and mathematics may not be contaminated by elements which are foreign to deductivity. If for instance the system commands us to start geometry with the affine plane, then physical space with geometry with its solid bodies, its distances, deplacements and fittings is only in obstacle to good mathematics.

This is one philosophy, and I think it is a good philosophy. As long as people are building a deductive system they are often obliged to close their eyes for the seductions of reality. But there are other activities, and there are people who will never build deductive systems of their own or even rebuild those of others though they need mathematics. We cannot expect them to espouse this philosophy.

Finally nobody can deny that we live in space, that we move in space, that we analyze space, to be better adapted to it. In fact, we are so much used to space that we easily forget about its importance for ourselves and for those we are educating. Let us ask a few haphazard questions:

Why does a piece of paper fold along a straight line?
Why does a rolled piece of paper become rigid?
Why does a tied paper ribbon show a regular pentagon?
How do shadows originate?
What is the intersection of a plane and a sphere, of two spheres?
What kind of curve is the terminator on the moon?
Why can the radius of a circle be transferred six times around periphery?

How come a beautiful star arises by this construction?

Why is the straight line the shortest?

Why do congruent triangles fit to cover the plane and why do congruent pentagons in general fail to do so?

How can people measure big distances on the earth, the diameter of the earth, and distances of celestial bodies?

What does a cube look like if viewed along a spatial diagonal?

What is larger, the superficies of a sphere cap, or that of the cylinder around it?

What is the shortest path for a light ray to travel from one point to another while touching a mirror?

How does a kaleidoscope work?

What is the biggest sphere in a tetrahedron?

Which kind of closed curves are equally as wide in all directions?

How does the liquid level in some vessel change if a certain quantity of liquid is added?

What is the relation between real and apparent size of a body?

If a cube is split into six square pyramids with their vertices in the center and these pyramids are turned outside upon the corresponding faces, why does a rhombododecahedron arise?

How can you measure the inclination of a line and a plane, or of two planes?

Is there a horizontal (a vertical) line in every plane?

What is the difference between a right and a left screw and why are not they equivalent under rigid motions?

What is a rigid motion on the sphere?

Why is a convex polyhedron rigid?

Why can a table with four legs wobble, and what is the difference with a table with three legs?

Why does a door need two hinges, and how can we add a third?

And finally the old question: why does a mirror interchange right and left though not above and below?

Notice that I did not ask any questions of practical use. My task would have been much simpler if I had done so. Notice, too, that there were no "puzzles" in the list, but only questions of principal importance and which nevertheless can be answered with little ingenuity.

Many mathematicians will consider my questions as irrelevant in teaching mathematics. With no regret they would leave them to physicists or other kind of people if need be that students get acquainted with space. Indeed, almost none of those questions can be fitted into a preconceived system of mathematics. To be sure, this is not only true for such questions of geometry, but for almost all questions which arise in reality. In particular, probability is an instructive case. All real problems of probability are irrelevant for an axiomatic probability fitted into a preconceived system of mathematics, and axiomatic probability in turn is irrelevant for solving real problems in probability.

I espouse the philosophy of teaching mathematics related to reality. Mathematics is important for many people because it admits of multifarious applications. I do not trust teachers of other disciplines to be able to tie the bonds of mathematics with reality which have been cut by the mathematics teacher. Moreover, I do not believe that mathematics, not tied to lived reality with strong bonds, can have a lasting influence in most individuals. We mathematicians retain the mathematics we learned, because it is our business. People usually forget what is not related to the world in which they live. For most people mathematics cannot be an aim in itself; if they have learned it in an unrelated way, they will never be able to use it.

CONCRETE MATERIAL THE CHILD CAN HANDLE

Tatiana Ehrenfest developed an introductory geometry course of exercises in space. Meanwhile many others designed concrete material, in our country for instance the van Hieles and van Albada, to stimulate a geometric activity of folding, cutting, glueing, drawing, painting, measuring, fitting, paving. People generally call it experimental geometry. I do not like this word. If the word means that the student is expected to experiment, then the greater part of his mathematical activity should be experimental, as is ours as far as we are creative mathematicians. If the term is chosen to remind of experimental physics, it is wholly mistaken. "Experimental geometry" suggests procedures like determining π by measuring a string laid around a circle periphery, a silly exercise you will never meet in such collections. In former times they called it mensuration, an activity which leads nowhere. The introductory geometry exercises by Ehrenfest and others, however, are zero level activities in that they prepare for the higher levels. The attitude of the children who work with this material greatly differs from the realism of the physical laboratory. To illustrate this I will give you a few examples.

Children in Dina van Hiele's class (12 year olds) got the task to design a sidewalk pavement with square flagstones. One of the children drew stones with cracks and sandgrains in the groves between the stones. One hint was sufficient to convince the child that these were inessentials in geometry.

Dina van Hiele gave the children bags with congruent cardboard triangles, quadrangles, pentagons, hexagons, which should be used as tiles to pave the plane. The pieces were not congruent enough to give a good fitting, but the children never complained; they built the patterns as though they were regular and understood them as regular. In general from the beginning of geometry they accept bad fittings and imperfect congruencies and symmetries as though it were perfect ones.

As the children get the task to draw a triangular pavement, they will first use the cardboard material but as soon as they have grasped the pattern, they will finish building and start drawing. None of them will add the drawn triangles piecewise but all of them will first draw long horizontal lines between which the triangles are to be fitted. Some of them use a second set of parallel lines; it is very rare that a child starts with drawing immediately three sets of parallel lines.

DINA VAN HIELE'S EXPERIMENTS

Before Tatiana Ehrenfest it was an axiom that geometrical instruction started in the plane. More precisely it started with points, continued with lines, and since the plane were the place to locate these objects, all happened in the plane. Tatiana Ehrenfest once told when she taught classes, parents complained that she discussed spheres with children before they had learned defining straight lines. In many countries plane geometry started in the 7th grade, geometry in space in the 10th or 11th grades. Children who were good in plane geometry, often failed in space geometry. Their spatial imagination had been killed by too much and too exclusive plane geometry. In our country the usual start in geometry is now in space. To give an example, I sketch Dina van Hiele's first course, with 7th graders.

She started geometry with the cube, but of course with no definitions. Cubes of different material were shown, cardboard, wooden, meccano cubes. The discussion led to constructing cardboard cubes. In this process the children became acquainted with the geometrical instruments and with fundamental notions such as right angles (defined by folding). In the meccano cube two kinds of diagonals were identified. After face diagonals had been drawn on the selfmade cube, the question arose how to find exactly the spatial diagonals. The teacher showed a cube with its upper face open to allow a look inside. There was a cardboard diagonal partition wall in it. The wall could be removed. It was a rectangle. The children understood this hint to construct the spatial diagonals. The children made a cardboard partition rectangle and noticed that it could be fitted in several ways into the cube. In how many? Other examples of double counting were dealt with. The children made more regular bodies.

The class then proceeded to studying symmetries first in space, then with plane figures. One half of a symmetric figure was completed to a whole. A mirror was used to check the result; the discussion led to an analysis of the concept of symmetry. A host of symmetric figures were considered and constructed. The discussion then focussed on the rhomb as a starting point for many constructions. This trimester of geometry finished with angles. Of course no definition of angles was given. Angles were made, estimated, measured. The principle of this first stage of geometry was, as Dina van Hiele put it: Rationally guided operations with concrete material.

A detailed report on a second trimester's course has been the main subject of Dina van Hiele's thesis. It is the wealthiest treasure of didactic observation and analysis in mathematics I ever saw.

The subject of this course was "tiles". After the children had become acquainted with congruency and regular polygons, they were given bags containing congruent regular polygons as a first concrete material to study the covering of the plane by congruent shapes. The children discovered by reasoning that the angles of a triangle sum up to 180°, the analogous facts for other polygons, and the interrelation between these facts. They discovered why certain shapes of tiles did not fit to cover the plane. (Not all pentagons, but are there pentagons that fit?) Logical relations were put into arrow patterns. In the triangular pavement the children discovered parallelograms, "saws", "stairs", equality of angles, "enlargements", symmetries and regularities; areas were compared by decomposition. Logical arrow patterns were combined into logical "trees".

The third trimester subject was again congruency, mainly in space. Solids were made, and volumes were compared by decomposition. The final piece of work and the acme of the course was the construction of the rhombododecahedron.

FITTING

I could only sketch the subject matter of Dina van Hiele's first course. The method and its analysis is more relevant. Though the plane provides easier opportunities of logical analysis in geometry, space with its solids is more concrete. It better meets the children's creative wishes. Figures are drawn, solids are made.

There is one exception in plane: paving a floor with congruent tiles. Here the leading psychological idea, I mean fitting, is the same as in space and it is realized as concretely. Fitting is a motor sensation. Psychologists can tell you how strongly the motor component of the personality is marked at a young age, how important motor apprehension and memory may be.

Things fit. Do children ask why? Apart from a rare exception young children do not. All these miracles of our space do not seem to make any impression. But they grind as millstones. The highest pedagogic virtue is patience. One day the child will ask why, and there is no use starting systematic geometry before that day has come. Even worse: it can really do harm.

The clue of geometry is the word "why". Only joy-killers will deliver the clue previously. The miracles of fitting are to prepare maturity for systematic geometry. But even if this has been reached, they will not become redundant. They are raw material for geometrical thinking. The child should recall and reconsider the treasure of old problems at every new stage.

Of course fitting as a leading idea requires that space is acknowledged as the home of solid bodies. There are no fitting problems in affine geometry. Affine geometry is a sophistication that needs a lot of deductive experiences to be invented. It can never be the start of geometry.

DEDUCTIONS

There are many ways of organizing the child's exploration of space. I quoted Dina van Hiele's because it has most profoundly been analyzed. The period covered by Dina van Hiele's course shows a steady though discontinuous growth of the child's ability to logically organize his own activity. On level 0 the child is thinking as it were with his hands, his eyes, his kinesthesis. To the viewer who knows mathematics the child's activity looks like mathematics. When the teacher shows the partition wall of the cube, the child succeeds in constructing its spatial diagonal, and this activity can be interpreted by the viewer as the same reasoning that leads himself to construct the spatial diagonal. In geometry the child can for a long time continue this 0-level activity until it starts reconsidering it consciously – there are quite a few who excellently perform on this level and never reach a higher one, but the reason that they fail often is that they are pressed too early and helped by algorithms to simulate acting on a level they did not reach.

Several times I stressed that in this approach no definitions are provided for geometrical objects when they are introduced. If the antididactical inversion is applied, a deductive exposition starts with definitions. (Traditional geometry texts even define what is a definition, which is still a higher level of learning.) In socratic didactics such an approach would be impossible since nobody can define a thing unless he knows what he should define. It is even worse in mathematics, where "definition" has a very special meaning. A definition in mathematics does not just tell people what a certain word is supposed to mean, but definitions become links in a deductive chain, and you cannot forge the link unless you know where it should fit.

If Dina van Hiele showed their students a cube, a rhomb, a parallelogram, or if she gave examples of congruent figures, the students never failed to understand which shape she meant, as well as they recognize what is a chair, or a bottle, or a doll though this has never been defined nor have they been confronted with all objects of these sorts. Of course there may be some uncertainties, for instance whether a square should be counted among the rhombs, and a rhomb among the parallelograms. Though the teacher can impose definitions that settle the questions, this means degrading mathematics to something like spelling, ruled by arbitrary prescriptions.

If the child knows what a rhomb or a parallelogramme is, it may visually discover properties of these shapes. There are a lot of them, and in the class discussion children will recite them. A parallelogramme has its opposite sides equal and parallel, its diagonals bisect each other, opposite angles are equal, adjacent angles are supplementary, the parallelogramme has a centre of symmetry, it can be split into congruent triangles, congruent parallelogrammes will cover the plane. There are a host of visual properties which ask for organization. Here starts deductivity; rather than being imposed it unfolds from local germs. The properties of the parallelogramme become deductively interrelated, and finally one emerges (maybe different ones for different children) from which the others can be deducted. This property can be taken as a definition, and now it can be understood why a square and a rhomb should be considered as parallelogrammes. By this activity, the child learns defining, too, and it experiences that in mathematics a definition is more than a description, that it is a means of organizing properties of a thing deductively. However, on this level the child would not yet be able to analyze the process of defining and to tell what is a definition. Indeed, this is again a higher level of the learning process.

To those who believe in a system, this method is a horror. A deduction needs a firm ground, and you cannot reason rigorously about things you have not rigorously defined before. This is a sound philosophy though it is not that of a creative mathematician. Definitions are not preconceived to derive something from them, but more often they are just the last element of the analysis, the finishing touch of organizing a subject. Children should be granted the same opportunities as the grown-up mathematician claims for himself. Telling a kid a secret he can find out himself is not only bad teaching, it is a crime. Have you ever observed how keen six year olds are to discover and reinvent things and how you can disappoint them if you betray some secret too early? Twelve year olds are different; they got used to imposed solutions, they ask for solutions without trying. Good geometry instruction would be a means to teach them a lot of things, to teach them organizing a subject matter, and what is organization, to teach them conceptualizing, and what is a concept, to teach them deducing, and what is a deduction, to teach them defining, and what is a definition, to have them to distinguish why some organization, deduction, definition is better than others. Traditional in-

struction is different. Rather than giving the child the opportunity of organizing spatial experiences, the subject matter is offered as a preorganized
structure, with all concepts, definitions, deductions preconceived by the
teacher who knows of every detail which aim it serves – or rather by the
textbook author who has carefully built in all his secrets.

THE SYSTEM

Traditional geometry was taught as a deductive system. It was a fake system
but by teaching it teachers used to indoctrinate themselves to believe in the
system. In that time geometry was the only part of mathematics that was
taught according to a system, while other parts were simply collections of
thumb rules. Things have changed. There have been rare attempts to save
the character of geometry by means of some Hilbert style axiomatics. The
other extreme is, rather than mathematics, to teach a system of mathematics
and to fit geometry into this system. Which pieces of geometry are really
fitted into the system and on which places, does not depend on geometry but
on the system. The usual frame is to begin with affine vector space on which
later on an inner product is imposed. The most conspicuous aspect of space,
I mean fitting, is disregarded and ignored by affine geometry. But it is still
worse. In such a rigid approach no opportunity is left for children to explore
space and the bodies in it, to organize the subject matter, to invent definitions and deductions. Among all the problems I quoted when I asked "what
is geometry?" there is none that can meaningfully fit into such a system.
Maybe you can compute the intersection of a sphere and a plane or of two
spheres by linear algebra but to discover that the result is a circle you have to
know what is a real sphere in real space. Of course you can figure out that
the radius of a circle is the side of the inscribed regular hexagon by means
of an analytic proof if you like proofs that obscure all essentials. To discover
that congruent triangles can cover the plane, while pentagons can in general
not, linear algebra is entirely unfit, but is even no appropriate means to *prove*
such facts. I would be able to continue this way with all examples I gave
earlier. The geometry allowed by linear algebra is an utterly dreary product.
Its highlights are to prove that two different straight lines can have none or
one intersection, that for circles these numbers are 0, 1, or 2. Maybe vector
algebra still provides a tasteless proof for the three bisectors of the sides of a
triangle passing through one point. The only geometrical subject that could
adequately be tackled by vector space, I mean barycenter and convex bodies,
is generally disregarded because it does not fit into the system of mathematics.
   Usually such a system of mathematics is not less ramshackle than old
geometry was, but today the authors learned how to make more successful

attempts to hide its deficiencies. This, however, is not the feature for which I blame system builders. It is rather the dogmatism of the system, and the didactical inversion. System building could be one of the aims of mathematical education and maybe even a realistic one provided that organizing a subject matter has been exercised on a lower level. An imposed system, however, prevents the student from learning organizing a subject matter on any level.

GEOMETRICAL AXIOMATICS

Some people have been anxious to save genuine geometry from denaturation by linear algebra. To do so, they tried geometrical axiomatics. In fact, it is another way to kill geometry. It is more or less Hilbert-style axiomatics, with a long list of axioms, most of them seemingly trivial, though there may be a few quite complicated ones among them. The deductional structure usually is quite sketchy; the greater part of argumentations consists of such formulas as "it is easily seen that ...", and "we omit the proof which would be lengthy though it would not offer any new insight". It is a frank confession and it is true; reasoning within such system is, indeed, a dreary occupation. What is interesting in an axiomatic system is constructing it and verifying whether it is complete, that is, checking whether its geometry can be coordinatized in the usual way. If somebody needs geometry in algebra or analysis, he never reasons within an axiomatic system, which would be much too complicated.

It is not because of its complexity that I blame such an axiomatic system. It is the way in which it is offered to the students. They should use it to perform mechanical deductions in it, an activity which the author honestly dismisses as irrelevant. The essential activities around such an axiomatic system are reserved to its author: first, organizing globally the geometric subject matter to get the axiomatic system, cutting the bonds with the organized subject matter, to afterwards restore them by checking the completeness of the system. Geometrical axiomatics cannot be meaningful as a teaching subject unless the student is allowed to perform these activities himself. Usually he is not allowed to do so. Either the author judges that axiomatizing is a business of grown up mathematicians rather than of learners, or he knows from his own axiomatizing activity or a thought experiment that it would be much too difficult for the student. Indeed, a student who never exercized organizing a subject matter on local levels will not succeed on the global one. Prefabricated axiom systems have merits of their own. They are an acceptable subject matter for people who are experienced in axiomatizing. If a student has learned axiomatizing with easier material, he will recognize in a complicated axiomatic system the same features he knows from his former experiences, and he will be able to disentangle this system and to understand it as though he built it himself. But if axiomatizing has never been exercized, such an axiomatic system of geometry is only one more piece of indigestible mathematics.

The method by which the axiomatician attempts to save geometry from the claws of linear algebrists, is: competing with the mathematical rigor of linear algebra. It is a mistaken idea.

RIGOR

System builders know one level of mathematical rigor and often they manage to stick to this one during a course of many years. All below this level, which is their own, they consider as fake, and all above as highbrow. Active mathematics, however, knows many levels of rigor, and good teaching should respect them. It should also respect the fact that rather than being imposed each of these kinds of rigor should naturally develop in the learning process. On a low level it can be rigorous figuring out $8+5$ by counting five more steps from eight. On the next level the learner is required to split it into $(8+2)+3$ and to use mental addition tables up to 10. Still later it is rigorous to know $8+5=13$ by heart and to use it as a technique of addition tables up to 20. On a certain level it may be rigorous to read $5-8$ from the number line, on a higher level rigor would mean to prove it by a kind of induction, on a still higher level rigor should require that it is understood as a definition, and finally it can be rigorous as a deductive consequence from an axiomatic system.

Every subject matter knows, on the level on which it is tackled, its own criteria of rigor. In the learning process they arise as tacit conventions between the learner and his surroundings, which are dropped or modified as the learning process continues. Reaching a certain level implies knowing by which rules of rigor it is governed, the main deficiency of an algorithmic activity on a level which has not actually been reached is that the learner does not master the rigor of that level.

This level theory of rigor is not to excuse sloppiness. Verbalistic definitions and arguments that in fact neither define nor argue anything, are as bad on every level. Circularity of arguments shows the same features on any level, though the criteria of circularity may vary. In a highly formalized system non-circularity can be checked by a computer, but you have to educate your mathematical sensitivity to feel on any level what is a circular argument. Axiomatics is not at all the highest level of rigor. Most axiomatics are poorly formalized, in particular those of geometry. This is no blame if the degree of formalizing corresponds to the level on which the axiomatic system is presented.

AXIOMATICS AND TRADITIONAL DEDUCTIVITY

Often the objection is raised: Traditional deductive geometry is also axiomatics though it is the bad one of the juggler who is hiding his axioms in his tophat and sleeves. Is not honesty a better policy?

The assertion that traditional deductivity and axiomatics proper are not essentially different is both right and wrong. Axiomatics from Pasch and Hilbert onwards up to our days, presupposes the syntax and semantics of of every day language. One defines "lines are parallel..." and continues speaking of "parallel lines", "parallel projection", "parallelism", and so on, one defines "intersecting lines" and continues with "lines that meet each other". Of course this is entirely justified on this level. But there are other levels. One can take offense of this linguistic unformality and try to formalize axioms and other statements. Formalizing, too, knows levels: on a very high level even deduction is being formalized. It is not true that axiomatics is the summit of honesty and rigor. Every level knows its own honesty and rigor, which cannot be enforced in teaching, if the student has not reached this level. It is again and again attempted, though with sham success. Geometrical axiomatics is particularly dangerous. High level tricks are so easy and captious to be built in. The author hopes that thanks to the automatism of the axiomatic method these built-in features would function automatically, that is, without the students' understanding of the trick. I know such policy is not unusual in teaching though conscientious educators fight it. I fear axiomatics may lead us deeper into this swamp.

If didactically viewed, axiomatics and traditional deductivity are much different. The difference can be characterized as that between global and local organization of a subject matter. Everybody knows how long it lasts to have an average student view a proof as a whole. It lasts still longer to have him oversee the connections with other propositions. If, with much trouble, he has reached this level, he is pressed upon the next, the global organization.

There is still another important factor. In the course of his instruction the student has become acquainted with two kinds of defining, the descriptive definition which fixes a familiar subject by reading characteristic properties from it, and the algorithmically constructive, creative definition which produces new subjects out of old ones. Axiomatics shows another feature by which it is distinguished formally and didactically from traditional deductivity: the implicit definition, which through descriptive, is creative but not in an algorithmic way.

LOCAL ORGANIZATION

How to save geometry if axiomatics cannot do it? Nobody would ask how to save physics or zoology, which have never been axiomatized. Students learn to figure out the price of 3 pounds of sugar if that of one pound is given, and the area of a rectangle if the sides are known, though the greater part of the notions involved have never been fitted into an axiomatic system. In applying mathematics you never move within an axiomatic system, and if by chance somebody would ask how to save applications, he probably would mean how to protect them against the axiomatic dogmatism. Probability is a most beautiful application which is gravely endangered in teaching to be killed by axiomatics.

Geometry can be saved if it is presented as a field in which the student can be active. As a prefabricated subject it is due to die by suffocation.

I already explained how in introductory geometry the student can learn to invent organizing the shapes and phenomena in space, by means of geometrical concepts and their properties. On the next level he would organize these concepts and properties by means of relations of a logical nature. On a still higher level this relational system may be a subject of investigation.

Let us take one example. The crucial mathematical childhood experience of quite a few mathematicians who wrote autobiographies, was the perpendicular bisectors of a triangle passing through one point. It is, indeed, a beautiful theorem. Children can easily find it, provided it is formulated in a less symmetric way: "Draw the bisectors of AB and BC, which intersect at M; look where the bisector of AC passes." Let us analyze the proof in the same way as the learner should do after he found it.

The proof rests on the property of the bisector of XY being the set of all points equidistant from X and Y, which may have been recognized by symmetry arguments. M is on the bisector of AB whence

$$MA = MB;$$

M is on the bisector of BC whence

$$MB = MC.$$

From both follows

$$MA = MC,$$

whence M is on the bisector of AC.

The proof is a combination of a few surprises. The first and second equalities are found by applying the bisector property one way, the third is a consequence of applying it the other way, the third is a consequence of applying it the other way. First and second, M is on the bisector, thus M is equidistant, third, M is equidistant, thus M is on the bisector. I think this is the first psychologically convincing occurrence of the logical complex characterized by such concepts as inversion, necessary and sufficient, if and only if.

The next surprise is the transitivity of the equality of line segments. The seemingly trivial property of transitivity should be made explicit to understand the proof. Again I think it is the first possible approach to the transitivity property and the first example of its productivity.

The third surprise is that such a symmetric statement as that about the three perpendicular bisectors must be tackled in an asymmetric way to prove it. It is an important step to learn that "three lines pass through the same point" means the same as "one line passes through the intersection of the two others". It is the first example of a methodological paradigm which serves up to the highest levels of mathematics.

The fourth and maybe the greatest surprise is that an incidence theorem (three lines passing through one point) is proved by metric arguments. It requires a not so easy analysis to understand the profound reason of this feature.

Finally it may be mentioned that the theorem leads to the fascinating construction of the circumcircle.

The proof of the theorem on the perpendicular bisectors is not only a marvelous piece of geometry, and a rich source of didactical ideas, it is a good example in geometry for what I have called, local organization. It can be dealt with as soon as children have understood the perpendicular bisector as a locus of equidistance. They need not be able to prove this crucial property of the perpendicular bisector. A proof of the locus property of the bisector cannot contribute anything to the understanding of the circumcircle theorem. I even doubt whether a proof of such an obvious fact as the locus property of the bisector can be recommended at an early stage. At a more advanced stage, it would make more sense to ask why the perpendicular bisector of XY is the locus of points equidistant of X and Y; then one may even ask why equality is transitive, which comes down on understanding what equality means. One can ask why the bisectors of different sides of the triangle intersect, why straight lines have at most one intersection. Of course such an arrangement contradicts the philosophy of mathematics as a prefabricated system. It is the way of exploration, in mathematics and in any science whatsoever, the way to understand and explain phenomena.

The question why the bisectors are concurrent is much alike that why the electric bell rings or why it does not, why the stomach does digest food and why it does not digest itself; why comets have a tail, and who committed the murder. It is questions "why?" asking for a reason or for a cause. And they have also in common that no answer is definitive. You can continue asking "but why is the bisector a locus of equidistance?" "why is equality transitive?", "why does the current activate the electromagnet?", and so on. The answer to any question contains the germs of fresh ones. It is an apparently infinite chain, although Aristoteles believed it could be suspended on the principles as the first reason and on God as the first cause. But does it really matter whether the chain is infinite or not, if at least I am able to avoid circularities? The practice of every day knowledge as well as of science, indeed, is, at a certain moment, to stop asking why. To repair the bell, I do not need Maxwell's theory, and if I study Maxwell's theory, I can dismiss field quantization.

I admit from this point onwards mathematics is a bit different. Mathematicians invented the axiomatic trick, that is accepting the vice of circularity as a virtue. You stop asking what are points and lines; instead of an explicit definition you tell what you are allowed to do with them. Actually every science behaves so, every science is based on implicit definition, but mathematics is the only one in which this is cultivated as the summum of deductivity, and even more, after a system has been axiomatized, its bonds with reality can be cut; by anontologization it becomes a self contained field.

We are again back at axiomatics, but I think we can now more clearly distinguish what axiomatics means. A mathematical text may start with axioms because it is ready made mathematics. Mathematics as an activity never does so. In general, what we do if we create and if we apply mathematics, is an activity of local organization. Beginners in mathematics cannot do even more than that. Every teacher knows that most students can produce and understand only short deduction chains. They cannot grasp long proofs as a whole, and still less can they view substantial part of mathematics as a deductive system. We are lucky if they can learn organizing locally a mathematizable field of reality or a piece of mathematics itself, because this is just what they will need in every day life and in their profession.

Organizing locally is not a deficient or illicit or dishonest activity in mathematics. It is a generally accepted attitude of the grown up mathematician in pure and applied mathematics even if he would never publish such exercises. None of the various axiom systems of geometry has ever been used to find or to prove geometrical statements if they are needed in algebra or analysis. It is much too complicated and the road from the axioms to important theorems is much too long. The proof is superseded by the firm conviction that it can be done but that it is hardly worthwhile. One is satisfied with the local organization up to a changing horizon of evidence. Coxeter's *Introduction to Geometry* is a marvellous demonstration of this attitude. The author knows in any case exactly where this horizon is lying, and which kind of rigor is adapted to the subject matter, for instance, that no recourse to the axioms is needed for Morley's theorem on the trisectrices, whereas a theorem like Sylvester's on collinear points which essentially depends on order properties, needs an axiomatic background.

OTHER VIRTUES OF AXIOMATICS

None of the axiomatic systems of geometry are such that they can be taught high school students as an experience, but maybe some could be offered to otherwise axiomatically experienced students (I mean students experienced in axiomatizing) as a subject to be analyzed. What would be the educational context of such an attempt?

We know that axiomatics is important, and if possible, the student should learn what axiomatics is. There are, however, so much easier subjects to be axiomatized than geometry, such as groups, measure, linear order, cyclic order, angles, and so on. I have already rejected geometric axiomatics as an expression of rigor. What is rigor, depends on the context. To justify axiomatizing geometry by arguments of rigor, one has to show a context in which local organization is not adequate. There are such contexts but usual axiomatics does not apply to them.

Another benefice of axiomatics is cutting ontological bonds. In the usual abstracting axiomatics it is too easy a business. In geometrical axiomatics, which is describing rather than abstracting, it is much more difficult. Linear algebra on the other hand, contains too little geometry to make the cut of ontological bonds effective. It would be worthwhile to show the student how geometry can be made independent of physical space but it is not easy. Of course, you can tell it with formulae like "it is easily shown", and "we omit the proof", but this is not more than verbalism.

Axiomatics is also an organizational form of foundational investigations. The axiomatical structure allows to investigate the scope of such and such an axiom by omitting or replacing it. This, too, can more easily be done with simpler axiomatic systems though it shows special aspects and charming features in geometry.

I would mention here an approach which is foundational though it has not been worked out to full fledged axiomatics. It is P. J. van Albada's:

Within spatial euclidean geometry one turns to geometry on the sphere, which by identification of the antipodes may be even transformed into the elliptic plane. Theorems known from euclidean plane geometry are investigated in elliptic geometry. Some remain valid, others cannot be transferred. To become familiar with elliptic geometry one tries to derive theorems of elliptic geometry from each other. Theorems that hold in both geometries ask for proofs that are valid in both of them. This leads to a problem which the student should have met in abstracting axiomatics: a common fundament of euclidean and elliptic geometry. I have not yet studied this problem sufficiently to tell you how it can be solved on high school level: I suppose the best way would be to try something like the approach of the Helmholtz-Lie space problem.

(In this proposal the elliptic geometry could not possibly be replaced by the hyperbolic one. The model of the hyperbolic plane is artificial and hardly accessible to a synthetic approach, and without a model, that is, as Bolyai and Lobačevski did it, it is hardly recommendable.)

GROUPS IN GEOMETRY

Transformations in geometry were long ago advocated by F. Klein as a consequence of his so-called *Erlanger Programm*. The breakthrough of transformations in geometry is of a rather recent date. How to explain this delay, even in Germany, where Klein had been the venerated master of a generation of teachers?

It is a complicated story. Foundations of geometry in the 19th century originally meant the Helmholtz-Lie foundational approach of defining a geometry by group axioms. Klein was never a foundationalist; he was even a complete stranger to the axiomatic idea. The *Erlanger Programm* is classifying given geometries by their groups. The slogan "geometry is the invariance theory of a group" is only intelligible within the restricted frame of a few classical geometries which may be derived from projective geometry by distinguishing certain subgroups of the projective group. Not before E. Cartan's homogeneous spaces has the *Erlanger Programm* been interpreted in a foundationalist sense. As to the technique Klein's geometry has always been algebraic rather than synthetic. On the other hand Pasch and Hilbert turned away from the Helmholtz-Lie group theory interpretation of foundations of geometry. Without doubt Hilbert's *Grundlagen der Geometrie* have contributed to lengthen the life of Euclid's methods for half a century, whereas Klein's algebraic approach was too little geometry to be interpretable in school mathematics. At their best Hilbert's and Klein's approaches were teacher's background knowledge which hardly influenced school instruction. In a 1956 report I stated that though transformations were progressing in school geometry, the exposition was classical and there was not any textbook based on the transformation idea. Even the problem how to introduce and to teach transformations had not received due attention. Though meanwhile many textbooks have been going to teach transformations, rather than being solved, the problem has been aggravated. Wrong concretizations to visualize transformations have become quite usual in experiments and textbooks. It is an old misapprehension in textbooks, and it is continuously made in

Piaget's work, to visualize a transformation as picking up a figure and laying it down elsewhere without changing its shape or after having applied some similarity or affinity. With the free mobility of figures no group can be formed. To get a group one has to pick up the whole plane (or space) and put it down elsewhere. Free mobility of figures is much more intuitive than geometrical transformations. The less intuitive is very likely to be blocked by the more intuitive one, especially if moving models are used.

A more recent, and even more mistaken concretization of transformations, say of the plane, is putting pawns or persons on selected spots of the plane and interpreting a transformation as a command to these figures to change their places. Of course, all is due to end up in nonsense.

Good teaching of geometrical transformation is not easy, indeed. Maybe this has been one more reason why transformations progressed so slowly in school education. In the 1956 ICMI report I stated there is still some hope left. There is one transformation that is immediately seen as a transformation of the whole plane, rather than as the movement of a figure in the plane or as a march of selected pawns. It is axial symmetry. Central symmetries and rotations are much more difficult; translations is the most difficult case. If the teacher starts with axial symmetry and if rotations and translations are introduced as products of axial symmetries, there is a real chance that the child will grasp the concept of transformation even in the introductory phase.

There is another approach to geometrical transformations. It is by means of the square lattice, which is the most natural, or even the only natural infinite figure. Free mobility of the square lattice can, indeed, suggest translations and rotations of the whole plane.

There are, however, more reasons, why to start with symmetries. They are more interesting than translations and rotations. To a young child congruent figures are the same. It will not hit upon the idea that something has happened if a figure has only been carried to another place. To an unsophisticated mind translating is no transformation. In this regard rotation is somewhat better than translation. If a cube is translated, nothing has happened; if it is turned and put on a corner, something has changed. But mirror reflection gives the strongest feeling of an important event. Symmetry as a transformation is more attractive, more abundant, and more problematic than translation and rotation. In almost all national ICMI reports of 1956 its usefulness in teaching was stressed. Meanwhile symmetry has vastly penetrated into textbooks as far as they teach geometry.

There is one Dutch course of modern geometry by transformations for 12 year olds, by Kuipers, Siepelinga, Troelstra and Tromp. They start with symmetries and introduce translations and rotations as products of symmetries; equiformities and affine transformations appear later. The books are written in a group theory spirit though groups are not explicitly mentioned. Deductive bonds are loose in the beginning to grow gradually stronger.

TEST PAPERS

Such new methods are more readily accepted by the students than by the teachers. Traditional geometry was difficult to teach, but its level of rigor was firmly ruled by a common opinion. Teachers complain that with deductively looser methods grading of testpapers becomes more problematic. "We cannot any more tell wrong from right", is a frequent complaint. They overlook that with old geometry it was easier only within a complex of mouldy and rusty conventions.

Old style geometry will not survive for more than a few years, but if teachers, indeed, choose the easy way to tell right from wrong when grading testpapers, geometry will be lost. Problems on sets and linear algebra are more reliable provided they are dull enough.

Geometry is endangered by dogmatic ideas on mathematical rigor. They express themselves in two different ways: absorbing geometry in a system of mathematics as linear algebra, or strangulating it by rigid axiomatics. So it is not one devil menacing geometry as I suggested in the title of my paper. There are two. The escape that is left, is the deep sea. It is a safe escape if you have learned swimming. In fact, that is the way geometry should be taught, just like swimming.

University of Utrecht, The Netherlands

PETER HILTON

TOPOLOGY IN THE HIGH SCHOOL

A vision of the future, dedicated to my great teacher and intimate friend
J. H. C. (Henry) Whitehead, on the tenth anniversary of his tragic and untimely death.

1. INTRODUCTION

There has long been a debate as to how geometry should be taught in the High School, and many papers contributed at this conference will doubtless be concerned with this important question. My own attitude is based on certain principles which should, I believe, inform one's approach to the choice of content in a High School mathematics curriculum and to the choice of methods of presentation. A topic is only deserving of inclusion if it enriches the student's experience by illuminating past or current interests and concerns, or if it is capable of future – but not too distant future – application. The experience referred to may be mathematical or non-mathematical; like-wise, the application may be outside mathematics, or it may be the application of a general mathematical procedure to a more specialized mathematical situation. On this criterion, geometry deserves its place as an application of algebra and as a model of the world of experience. It is not, however, obvious that, in an inevitably crowded curriculum, one can justify the inclusion of geometrical material (for example, problems of coordinatization or Hilber-tian axiomatics) which, while elegant and deep parts of mathematics, are somewhat *sui generis* and do not make a strong impact on the rest of mathe-matics or on the world of scientific knowledge served by mathematics.

However, I would maintain that topology does merit inclusion on the basis of the criterion adduced above. One does not have to follow Piaget all the way to agree that topological properties are among those most imme-diately apprehended by our intelligence when coupled through our senses with the world of experience. It is relatively unimportant whether a rectangle is really a square; but we are immediately alerted if, by removing a subset, a space becomes disconnected. Even if it remains connected, the existence in that space of essential holes is a matter which claims our immediate attention and demands our concern. A recent approach to structural linguistics eluci-dates as the basic 'action verbs' those corresponding to topological singula-rities of vector fields – "start", "stop", "cut", "meet", "give", "take", "cross", "exchange", etc. Thus topology does deal with the world of experience, and seeks to answer questions which seem basic, palpable, and natural. Moreover, topology has immensely strong interaction with other mathematical disciplines, in particular, abstract algebra, geometry, and real and complex analysis. Thus, in a carefully articulated course, there could be constant reinforcement of basic ideas and rekindling of interest through the interplay of these disciplines.

Geometry is a vast subject, and so, too, is topology. Thus it is not particularly helpful merely to advocate that some topology find its way into the High School curriculum. However, one remark of a global nature, relevant to the choice of topics within topology worthy of a place in the syllabus, should be made – and may arouse controversy. Topology has been described by some as "rubber sheet geometry", and the impression has been given that it is essentially a "fun" subject, in which one tries to turn bicycle tires inside out, plays with pretzels and trefoil knots, and constructs Möbius bands and Klein bottles. I agree that these examples of topological spaces should be included as illuminative examples; but I think it very important to treat the subject, in essence, with the seriousness it deserves. For, if we do not, then there is clearly no case for preferring topology to more conventional geometry as a component of a well-constructed syllabus. Of course, I advocate that the student learn, by reading and experiment, of the fun to be had from these elementary topological configurations. But I emphasize that I am concerned primarily to exploit the intrinsic interest, to the observant and scientifically curious, of topological notions; to develop its interconnections with other mathematical disciplines; and to teach it in a way consonant with the student's later contact with the subject.

In this paper it would be unreasonable for me to attempt a comprehensive syllabus for High School topology. This would, in any case, be premature in view of the actual state of my own thinking on the subject. A good account of the possible structure of such a syllabus is outlined in [1] and is incorporated as Chapter 25 of [2]. I will content myself in this paper with only one fairly detailed description, namely that of the way I might seek to develop the concept of the *fundamental group* of a pointed topological space. My choice of this particular topic is based on the following considerations. I would suppose that it is generally conceded that a High School curriculum in mathematics should include some group theory. This group theory will itself have arisen by abstracting from certain mathematical situations (many of them geometrical) in which groups naturally arise. It is, therefore, not introducing any new way of thinking to educe a group to measure the number of holes (or *holiness*) of a topological space. However, we have here the possibility of doing two very exciting things not previously available to us.

First we can bring out the *functorial* nature of the fundamental group, which appears as a (homotopy invariant) functor from the category of pointed (= based) topological spaces and pointed maps to the category of groups and homomorphisms. Why should we wish to do this? Not, certainly just to show that we are "with it"! There is, I maintain, a serious reason for seeking to familiarize a High School student with the notions of category and functor, and this resides in the fact that he is thereby acquiring first-hand experience of current mathematical language and methodology. It is a frequent and well-justified complaint of the student that his early mathematical experience does not give him a clear picture of what sort of thing he will be doing later in his mathematical education; for example, he becomes extraordinarily adept at differentiating and integrating familiar functions and finds that this skill is not highly rated among university mathematics majors, although it stood him in excellent stead in High School. Of course, I do not claim that the language of categories and functors is exclusively the language in which modern mathematics is conducted; but it pervades so much of modern algebra and topology and is beginning to penetrate so many other disciplines that it must be learnt by any one wanting to become conversant with modern mathematics in the large.

A second exciting consequence of studying the fundamental group depends on the possibility of realizing a group as the fundamental group of a polyhedron, which may be taken to be two-dimensional and will be compact if the group is finitely presented. This leads to the topic of combinatorial group theory and to the idea of doing group theory through the fundamental group representation. The significance of subgroups, normal subgroups, and quotient groups would then emerge very naturally through the introduction of the notion of covering space, but it may be that this concept would have to await the reappearance of topology the second time round the "spiral" curriculum.

It is certainly a matter for argument just how complete a treatment of the fundamental group, at this level, should and could be. Clearly there are many results which can be stated meaningfully and appropriately on first introduction to the fundamental group, but whose proof would be probably too difficult for most students; as an example I would instance the simplicial approximation theorem. However, it is almost certainly true that there are fewer such results in elementary topology than there are in the differential calculus. For the latter rests, logically, on deep properties of the real number system and continuous functions and on limiting notions of considerable sophistication. Of course, one cannot do topology without the notion of continuity, and one cannot introduce the fundamental group without mentioning the unit interval, but the properties one requires of continuous functions are far more accessible than, for example, those required in a careful treatment of maxima and minima. Moreover, one can give a purely combinatorial definition of the fundamental group of a simplicial complex and then the entire analytical sophistication of the subject resides in the key theorem, based on a fairly elementary application of the simplicial approximation theorem, that this combinatorial gadget coincides with the topological fundamental group of the underlying polyhedron of the simplicial complex. I state as my firm opinion that quite as much significance, for the student's present and future understanding of mathematics, lies in the statement of this key theorem as in its proof. For here he meets a good example of a basic aim of mathematics – to show that a certain notion, derived from a somewhat complicated mathematical structure, is in fact independent of certain elements of that structure. The combinatorial definition of the fundamental group requires the explicit description of the simplicial structure overlaying the polyhedral space, yet the fundamental group depends only on the underlying space – indeed, only on its homotopy type.

This result, typical of so many invariance theorems in mathematics, by no means constitutes the only reason, even at this elementary level, for introducing simplicial complexes. It is an agreeable fact about the fundamental group of a space, which distinguishes it, for example, from the homology groups, that it may be defined with a minimum of machinery. The two notions involved – composition of loops and equivalence classes of loops under deformation – are both very intuitive, and thus the fundamental group requires little sophistication for its understanding. However, it is a serious defect of many elementary presentations of this concept that no clue is given as to how the fundamental group may be calculated. Again there is an important message here; one may have an excellent tool for theoretical investigation – that is, for the development of mathematical theory – but one should also know how to utilize it in special cases. It is an arid satisfaction to know that two spaces X and Y cannot be homeomorphic if their fundamental groups are non-isomorphic, if one has no means of computing their fundamental groups, even in the sense of being able to write down generators and relations. The introduction of the simplicial structure is thus well-motivated, once the fundamental group has been discussed in its topological form, by the demand to be able to give a *presentation* of the group (in the technical, group-theoretic sense). Further theory, of an entirely combinatorial nature, then enables one to produce more economical presentations, and a natural goal of this sequence of ideas is the realization of any group as the fundamental group of an appropriate polyhedron.

Apropos the argument in the foregoing paragraph, a remark may be relevant relating to a current controversy about the training received by mathematics Ph.D'.s. There is no love lost between the new Ph.D. and industry in this country. The former finds the latter, as an employer, restricts his freedom and fails to provide a good atmosphere for his work. The latter finds the former quite unsuitable, primarily because he is trained as a theory-builder rather than a problem-solver. This complaint is also often made by university departments in the sciences (physical, biological, social) who seek the services of the mathematician in connection with their own teaching and research program. I mention this here because topology is often cited as typical of those mathematical disciplines encouraging the building of structures rather than the solution of problems. I believe this charge to be wholly unfounded; many beautiful problems in topology have been solved in recent years, and it is a function of the instructor's (or researcher's) approach whether he gives due emphasis to both aspects of the subject. I would hope that the treatment of the fundamental group suggested in this paper presents a balanced description of both aspects – and does so explicitly.

It is in Section 3 of this paper that I present detailed recommendations for a study of the fundamental group. Section 2, on the other hand, contains a summary of the other topological notions that I believe appropriate in the High School. It is partly due to lack of space that I present only an outline of these recommendations, and partly due to the fact that my own thinking on this problem has not yet reached a definitive stage. Partly, too, as references [1, 2] suggest, the material described in Section 2 is much less innovative than that described, in greater detail, in Section 3. I would hope that it is not really in dispute today that some topology should feature in a High School mathematics program. This particular recommendation of the Cambridge Conference on School Mathematics, aroused no overt opposition and was, indeed, regarded as somewhat conservative by those familiar with the European educational scene.

Inevitably, however, one must meet the objection that one is over-crowding the High School syllabus. This objection can only be met in part by pointing out that one is not insisting, first time round, on a thorough, comprehensive treatment of the topics named, with all proofs provided. Proofs may be explicitly omitted or references can be given; the only incontrovertible desideratum is that no false statements should be made, since these produce disillusion and discontent among the more gifted students – and, in any case, no satisfactory unlearning procedure has yet been devised. However, I should face, already at this preliminary stage, the necessity of making some suggestions as to how room can be found in the curriculum for the topics recommended. I have already averred that I do not believe that other kinds of geometry are really more appropriate or more useful at this level, so that I would be happy if, the foundations of analytic geometry having been laid, a topology syllabus were regarded as fulfilling the geometry requirement. (Of course, I would be very happy in a situation in which *alternative* geometry options could be offered, including topology; but it does not seem realistic to expect this for some time, except in rare, privileged High Schools. Perhaps CSMP may show us the way here.) Beyond this recommendation, there is one proposal which, it seems, must be considered seriously, and that is to postpone the introduction of the differential calculus. To argue the merits of this proposal would surely require a paper quite as exhaustive as any at this conference. Here I must be content to repeat that the theory of continuous functions and limits on which the differential calculus is built is quite sophisticated, certainly more sophisticated than the foundation of elementary topology. On the other hand, to postpone the presentation of the calculus to the college freshman (or sophomore) year may well mean that many students never encounter this important topic, and may also produce difficulties with regard to the teaching of those subjects like physics, engineering and biology in which the calculus is required. I believe there are answers to these objections to delaying the calculus, but I prefer to leave the issue open. Of course, to postpone the calculus is to weaken one of the arguments given for introducing topology, namely, that it interacts so naturally with real and complex analysis. However, the interaction with algebra remains; and the argument is not entirely eviscerated since the concept of a *continuous* function remains central to topology and will, of course, have been met by the student in connection with his study of polynomials and other real-valued functions. Moreover, there is much insight into the nature of geometry to be gained by regarding the group of self-homeomorphisms of a space as a generalization of groups of motions in Euclidean geometry.

I must conclude this introductory section by reiterating my disclaimer to the right or intention to make prescriptive judgments on what is appropriate to a topology sequence in the High School. This paper is intended to form a basis for discussion; my only real evidence for the validity of my tentative recommendations is based on introspection. When I recall the hours spent on proving recondite geometrical propositions, or on establishing esoteric trigonometrical identities, I cannot but wish I could earlier have been exposed to the basic geometrical notions inherent in topology. This is not mere hindsight, insofar as my own critical attitude towards my High School syllabus antedated my own discovery of topology. Moreover, I was fortunate to have good teachers who succeeded in extracting some inspirational juice from even the driest of topics. However, it was not until I had the good fortune, during my freshman year at university, to attend a course by Henry Whitehead that I realized what projective geometry was really about, and that I first heard the word "topology" mentioned. Perhaps the inspiration provided by Henry Whitehead has carried me away – but it is a journey I would not have missed.

2. ESSENTIAL COMPONENTS OF A TOPOLOGY PROGRAM

In this section I briefly outline what I regard as the basic constituents of a High School topology program. That these constituents are suitable for the consequent work on the fundamental group is readily demonstrated; but I would prefer them to be judged in a broader context as providing the link between earlier geometric experience and the mathematical ideas the student will meet in his university courses. As indicated in the previous section they should also be judged for their intuitive content, since, I claim, they embody very natural and immediate concepts drawn from experience, and thus exemplify in the best possible way the proper role to be played by generalization and abstraction in mathematics.

The starting point must be Euclidean space $\mathbf{R}^n$ and its subspaces; continuity is then defined by means of *balls*. That is, $\mathbf{R}^n$ is endowed with a metric ϱ and a ball in $\mathbf{R}^n$ of *radius ε center x* is the set of points y with $\varrho(y, x) < \varepsilon$. If X is a subset of $\mathbf{R}^n$ and $x \in X$, then a ball in X, center x, is the intersection with X of a ball in $\mathbf{R}^n$. A function $f: X \to Y$, where X, Y are subsets of Euclidean space (not necessarily of the same dimension) is *continuous* if, for each $x \in X$, the counter image of a ball in Y, center $f(x)$, contains a ball in X, center x. The student should observe that this really is the "$\varepsilon\delta$"-definition – if he has already had that inflicted on him! He should also observe that we have here a natural category since the composite of continuous functions is continuous. Inclusions of subsets are obviously continuous; hopefully the student has not previously been led into confusion by confounding the codomain of a function with its image, so that he is ready to consider this important class of continuous functions.

The first generalization should be to *metric* spaces; the generalization is smooth and easy and there are Hilbert spaces available for motivation. Another motivation is, to be sure, the desire to find out which of the ingredients of a mathematical situation are essential for the conclusions currently under consideration. An *open* set in a metric space M is then defined as a subset S of M such that, if $x \in S$, then some ball, center x, is contained in S. One can then readily establish the axioms for the system of open sets; one also shows that the balls are themselves open sets, and one characterizes continuity in terms of open sets. Further a subset of a metric space M is obviously a metric space with the same (or induced) metric and then the open sets of a subspace are just the intersections of the open sets of M with the subspace. *Closed* sets are then defined as complements of open sets; it is important to show here that, though a closed set is the complement of an open set, the *set* of closed sets is not the complement of the *set* of open sets.

At this point the students should specifically remark that, in a sense, the metric has disappeared. Attention has been concentrated on the open sets and the closed sets, and continuity is defined most simply by means of those basic concepts. He should thus be ready for the notion of a *topological space*, in which a topology is specified by means of a family of distinguished subsets (the *open sets*) satisfying the axioms already described. The definition of continuity is immediate and we thus have a functor from the category of metric spaces to the category of topological spaces. This functor is neither one-to-one nor onto with respect to the objects of the categories. Each fact can be easily illustrated. The notion of a *metrizable* topological space emerges naturally, The *Hausdorff* property should be brought to the fore at this point as a property of metrizable topological spaces which is not shared by all topological spaces; of course, this property will reappear frequently in the sequel. The *lattice of topologies* on a set deserves mention; but, more important are the universal characterization of the subspace and quotient space topologies, as well as their characterizations in terms of *coarseness* and *fineness*.

It is convenient to have the notion of *neighborhood*; this should not be required to be open, so that one may meaningfully talk of open and closed neighborhoods. Attention should be very much concentrated on metrizable spaces – and, indeed, on subsets of Euclidean space – but it is reasonable to refer here to countability axioms.

The *topological product* is an important notion which it would also be convenient to have here since it provides a context for many future discussions and an illustration of concepts to be introduced later. It is first defined explicitly by specifying the topology on the Cartesian product, but the universal mapping property should be made explicit. The product of metrizable spaces is, of course, metrizable. More precisely, there is an obvious product of metric spaces and the functor from metric spaces to topological spaces is product-preserving.

The next big idea must surely be that of *compactness*, defined by means of open coverings. *Sequential compactness* should also be mentioned and their equivalence for metric spaces at least asserted. That "compact=closed" for Hausdorff spaces and that the continuous image of a compact set is compact are easily proved; even the property of the Lebesgue number is not difficult to prove and its importance warrants its inclusion. One conspicuous pay-off is that the theorem on uniform continuity becomes a trivial consequence.

From compactness one should probably move to connectedness, pointing out how this definition derives from an attempt to give precision to a very obvious intuitive concept (the same is, of course, true of the concept of continuity). Again its preservation under continuous functions is a crucial property and this can be seen to be at the heart of some fundamental properties of continuous real functions. The idea of disconnecting a space (e.g., the Jordan curve theorem) finds a natural place here, though there are pedagogical problems of "belaboring the obvious". An example should be given to contrast connectedness with path-connectedness; but again it should be pointed out that the concepts coincide for nice spaces, and the emphasis should shift to the latter concept. The study of paths leads naturally to the concept of homotopy and deformation. This is the opportuinity for a contrast of the topological classification by *homeomorphism* with the homotopy classification by *homotopy type*, and of the methods used. It also suggests that other classifications might be considered, so that the way is prepared for the introduction of *combinatorial structure*.

I believe that attention should be concentrated on *finite simplicial complexes* covering compact *polyhedra*, though perhaps, non-compact polyhedra could be mentioned. My reason is that the complexes should, as I say, be geometric, in the sense that they are simplicial subdivisions of topological spaces which are themselves subsets of Euclidean space. I would suppose that the case for considering abstract simplicial complexes would best be made in the context of the study of the fundamental group; and it is, clearly, the abstract complexes which lend themselves most easily to generalization to the infinite case.

The preparation, then, would be through the affine structure of simplexes, and would early concern itself with barycentric coordinates. It would probably be worth discussing barycentric subdivision and proving that the mesh of a complex decreases to zero with successive subdivision. This would probably require some discussion of convexity, but this is, in any case, in place in any geometry syllabus. (It is also in place in elementary function theory, though it is often neglected.) Simplicial maps are easily defined, leading again to a good and important categorical notion; and by introducing the locally-linear continuous map of polyhedra induced by a simplicial map, one arrives at one of the most important functors of topology. The *simplicial approximation theorem* then measures the "faithfulness" of this functor and is the key to establishing the relation between the combinatorial and topological theories.

I have refrained from making any specifically pedagogical recommendations in this article, confining myself to a description, in outline in this section, of the proposed content of a topology course. Any such recommendations would, of course, have to take into account the entire educational context. If one were in a position to provide the student with an "activity package", then I would propose that the simplicial approximation theorem be the subject of such a package. Failing this I would hope a pamphlet were available enabling the most "involved" students to get an idea of how its proof runs.

Subcomplexes should be discussed; from these the notion of subpolyhedra immediately follows, but even more important, one can discuss the whole question of *relativization,* another key concept of mathematical methodology. Contiguity might be introduced as the combinatorial analog of the homotopy relation and the simplicial approximation theorem extended to cover a homotopy between continuous maps. Special topics which might be introduced at this stage should include the *Euler characteristic* and the detection, through the simplicial complex, of the *components* of a polyhedron.

There are many further notions the student might well be given "if time permits". I would myself be happy to see some reference to *manifolds* in the course I am outlining. I would also hope that the notion of *localization* of topological properties and concepts could be aired with special mention of local *compactness* and local homeomorphism. The latter would be most useful in describing the topological structure of a manifold; it would also be reasonable to treat covering spaces and covering-space projections as a rich source of important illustrations of local homeomorphism.

It goes without saying that the entire treatment, from beginning to end, must be accompanied by a wealth of well-chosen illustrative examples and exercises for the student.

3. THE FUNDAMENTAL GROUP

After this rapid summary of what, at my present stage of thinking, I would regard as proper to an elementary course in topology, I come to my more specific proposal relating to a course devoted to the fundamental group. Prerequisite to this course would be the content, at least approximately, of Section 2, together with an introduction to group theory. The latter should, hopefully, have included the notion of presenting a group by means of generators and relations.

Right at the outset I would suppose that we are dealing with path-connected spaces X furnished with a fixed base point x_0. The notion of being path-connected has already been introduced; we now define a *loop on X at x_0* to be a continuous map

$$f : I \to X$$

from the unit interval $0 \leqslant t \leqslant 1$, which we write I, to X such that $f(0) = f(1) = x_0$. We may write $\dot{I}$ for the frontier $\{0, 1\}$ of I in $\mathbf{R}$.

We are going to *compose* loops and we are going to *deform* loops. After an informal introduction to these basic notions, we prepare the way for the precise treatment through the following lemma.

LEMMA 1. *Let X be given as the union of a finite number of closed subsets,* $X = \bigcup_{i=1}^{n} F_i$, *and let $f: X \to Y$ be a function from the topological space X to the topological space Y such that $f \mid F_i$ is continuous for each i. Then f is itself continuous.*

This lemma should be proved; it would also be advisable to take time here to show that the words "finite" and "closed" cannot be omitted from the enunciation, though one can omit "finite" if one replaces "closed" by "open". The lemma should then be immediately applied to validate the definition of composition,

$$(l*m)(t) = \begin{cases} l(2t), & \text{if } 0 \leqslant t \leqslant \tfrac{1}{2} \\ m(2t-1), & \text{if } \tfrac{1}{2} \leqslant t \leqslant 1 \end{cases} \tag{2}$$

of loops l and m (on X at x_0) and to show that (2) is compatible with the homotopy relation. Here we say that $l_0 \simeq l_1$ (the loop l_0 is homotopic to the loop l_1) if there exists $F: I \times I \to X$ such that $F(t, i) = l_i(t)$, $i = 0, 1$, and $F(i, u) = x_0$, $i = 0, 1$ (that is, the homotopy is *rel $\dot{I}$*). Then (2) is compatible with the homotopy relation in the sense that if $l_0 \simeq l_1$ and $m_0 \simeq m_1$, then

$$l_0 * m_0 \simeq l_1 * m_1. \tag{3}$$

It follows from (3) that (2) induces a composition in the set, $\pi(X, x_0)$, of homotopy classes of loops on X at x_0; if $[l]$ is the homotopy class of l then we set

$$[l][m] = [l*m], \tag{4}$$

and (3) guarantees that this composition in $\pi(X, x_0)$ is unambiguous. At this point analogies may well be drawn with other familiar situations in which one defines an operation on sets of equivalence classes by "picking representatives" – for example, modular arithmetic.

Next one defines the reverse of the loop l to be the loop $\bar{l}$, given by

$$\bar{l}(t) = l(1-t) \tag{5}$$

and shows that $[\bar{l}]$ depends only on $[l]$. One is ready for the first big theorem.

THEOREM 6. *The set $\pi(X, x_0)$ is a group under the operation (4). The neutral element is the class e of the constant loop $c(t) = x_0$, $0 \leqslant t \leqslant 1$; and $[l]^{-1} = [\bar{l}]$.*

The proof should not be given in tedious detail. One verification may be carried out very explicitly, for example, the demonstration that $[\bar{l}]e = [\bar{l}]$.

Thus we define $F: I \times I \to X$ by

$$F(t, u) = \begin{cases} l\left(\dfrac{2t}{1+u}\right), & \text{if } 0 \leqslant t \leqslant \tfrac{1}{2}(1+u) \\ x_0, & \text{if } \tfrac{1}{2}(1+u) \leqslant t \leqslant 1, \end{cases}$$

use Lemma 1 to verify that F is continuous, and then observe that F is a homotopy from $l*c$ to l. One should then point out what is really going on geometrically; one is taking the unit interval and, keeping its endpoints fixed, one is stretching $\langle 0, \tfrac{1}{2} \rangle$ to cover the whole interval and compressing $\langle \tfrac{1}{2}, 1 \rangle$ onto its endpoint 1. It is useful to have the informal language so flexible that one can think of the t-coordinate as representing either a space-dimension or a time-dimension. Thus the difference between $l*c$ and l is that, in $l*c$, one rushes round the loop l in half the time and rests at x_0 for the second half. Plainly $l*c$ and l should be homotopic and so should $c*l$ and l. Likewise one points out that composition of loops is not strictly associative, but that the difference between $(l*m)*n$ and $l*(m*n)$ amounts to the time allocated to l, m, n in executing the entire loop. One then passes to Figure 1, and finally invites the student to provide a formal proof.

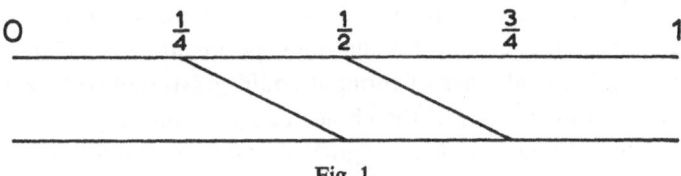

O $\qquad$ $\dfrac{1}{4}$ $\qquad$ $\dfrac{1}{2}$ $\qquad$ $\dfrac{3}{4}$ $\qquad$ 1

Fig. 1.

It remains to show that $l*l \simeq c$. Here again we may use both "space-language" and "time-language" to make the idea of the argument clear.

Examples can only be treated informally at this stage, since no methods have been given for calculating *fundamental groups* $\pi(X, x_0)$. However, plausible reasons can be given to suggest that the group does, in some sense, count the number of holes at least when circular holes are cut out of a rectangular piece of cardboard.*

One next develops the idea of the *induced homomorphisms*. Given a map $f: X, x_0 \to Y, y_0$ (that is, f is continuous and $f(x_0) = y_0$) then one may asso-

---

* Of course, a group does more than just enable one to *count* holes. It is a triumph of algebraic topology that it endows a space with natural group structures, like the fundamental group and the homology groups, which penetrate deeper into the topology than do the numerical invariants derived from these groups. When the student can calculate fundamental groups, he should cetainly be offered the example of the real projective plane, whose fundamental group is cyclic of order 2. Thus there is a loop which is not contractible; but if executed *twice*, it becomes contractible!

ciate with each loop l on X at x_0 the loop fl on Y at y_0. Moreover

$$fl_0 \simeq fl_1 \quad \text{if} \quad l_0 \simeq l_1, \tag{7}$$

and

$$f(l*m) = fl*fm. \tag{8}$$

Relations (7) and (8) ensure that f induces a homomorphism, which we write $\pi(f)$, from $\pi(X, x_0)$ to $\pi(Y, Y_0)$; we merely define

$$\pi(f)[l] = [fl]. \tag{9}$$

Further if $f \simeq g : X, x_0 \to Y, y_0$, the homotopy keeping x_0 at y_0, then $[fl] \simeq \simeq [gl]$. This observation, together with some trivialities, establishes the second major theorem.

THEOREM 10. *π is a functor from the category T_0 of based topological spaces to the category of groups. The functor π is an invariant of based homotopy type.*

It is true that we have explicitly defined π only for path-connected spaces, but there is no difficulty in extending the definition. There is also no real point in so doing. However, the question does arise as to the actual nature of the dependence of $\pi(X, x_0)$ on the choice of x_0. The situation should be explained and can easily be made plausible – it is not even too difficult to make it precise. It would be a fascinating experiment to see if the students trained in mathematical ways of thought could grasp that $\pi(X, x_0)$ is independent up to isomorphism of the choice of x_0 within its path-component, but that π is not a functor on the category of path-connected spaces since the isomorphism is not canonical.

One should now feed in enough facts to illustrate the methodology of exploiting a functor. If the students will accept (what will be established later) that the fundamental group of a circle is not trivial, then one may prove that a circular disc may not be retracted onto its boundary and hence that every map of a circular disc into itself has a fixed point. Likewise one cannot deform the surface of a sphere into a bicycle tire (torus).

This stage of the treatment of the fundamental group should close with the elementary demonstration that the fundamental group of a topological product is the direct product of the fundamental groups of the factors. This yields the precise value of the fundamental group of a torus, or of a cylinder, given the fundamental group of a circle.

At this point it is to be hoped that the student is ready to consider the question of how a fundamental group can be calculated. First, the meaning of this problem must be made more precise-when can we be said to "know" a group? Clearly, it will be necessary to talk a little about the presentation of a group by means of generators and relations – geometrical examples of symmetry groups are particularly appropriate here. An ambitious experiment would involve some digression on the undecidability of the isomorphism question for groups.

Next, the plan of attack should be explained. We define the fundamental group of a simplicial complex by a purely combinatorial procedure – this part of the procedure has nothing, in principle, to do with topology. Then, however, we establish, by the method of simplicial approximation, that this combinatorial fundamental group is isomorphic to the fundamental group of the polyhedron underlying the simplicial complex. This is, of course, the main result of the entire theory. The fundamental group of a simplicial complex is, indeed, very easily described in terms of generators and relations. It is not too difficult to reduce this presentation to an extremely economical one. We are then in a position to calculate fundamental groups and even to prove some more theorems about them.

In greater detail the program might unfold as follows. We start with a simplicial complex K with a preferred vertex a. An *edge-loop*, or loop, on K at a is a finite sequence of vertices of K,

$$a^{i_0}a^{i_1} \dots a^{i_n}, \ a^{i_0} = a^{i_n} = a \,, \tag{11}$$

such that successive vertices belong to the same (closed) simplex of K (repetitions are allowed). Obviously the edge-loops stand in close analogy with the loops on a topological space and it is easy to associate a loop running round the appropriate edges of the polyhedron $|K|$ underlying K with the edge-loop (11). We compose edge-loops by setting

$$l * m = aa^{i_1} \dots a^{i_{n-1}}aa^{j_1} \dots a^{j_{m-1}}a \tag{12}$$

if $l = aa^{i_1} \dots a^{i_{n-1}}a$, $m = aa^{j_1} \dots a^{j_{m-1}}a$. Again, this composition is very natural, especially in view of the analogy. The reverse of l, above, is given by

$$l = aa^{i_{n-1}} \dots a^{i_1}a \,. \tag{13}$$

It remains to imitate homotopy. This, of course, is the subtlest step. The *allowed moves* are (i) the deletion of a^{i_r} from edge-loop $\dots a^{i_{r-1}} a^{i_r} a^{i_{r+1}} \dots$ if $a^{i_{r-1}}$, a^{i_r}, $a^{i_{r+1}}$ belong to a single simplex of K, or the opposite of this move (i.e., the insertion of a^{i_r}), together with the very special move (ii) replacement of aa by a or the opposite of this move. These moves must be carefully exemplified. It should be pointed out that the substantial move (see Figure 2) is that of replacing $\dots ac \dots$ by $\dots abc \dots$, where we have a triangle abc in K (or the opposite). The other moves are just "tidying up operations" associated with repeated vertices in an edge-loop (or the opposite). We say that two edge-loops are homotopic if one may be obtained from the other by a finite sequence of allowed moves. This is plainly an equivalence relation. To see that it is compatible with the composition (12) we invoke the import methodological principle that, since the homotopy relation is *generated* by the allowed moves, it is sufficient to check that the allowed moves are compatible with composition of edge-loops. The following theorem, analogous to Theorem 6, is, however, quite trivial by comparison. We write $[l]$ for the class of l.

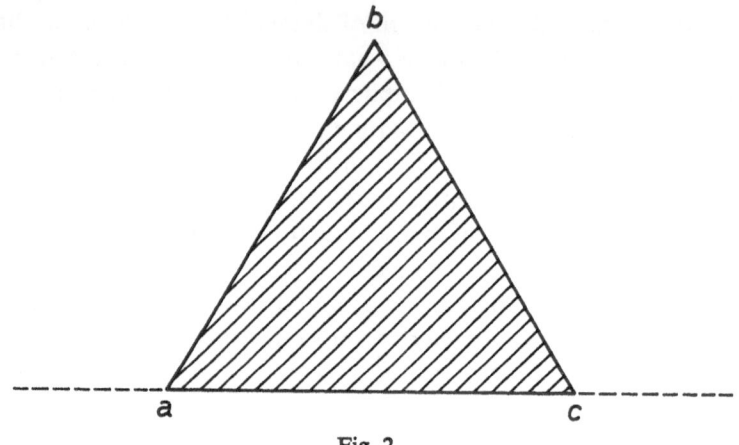

Fig. 2.

THEOREM 14. *The set $\pi(K, a)$ of homotopy classes of edge-loops is a group under the operation induced by* (12). *The neutral element is the class of a, and* $[l]^{-1} = [\bar{l}]$.

It is also an easy matter to develop the idea of the homomorphism $\pi(f)$: $\pi(K, a) \to \pi(L, b)$, induced by a simplicial map $f: K, a \to L, b$, and thus to show that π is a functor from the category of (based) simplicial complexes and (based) simplicial maps to the category of groups and homomorphisms. It is probably inadvisable to pursue too explicitly the homotopy invariance of f as this would require a review of the contiguity relation for simplicial maps, but this aspect should, of course, be available to any student who raises the question.

We have already indicated how an edge-loop on K at a gives rise to a loop on the underlying polyhedron $|K|$ at a. It is again obvious that an allowed move on an edge-loop leads to a deformation of the associated loop and that the operations of composition in $\pi(K, a)$ and $\pi(|K|, a)$ match under the given association. Thus we have a homomorphism

$$\varrho: \pi(K, a) \to \pi(|K|, a) \qquad (15)$$

and the main theorem of the theory is the following.

THEOREM 16. *The homomorphism $\varrho: \pi(K, a) \to \pi(|K|, a)$ is an isomorphism.*

Probably one should only sketch the proof, giving the diligent student a reference or a specially prepared pamphlet for the details. The ideas are, however, so important that it would be wrong to give no indication of proof. Also we have in the homomorphism ϱ a nice example of a *natural* transformation and this should be drawn to the student's attention, though not in a heavy-handed way.

The force of Theorem 16 is that, to calculate the fundamental group of a polyhedron, one may make a purely combinatorial construction based on the simplicial structure. This is the *practical* implication and leads immediately to the calculation of the fundamental group of a sphere, for example. (One should, of course, point out that $\pi(K, a)$ depends only on the 2-skeleton of K.) On the other hand, there is the *theoretical* implication, that $\pi(K, a)$ does not depend on the particular triangulation of $|K|$, but depends only on $|K|$ itself, indeed only on its homotopy type. This is far from obvious *a priori*.

There remains the question as to how one can calculate $\pi(K, a)$. We say that a simplicial complex L is 1-*connected* if it is connected and $\pi(L, b)$ is the trivial group for any vertex b of L – we know that $\pi(L, b)$ is independent, up to isomorphism, of the choice of b. Now let K be a (finite) connected simplicial complex. Then there is certainly a 1-connected subcomplex L of K containing all the vertices of K; for example, we may take a *tree* (1-dimensional with no non-trivial loops) in K containing all the vertices of K, and the underlying polyhedron of a tree is plainly contractible. Having chosen L, we define a group $G(K, L)$ as follows:

$G(K, L)$ has one generator g^{ij} for each edge $a^i a^j$ of K (we allow $i=j$); the relations are

(I) $g^{ij} = 1$ if $a^i a^j \in L$ (allow $i = j$)

(II) $g^{ij} g^{jk} = g^{ik}$ if $a^i a^j a^k \in K$ (allow any of i, j, k to coincide).

Notice immediately that $g^{ii} = 1$ and $(g^{ij})^{-1} = g^{ji}$. Then the final theorem to achieve our program is

THEOREM 17. $G(K, L) \cong \pi(K, a)$.

The proof of this theorem requires that one introduce *edge-paths* and composition of edge-paths. The definitions are obvious and will surely be supplied by the student who understood the theory of edge-loops. Of course, one can even apply the allowed moves to edge-paths to generate a good notion of homotopy of edge-paths. However, the purpose here is to set up homomorphisms $\varphi : G(K, L) \to \pi(K, a)$ and $\psi : \pi(K, a) \to G(K, L)$ and prove them mutually inverse. To define φ one first associates with each vertex a^i a path x^i in L from a to a^i (associating the path a with the vertex a). Then if

$v^{ij} = x^i * a^i a^j * \bar{x}^j$, v^{ij} is an edge-loop and one defines $\varphi(g^{ij}) = [v^{ij}]$. At this point the whole concept of setting up a homomorphism by assigning values to the generators and checking the compatibility with the relations comes under discussion; but in a context in which it is clear by drawing pictures that all is well. It is also easy to see that φ is onto, but less easy to see that φ is one-one. We recommend that this last step be achieved by introducing ψ. We first define a function θ from the set of edge-paths of K to $G(K, L)$ by

$$\theta(a^{i_0}a^{i_1} \dots a^{i_n}) = g^{i_0 i_1} g^{i_1 i_2} \dots g^{i_{n-1} i_n}.$$

Then relation (II) shows that if the path u' is obtained from the path u by an allowed move, $\theta(u') = \theta(u)$. Also $\theta(l*m) = \theta(l)\theta(m)$, obviously. Thus, θ, restricted to the edge-loops, induce $\psi : \pi(K, a) \to G(K, L)$. However, the verification that $\psi\varphi$ is the identity, easy as it is, requires that θ was defined for edge-paths and not merely for edge-loops. Again, this easy verification allows the student to concentrate on the important mathematical principle that the relation $\psi\varphi = 1$ is established over the whole of $G(K, L)$ once it is established on the generators.

Before proceeding to give the host of examples which now becomes available, it is as well to show just how economical $G(K, L)$ really is. In fact, in view of the given relations, we really only need a generator g^{ij} if $a^i a^j \in K - L$ and $i < j$, and we only need the relation $g^{ij} g^{jk} = g^{ik}$ if $i < j < k$, the appropriate g^{rs} to be replaced, of course, by 1 if $a^r a^s \in L$. This leads, provided L is taken fairly big, to a very simple system of generators and relations. For example if K is a bunch of k circles (triangles) then $\pi(K)$ is *immediately* seen to be free on k generators.

This description of how a first course in the fundamental group might develop has been concerned to stress the characteristic mathematical methodology that enters into the development. Of course, there are other features of the course which are dictated by considerations of sound pedagogy and I do not believe it necessary to stress yet again the importance of constant exemplification and application. The one point which I have not developed in this description, so far, but which was referred to in the introductory section, is the possibility of doing group theory via the fundamental group. There is, first, the argument that every group may be realized as the fundamental group of a polyhedron and the polyhedron may be taken to be compact if the group is finitely presented. This theorem is not so easy to establish rigorously but, as with other aspects of this subject, it is not difficult to render it plausible. Given this theorem, the rest is again readily available using the combinatorial definition of $\pi(K, a)$. One may construct a *covering complex* $\tilde{K}_Q$ corresponding to any subgroup Q of $\pi(K, a)$. Moreover if Q is normal in $\pi(K, a)$ than the quotient group $M = \pi(K, a)/Q$ acts as a group of cover-transformations of $\tilde{K}_Q$. The theorems that a subgroup of a free group is free and that a subgroup of index d of a free group on n generators is free on $d(n-1)+1$ generators follow immediately – of course, there is no problem about realizing a free group as a fundamental group. This type of material may be hinted at at the end of the course; it is probably better regarded as a new packet in the topology matrix. An excellent treatment of the ideas presented in this suggested course, including those outlined above relating to covering complexes and covering spaces, is to be found in [4], except that Massey does not treat simplicial complexes, which are conceptually far simpler than the CW-complexes of current homotopy theory.

Cornell University, Ithaca, N.Y., U.S.A.

REFERENCES

[1] H. B. Griffiths, *Topology, Mathematics Teaching*, Association of Teachers of Mathematics, 1967.
[2] H. B. Griffiths and P. J. Hilton, *Classical Mathematics*, Van Nostrand, 1970.
[3] *Goals for School Mathematics*, Cambridge Conference on School Mathematics, Houghton Miflin, 1964.
[4] W. S. Massey, *Algebraic Topology: An Introduction*, Harcourt, Brace and World, 1967.

M. A. JEEVES

SOME EXPERIMENTS ON STRUCTURED LEARNING

INTRODUCTION

When an experimental psychologist accepts an invitation to share in a conference of distinguished mathematicians and mathematical educators it is most important that when he speaks he makes clear his standpoint and presuppositions and the limitations of what he has to say. With the widespread interest in and enthusiasm for 'new mathematics' of one sort or another, it is inevitable, and indeed highly desirable, that educational psychologists should be closely involved in attempts to evaluate the efforts and results of the different innovators in this field. In general, the educational psychologists' concern tends to be with the gross or overall effects of what the different materials and methods achieve. They therefore tend to be concerned with things like success and failure rates, and rates of progress in developing various mathematical skills. To put it another way, they tend to be more concerned with *what* is achieved than with *how* and *why* it is achieved. By comparison, and from the standpoint of an experimental psychologist interested in understanding more about complex learning and thinking, my primary interest is in understanding the *processes* whereby the end products are achieved and only secondarily in success or failure rates. To put it another way I want to know what factors are responsible for success or failure not just how many subjects succeed or fail. Perhaps by way of illustration I may compare an approach used extensively by the late Sir Frederic Bartlett and his students. He repeatedly pointed out that in studying perceptual-motor skills it was more important to know *how* the various component parts of a complex skill were built up and integrated into a whole, than to know whether the skill was successful or not. Though of course, this latter is also very important. What I am suggesting is that an approach which has been tried and clearly demonstrated to be successful in twenty five years of intensive study of perceptual-motor skills should at least be given a fair trial in the study of mental skills. That then is the framework and starting point of the studies I shall be reporting in a moment. As will soon become evident, the other main formative influence on the work I shall be reporting arises from the early studies of human thinking by Bruner and his colleagues.

One further introductory word is called for as regards the psychological theoretical standpoint of the work reported here. As you know there are still two strong but very different theoretical traditions in psychology, usually labelled stimulus-response type and cognitive type. The stimulus-response theorists, or perhaps better called associationist theorists, speak in terms of associations between discrete events in the environment (the stimuli) and discrete events associated with the organism (the responses). The cognitive theorists are more likely to talk about formal categories of behaviour, giving relatively less attention to the topographic aspects of behaviour classed within those categories. As Mandler (1962) sees it "The important questions" as regards this theoretical controversy "are: Do organisms learn generalizable, but discrete, responses in specific situations or are rules of behaviour, maps or schemata laid down which connect various behaviours and environmental inputs? Do organisms learn what to "do", or do they learn "what leads to what"?" And commenting on the cognitive theories Mandler adds "... they belong to a much larger class of theories which have claimed to be "structural", including not only the gestalt school and its heirs but also the speculations of Piaget, Bartlett and Hebb among others. What these positions have in common is a postulation of the organization of behaviour not derivable from any combination or association of stimulus-response links." My own view, about which I shall say more at the end of this paper, is that to assert that we must be *either* stimulus-response type theorists *or* cognitive type theorists is to pose a false dichotomy. I shall argue that there is now enough evidence to suggest that cognitive processes may in fact use as building blocks smaller chunks of behaviour built upon stimulus-response type principles.

These opening remarks are, I suppose, more by way of an apology, for whilst you mathematics educators are making leaps and bounds in the development of new and existing approaches to the teaching of mathematics, we are slowly plodding along behind trying to find out *why* this or that method is more successful than another. In justification of our approach we should argue that as scientists interested in behaviour we are not only concerned to be able to *predict* accurately but also to *understand why* we are able to predict accurately. Of course, there is more to it than this, because we also believe that when we understand *the processes* involved in accelerated learning of one part of mathematics we may then be in a better position to generalize effectively to other areas.

SOME PRESUPPOSITIONS

I said earlier that I must lay bare my presuppositions. Here are some of them. First, that structural learning is a complex activity and that we are not likely to understand it by studying the performances of subjects on tasks which last a few seconds or even a few minutes. The second is that to understand structural learning we must devise tasks which will enable us to externalize as much of the subjects thinking processes as possible and thus rely on subjects introspections no more than is essential. Thirdly, our experimental tasks must be of such a kind that they will provide us with information about the way in which component parts of the subjects performances, leading to their ultimate success or failure, follow one another and are interrelated. Two further considerations guided our choice of experimental tasks. First that in order to study transfer effects we must use tasks which are as far as possible totally unfamiliar to the subject prior to the start of the experiments, and second, that in order to study aspects of cognitive development our tasks must not only extend adult subjects but also be capable of performance by children.

Within the self-imposed restrictions of the approach I have just outlined we concentrated on a limited number of problems and issues. These can be best summarised in the form of questions we were asking. These included: (i) How can we study structural learning with a sufficient degree of control to allow some quantification of performance? (ii) Can we examine the more marked individual differences in performance in terms of strategies employed in structural learning? (iii) How can we maximise transfer effects not only in terms of levels of performance attained on later tasks, but also in terms of degree of efficiency and sophistication of strategies which are developed? (iv) Can we identify limiting factors in human information processing which affect rate of structural learning?

In the remainder of this paper we shall first describe the typical experimental set-up and then give examples of results of different experiments to illustrate answers to the questions just posed.

THE EXPERIMENTAL TASK

Pilot studies using simple mathematical groups made it clear that it was possible to gain considerable insight into the processes assumed to underlie structured thinking by using embodiments of such groups. A particular advantage of mathematical groups is that they stand to each other in certain definable relationships, such as embeddedness, overlap, recursion etc. In our early experiments the actual embodiment of the structures used was effected as follows. The subject sits facing a board with a window in it that can be opened or closed by the experimenter from behind. Cards containing appropriate symbols (such as coloured shapes) can be exposed in the window, and the subject is given duplicates of these cards. The first event in the subject's experimental environment is the appearance of a card in the window and this is the "dependent" variable. There are two independent variables, the card in the window previous to the card appearing as the event, and the card played by the subject. For the two-group game the subject is told: "We shall play a game and your job is to try to find out its rules. I can put either of my cards in the window here. Now look in the window and then play one of your cards, so that I can see which one it is. I shall then close the window and reopen it, perhaps with the same card showing again, or perhaps with the other card. What will appear in the window after you have played your first card will depend *only* and *wholly* on what was in the window before *and* on the card you played. This is a game between you and a mechanism. Remember that this really is a mechanism. What it does once, it will do again under the same circumstances. If one picture or symbol in the window once changed to a certain picture or symbol after you played a certain card, the same will happen again whenever you play the same card, providing the same picture or symbol was in the window previously. The mechanism has no sequence of its own, for the card you play is your own choice, and the mechanism cannot respond until you have played your card. This is in no sense an intelligence test. My only interest is to find out how you set to work to discover the rules of the game we are to play". Before commencing the task the subject is encouraged to ask questions and great care is taken to ensure as far as possible that he understands what he has to do. He is then given his two cards (if it is a two element group) and again shown where the first card to be exposed will appear. Records are kept of the cards exposed and the cards played by the subject and the predictions made by the subject throughout the game. Before the prediction stage of the game begins the subject was given four preliminary trials. After this he was told: "Now look at the window, play one of your cards and say which card you expect to see exposed next. We will go on like that until your expectations turn out to be correct every time."

In these early experiments only two element and four element groups were used. However, it was already evident that the task of recording the state, the play, the prediction and whether right or wrong, was already becoming sufficiently burdensome to necessitate the construction of a machine embodiment of these games. Accordingly, for later experiments an electronic machine was built on which it was possible to program mathematical group structures and non-group structures of up to ten elements. The subject now played a game against the machine, at each stage knowing the state of the machine, indicating his own play by moving the '*play*' dial and indicating his prediction after each play and before pressing the 'go' button, which would cause the machine to carry out the play indicated by the subject. The state, play and prediction panels are as indicated in Figure 1. The output

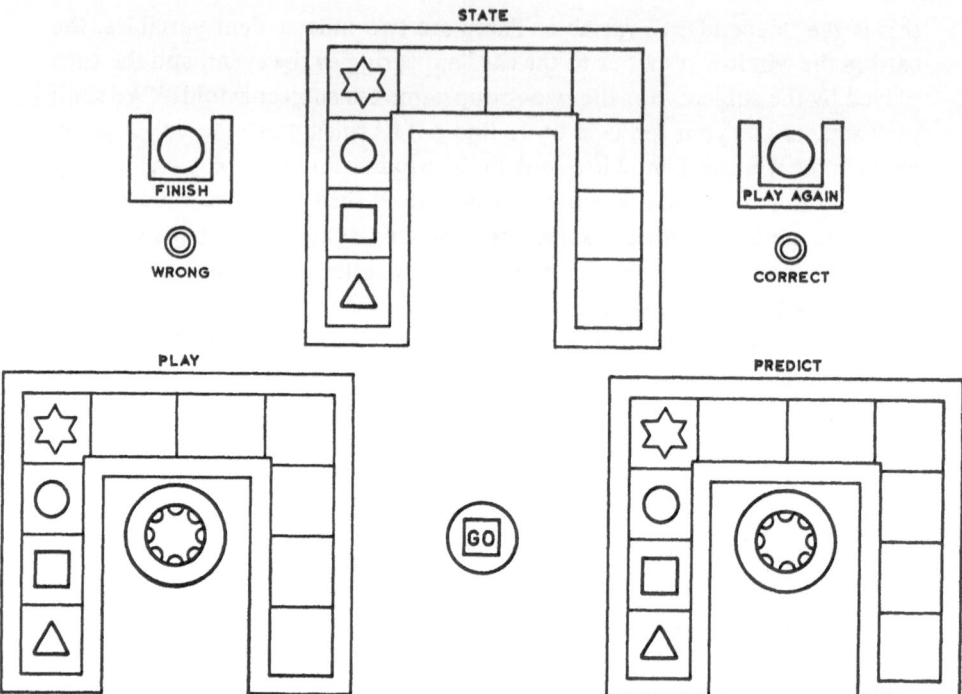

Fig. 1. All the symbols were dimly lit throughout the game. The symbol representing the state of the machine at any time was brightly lit, as also was the symbol the subject wished to play and the one that was his prediction of the next state of the machine. If the subject's prediction was wrong, the WRONG light showed red, if it was correct the CORRECT light showed green.

from the machine fed directly into an IBM card punch machine so that a complete record of the subject's performance was recorded automatically and in a form appropriate for computer processing. In the first series of experiments, subjects played each game until they reached a predetermined criterion of success, such as 10 correct predictions in a row, at which point they were tested systematically on all possible state and play combinations, and if they reached a criterion of 13 out of 16, or whatever it may be, correct predictions, that completed the game. In the second series of experiments, where there was particular interest in problems of transfer, it was considered essential to ensure that all subjects had the same number of plays (e.g., 140, on the 5, 6, 7, 9 element groups) and after this they were tested on all the possible combinations of state and play for that particular game. Further details concerning differences between early and later experiments will be given below when some of the results are being considered.

MEASURES TAKEN

As examples of the kinds of measures taken the following may be considered:

(1) In the early experiments only, the number of instances which the subject required in order to reach criterion performance.

(2) The number of erroneous predictions made in the course of either reaching criterion or in the course of the predetermined number of plays. This will hereafter be referred to as the "A operation score".

(3) The number of erroneous predictions made in the test phase of the game. For example, in a four element group there are 16 possible combinations of state and play and outcome which must be tested. The number of errors made in this test would represent the "B operation" score for a subject on this game.

(4) Equation scores. At the end of each task a subject was asked a series of questions of the general form: "What would you play with A to get a B in the window next?" This was a way of representing in the familiar terms of the symbols used in the display on the machine, an equation of the form "Where $ax = b$ what is the value of x?". Answers given to these questions enabled an equation score to be computed for each subject on each game.

These were the basic measures taken but in addition there were other scores such as extrapolation scores and many hypotheses scores, as well as strategy indices, details of some of which will be given below when their significance is discussed.

EXPERIMENTAL DESIGNS

In all experiments subjects attempted at least two tasks. In many of the later experiments concerned with the transfer effects of different structural relations subjects attempted four different taks on consecutive days. In most of our experiments the same tasks were given to groups of first year university students and 10 and 11 year old school children. Apart from that the subjects of the experiments were not specially selected in any way. Formal measures of intelligence were not taken and we can only express it as our opinion that the mean and range of the intelligence of the students was only slightly, if at all, greater than that of the children. It could be argued that we should have used carefully matched groups of subjects on a variety of personality and attainment tests. In reply we can simply offer it as our opinion that in the early stages of investigations developing new techniques for the study of structural learning we should concentrate on robust phenomena and differences which would stand out in spite of the variance introduced by using relatively unselected groups of subjects.

SOME EXAMPLES OF ISSUES STUDIED AND FINDINGS WHICH EMERGED

1. *Evaluations and Strategies*:

At the end of each game each subject was asked what he or she thought the "game" had been about. These evaluations fell very easily into three recognisable categories:

(a) *The Operator evaluation* regarded the card played as an operator acting on the card in the window. For example, in the two-group a subject might have said that the card *a* left the window unaltered, and the card *b* altered it.

(b) *The Pattern evaluation* split the number of different possible combinations into distinct groups and dealt with such groups as subwholes of the whole game. For example, in the two-group a subject might have said that if the cards played and in the window were the same then an *a* would follow, if they were different then a *b* would follow.

(c) *A Memory evaluation* would be one in which the subject stated either that he had found no rules, or that he had simply memorised the results of the combinations separately.

There were, particularly in the four-group, mixtures of these "pure" types and this made it possible to distinguish between six different types of evaluation.

Corresponding to these evaluations, operator, pattern and random strategies were found and measured. The operator strategy was the continued playing of the same card in order to find out its effect on the window. This was scored by counting the number of times a subject played cards in runs of three or more of the same card played, and dividing this number by the total number of cards freely played by the subject. For the pattern strategy, most subjects on the four game (where the four elements are *a, b, p, q*) appeared to split up the combinations into the following three groups (i.e., those subjects who did any such consistent splitting up):

> (*aa, bb, pp, qq*) referred to as the "square" or "*S* section";
> (*ab, ap, ba, pa, qa*) referred to as the "constant" or "*C* section";
> (*bb, bq, pq, pb, qb, qp*) referred to as the "triangle" or "*T* section";

Runs of cards played which allowed the subject to remain as far as possible within one such section were counted and this number again divided by the total number of cards played freely by the subject. As it was impossible to keep within the same section for long by the rules of the game, re-entry cards were also counted as parts of the "runs" if the outcomes of such re-entry cards were correctly predicted. For example, in the cyclic four-group there could have been a "run" like this:

$$(bp), (qb), (pq), (ab), (bq), (pp) \dots \dots \dots$$

This would have been counted as a run of five, provided that the fourth member of the run, i.e., the playing of b against a window a, was correctly predicted as b. Clearly (pp) is already out of the T section and so the run is broken there.

It was conjectured that the operator evaluation indicated a deeper insight into the structure than the pattern one, and that the pattern one was superior to the memory one. This was tested in a number of different ways and found to be the case. As an example, consider the number of cards the subjects had to play before they learned the structure. Subjects were subdivided into operator or partial operator evaluations; purely pattern evaluations; and pattern-cum-memory or purely memory evaluations.

The mean numbers of cards played by these groups were 101, 120 and 151 respectively (Table I). The first of these means differed significantly from the last at the $2\frac{1}{2}$ per cent level on a one-tail t test. Giving the groups as listed evaluation scores of 1, 2 and 3 respectively, a product moment correlation between these scores and the numbers of cards played came out as 0.5 with $p < 0.005$. It seems therefore that subjects do tackle the task more efficiently if they give 'higher order' evaluations as defined by the above list.

TABLE I

Number of instances or trials required to complete
the task as a function of evaluation

	Evaluations		
	Operator, or partial operator	Pure pattern	Memory, or pattern memory
Means	101	120	151

2. Transfer Effects

(a) Effect of order of presentation

There is considerable evidence from studies of perceptual motor skills that overall performance on both tasks taken together is in general better when the more difficult task preceded the easier task than with the reverse order of presentation (e.g., Szafran and Welford, 1950). It was therefore of interest to discover that with the two tasks used in the first series of experiments, the same general result was observed with these high level cognitive skills. To illustrate this point, consider the percentage of different types of evaluations given by subjects depending upon the order in which the two tasks were presented. As Table II indicates, there was a consistent trend for higher order evaluations amongst those subjects who had their tasks in the order 4 followed by 2, than those who had them in the order 2 followed by 4. This

TABLE II

Effect of order of presentation upon evaluation of the tasks

Game	Percentage		
	Operational	Pattern	Memory
Two-game			
2–4 subjects	35	18	47
4–2 subjects	42	54	4
Four-game			
2–4 subjects	24	26	50
4–2 subjects	17	66	17

trend applied both to the simpler game and to the more complex game. It seems therefore, that the practice of "throwing subjects in at the deep end" has paid off to a considerable extent both in the simpler and in the more complex tasks.

(b) *Effects of Structural Relations*

In one experiment all subjects were given four different groups administered on four successive days. The groups used and their structural relations with other tasks are summarised in Table III.

A full account of the results of this experiment which included a complete replication of the design in Table III but using eleven year old children as subjects, will be found in Dienes and Jeeves (1970). As examples of our findings we may cite the following:

(1) Children consistently found it easier to particularize than to generalize, and more difficult to generalize than adults. Thus, whilst the performances of adults and children were very similar on the five-group and the three-group taken together, given in that order (i.e., transfer based on recursion by particularization), they differed considerably when the tasks were presented in the converse order (i.e., transfer based on recursion by generalization) as Table IV indicates.

(2) Children and adults both found generalization by a factor more difficult than simple generalization. Since both adults and children also gave

TABLE III

Structural relations between groups

Task I	Task II	Task III	Task IV
Klein 4	3 ⟶ Recursion (Generalization)	5 ⟶ Recursion (Generalization)	7
Klein 4	5 ⟶ Recursion (Particularization)	3 ⟶ Recursion (Generalization)	7
Klein 4	3 ⟶ Recursion (Generalization) plus embeddedness (simple)	6 ⟶ Recursion (Generalization) plus embeddedness (simple)	9
Klein 4	3 ⟶ Recursion (Generalization) plus embeddedness (simple)	6 ⟶ Overlap (multiple)	9A
Klein 4	6 ⟶ Recursion (Particularization) plus embeddedness (simple in reverse)	3 ⟶ Recursion (Generalization by a factor plus embeddedness (simple)	9
Klein 4	6 ⟶ Recursion (Particularization) plus embeddedness (simple in reverse)	3 ⟶ Embeddedness (multiple)	9A

TABLE IV

The effect of order of presentation upon subjects' performances on two tasks taken together

	Adults			Children		
Tasks	Errors on "A" Op.	Errors on "B" Op.	Equation score	Errors on "A" Op.	Errors on "B" Op.	Equation score
3+5	64	4.7	13	88	10	9
5+3	59	5.4	14	60	5	13

better performances with embeddedness than overlap, this became more
striking on comparing performance in moving from the six-group to the
nine-group (simple generalization by a factor plus simple embeddedness).
Table V compares the performance on the nine-group depending on whether
it is immediately preceded by the three-group or the six-group.

TABLE V

The effect of order of presentation upon final task in a series of four tasks

Order	Adults			Children		
	"A"	"B"	Equation	"A"	"B"	Equation
9 preceded by 4-3-6	45	9	8	77	42	5
9 preceded by 4-6-3	74	20	6	95	37	2

(3) Where simple embeddedness plus recursion (3 followed by 9) was
replaced by multiple embeddedness (3 followed by 9A) the margin between
performance of adults and children narrowed significantly (Table VI).

TABLE VI

The effect of structural relations between tasks upon transfer

Order	Adults			Children		
	"A"	"B"	Equation	"A"	"B"	Equation
9 preceded by 3	74	20	6	95	37	2
9A preceded by 3	68	30	3	71	46	3

3. Individual Differences

Since our groups of subjects were not selected on the basis of intelligence
test or personality inventory performance we can only make two major
comparisons and these are between male and female subjects and between
younger and older subjects. Since some reference to age differences has
already been made in the preceeding section we shall here comment only on
sex differences.

Sex differences on selection and reception tasks. In one of our early experiments whilst half the subjects freely selected their plays, as already explained,
the other half had their plays selected for them by the experimenter. These
two groups will be referred to as the "selection" and "reception" groups

respectively. The reception subjects were divided into four groups: operator group, pattern group, mixed group, random group. In the operator group the same card was played in succession 8 to 12 times, all the cards being taken in turn. It was thought that this procedure would lend itself to the card played being regarded as an operator on the window. In the pattern group, combinations of a certain kind were repeated for some time, such as, for example, the card being played with the same symbol as the one on the card in the window. The mixed group were given the first half as operator, and the second half as a random sequence the whole time. The part of the experimental design with which this discussion is concerned is given in Table VII.

TABLE VII

Part of the experimental design used in "Thinking in Structures"

Selection Subjects	2-4			4-2
Men Women				
Reception subjects	Operator	Pattern	2–4 Mixed	Random
Men Women				

Since all the adult subjects in the experiment were university students, it was not expected that performance on a task of this kind would vary with the sex of the subject. However, in order to cover this eventuality, subjects were selected so that equal numbers from each sex were allotted to each experimental group. Surprisingly, it was discovered that amongst those subjects who were permitted to freely select their own orders of play, there were more males in the top half of the performance distribution than in the bottom half, and conversely there were fewer females in the top half than in the bottom half. Table VIII summarises this finding for the male subjects.

TABLE VIII

Performance of males on selection and reception tasks compared

Half	Selection	Reception
Top	12	4
Bottom	3	11

$\chi^2 = 6.56, p < 0.05$.

This comparison is based upon the evaluations given by the subjects but a similar distribution is obtained if subjects are ranked according to their error scores. This finding may be summed up by saying that women give higher order evaluations than men in the reception situation, whereas men give higher order evaluations in the selection situation. The same trend is observable in strategy measures, as well as in the total error scores. It might appear therefore that women are favoured by having their strategies imposed upon them whereas men are favoured if they are left free to select their strategies. Since this finding does not apply to a similar group of 11 year old children, it is interesting to speculate on the extent to which it is culturally determined. This finding may be cited as a particular example of a more general use to which this method of studying thinking may lend itself, namely relating thinking and personality. Bartlett (1965), writing about this aspect of our results has said: "The second matter concerns the possibility of using the ideas that lie behind the thinking-in-structures approach to throw some light on the little understood questions of thinking and personality. In this research, attention is called to the possibility of relationships between personality characteristics and the persistence and frequency of particular kinds of error, and the speed with which threatened errors may be recognised and corrected. These and similar matters all mark persistent personal and temperamental characteristics that can exercise a great influence on the course and efficiency of thought processes. With the procedures described in this section, these characteristics could easily be given the effective and experimental study they deserve".

4. *Analyses of results which focus on psychological issues of the mechanism of structural learning.*

(i) *The importance of short-term memory*

In looking for an answer to the question of *why* operator-type strategies were correlated with goodness of performance as measured by errors made, number of trials in achieving criterion or equation scores, it seemed at least partly because the more systematic processing of information made possible by operator type play reduced the load on short-term memory. This reduction of the load on short-term storage was made possible by an information reduction process of extracting a rule. Once the information could be coded in this way it would mean that instead of occupying one's limited short-term storage capacity by remembering a string of seemingly disconnected instances in a rote fashion, the important information, namely the rule, could either remain in short-term store and the items from which it was deduced could then be cleared, thus making room for holding a new array of instances from which further rules could be extracted, or possibly the rule could move into long term storage thus completely clearing the short-term store for repeating the exercise of further rule extraction and information reduction. Which of these two alternatives is the correct one does not concern us now and nothing in the present argument depends upon which one, if either, eventually turns out to be correct.

It would seem to follow from this emphasis upon the limiting effect of short-term storage that we should look again at our data and see how widespread was the tendency of goodness of performance to go with operator type stategies or other systematic strategies for the acquisition and short-term storage of information. Furthermore it would seem to follow that if we could modify the nature of the task to reduce the load on short-term storage (either directly by leaving the evidence displayed for longer or indirectly by putting the task of instance selection more under the subjects control thus further facilitating the ordered choice and selection of instances of state, play and outcome) then again this should result in improved performance. One further approach to this issue seemed also worth trying, namely studying performance on groups which on a rote learning hypothesis would be equal in difficulty, but which on a structural learning hypothesis, stressing the importance of short-term storage, should lend themselves differentially to breaking up into small chunks from which sub-rules could then be more easily extracted. We shall now briefly consider the evidence from each of these approaches in turn.

(a) *Strategies which reduce the load on short-term memory*
Reference back to Table I indicates the relationship between type of evaluation given and performance on the task. Further analysis showed that what applied to evaluations applied equally to strategies. Since it may be that the most important feature of an operator-type strategy is the systematic exploration that it affords of the function of each element in the group, together with the opportunity to reinforce at once what was being learned, we looked at the relationship between the tendency to use runs of four or more plays of the same elements and success or failure in reaching criterion on the task. As an example of this type of analysis on a different group we may compare the subjects who succeeded on the cyclic eight group with those who failed. Of the successful subjects, when they played the same element more than once successively, on an average 70% of such plays were in runs of 4 or more. By comparison those who failed played on the average only 40% of their total plays in this way. This difference is significant at the 1% level. Lest it be thought that a stimulus-response type reinforcement theory could equally well explain this relation between runs and success we should point out the

following. An S-R interpretation would presumably argue that reinforcing the same stimulus-response-outcome combination four of more times would consolidate learning of that combination. But, of course, unless the neutral was being played in this way, each time any element was played against a different outcome in the window it would be a *different* stimulus-response-outcome combination that the subject was faced with. On the other hand if he had learned to pay attention to the *role* of the element he was playing several times in succession, then he would indeed be confirming or infirming hypotheses he was developing concerning the *role* of that element. But S-R type theories do not to the writer's knowledge usually extend their formulations to the consideration of *roles* but simply to associations.

(b) *The effects of task modifications designed to reduce the load on short-term memory*

Remarks made by subjects after taking part in experiments on the machine described earlier, suggested that if changes were made to two features of the task, it would be easier to do. Firstly, they complained that the state against which they had played their selected element was no longer displayed once they had pressed the 'GO' button. (The exception to this would be when they played the neutral element.) Thus, if they wanted to check which state and play combination had given the present state of the machine they were unable to do so *unless* they had made a practice of *remembering* what had previously been the state of the machine. To do this they had to hold this information in short-term storage, along with any other information about earlier goes that they were retaining in order to discover regularities in the operating of the machine. Secondly, some subjects reported that at times they wanted to have the machine in a particular state in order to test out an hypothesis they had formed on the structural rules of the task. However, since the state was not directly under their control, but only indirectly controlled through playing elements against the machine, they could not do this. Accordingly a second version of the machine was constructed which left the previous state in view and enabled the subject to control both state and play. It was thought that both these modifications would reduce the load on short-term storage and the second would also give the subjects much more control over the task. Precisely which of these factors would be responsible for any observed changes in performance it is not yet possible to say. However, what is clear is that when these changes are made there is a dramatic change in rate of learning of the cyclic 4 group as Table IX indicates.

For reasons which are not relevant to this discussion there were slight procedural differences in running the experiments on the two machines. When the experiments were run on the Mark I we were giving all S's the same number

TABLE IX

Machine	Number of subjects	Mean error score on test phase	Mean number of plays either to criterion (on Mk. II) or after no further errors were made (Mk. I)
Mark I	58	5.66	108.7
Mark II	70	2.71	49.0

of plays, namely 120 and then testing them on the 'B' operation. Our data analysis however allowed us to say after which play no further errors had been made if the task had been learned in less than 120 goes. On the experiments on Mk. II subjects were tested in the 'B' operation as soon as they made ten errorless predictions. Thus from Table IX we see that whilst the subjects on Mark I made twice as many errors as the Mk II subjects (a difference significant at the 1% level) they took 108.7 plays to reach a level of performance reached by the Mk. II subjects in 49.0 plays (this difference is significant at the 1% level). This comparison suggests that if we are correct in thinking that one of the major psychological changes produced by the change in machine embodiment is in reduction of load on short term memory then such a change certainly produces a striking improvement in rate of structural learning.

(c) *Different Group structures and changes in load on short-term memory*
We may compare the proportion of subjects reaching criterion within 200 trials on the 2×4 eight group with that on the $2 \times 2 \times 2$ eight group. In a repeat of this experiment, in which subjects were not required to predict after every play and before pressing the 'GO' button, the proportions of those succeeding were $\frac{2}{15}$ on the 2×4 eight and $\frac{8}{15}$ on the $2 \times 2 \times 2$ eight group. In this latter experiment the difference between proportions is 0.40 and the standard error of the difference is 0.17. Since the difference is 2.35 times the s.e. of the difference and we had predicted a trend in this direction when designing the experiment, we may accept this difference as significant at better than the 5% level. There is thus some support for the suggestion that when we compare performance on two groups with equal numbers of elements, but which on the basis of their internal structures differ in suitability for breaking down into smaller chunks of learning then the group which can be more readily handled in terms of small chunks is learned more easily. We have argued above that one possible reason for this is that whilst the human short-term memory store may be able to hold sufficient information about a small subgroup to extract its rule structure, the same short-term memory store may be insufficient to hold enough information about a larger sub-

group without pencil and paper or some similar aid, and these were not
permitted in any of our experiments.

We may summarise this section by pointing out that evidence from three
different directions can be interpreted to support the view that where load on
short term storage is reduced, rate of structural learning improves.

(ii) *What is being learned?*

It was asserted earlier that from the psychological viewpoint it is more important to know something about *how* the subjects learned their tasks than
whether or not they succeeded. More specifically we could ask does analysis
of the process of learning support the view that an S-R associationist type
model can adequately describe what is taking place in the course of learning?
This is a complex issue which we have discussed in detail and with supporting
evidence in our second monograph. Here I can do no more than exemplify
aspects of the analytical approach we adopted when studying our data and
indicate how they supported the view that in addition to learning stimulus-
response-outcome combinations in a rote manner, something more was being
learned which we called 'structural learning'.

Consider, for example, the learning of the 3×3 nine group. Table X
highlights the way in which the sub-groups are embedded within the overall

TABLE X

9A break-up (showing subgroups)

structure and indicates the symbols used to represent the elements of the group on the machine. In designing the symbols which would represent different elements in the mathematical groups used as our experimental tasks, in some instances we built in certain perceptual cues, which, if used might assist in discovering how earlier tasks were structurally related to later tasks. For example, there are two "obvious" subgroups in the 9A task, one was represented by keeping the colour yellow constant and varying the shapes as between triangle, circle and square. This we shall call the *inner subgroup*, as its elements are clustered around the neutral, which was represented by the yellow circle. The other subgroup was represented by keeping the shape constant, in the shape of a circle, and varying the colours, i.e., by taking orange, yellow and red to form the three-cycle. One might ask about the differences in the amount learned in these two subgroups and other "comparable" parts of the matrix: one might also consider those subgroups where no perceptual constant was provided. The remaining parts could, for example, be the following:

(r) where red responses were given to red stimuli
(o) where orange responses were given to orange stimuli
(tr) where triangle responses were given to triangle stimuli
(sq) where square responses were given to square stimuli
(res) parts of the matrix not covered by any subgroup or other category.

The (r) and (o) parts of the matrix could be used as controls to check against the learning of the "yellow" subgroup, the (tr) and (sq) parts of the matrix could be used as controls to check against the learning of the "circle" subgroup, and the (res) part, which is the residue.

In order to study these effects quantitatively we shall look at the differential results on the B-operation scores based on different parts of the matrix to see whether our attempt to bring out the structure of the tasks by suitable symbolisation as indicated above has had any appreciable effect on the learning of it.

Table XI shows the break up of the 9A matrix into sub-groups, constant colour sections, constant shape sections and the non-group section: the numbers are the probability of making an error on the 'B' operation in the part of the matrix indicated.

We can group together the inner yellow and the outer circle, the rationale being that in each of these there is a perceptual invariant together with a subgroup with which it coincides. In the yellow subgroups this invariant is the colour yellow, in the circle subgroup it is the circular shape. Next we can take the "hybrid" subgroups i.e., where we have the structure of a

TABLE XI

	Inner (yellow)	Outer (circle)	Outer (1) (2) (hybrid)	(o)	(r)	(tr)	(sq)	res.
Children 9A preceded by 6 and 3	0.383	0.333	0.494, 0.408 $\bar{x}=0.451$	0.581	0.618	0.778	0.815	0.608
Children 9A preceded by 3 and 6	0.432	0.346	0.566, 0.496 $\bar{x}=0.531$	0.667	0.593	0.742	0.729	0.732
Adults 9A preceded by 6 and 3	0.32	0.28	0.35, 0.39 $\bar{x}=0.37$	0.45	0.38	0.49	0.20	0.456
Adults 9A preceded by 3 and 6	0.148	0.198	0.297, 0.259 $\bar{x}=0.278$	0.296	0.309	0.322	0.296	0.302

TABLE XII

	Subgroup perceptual invariant	Subgroup without perceptual invariant	Perceptual invariant, but no subgroup (Colour)	Perceptual invariant no subgroup (shape)	Residue
Children 9A preceded by 6 and 3	0.358	0.451	0.600	0.797	0.604
Children 9A preceded by 3 and 6	0.389	0.531	0.630	0.735	0.82
Adults 9A preceded by 6 and 3	0.30	0.37	0.44	0.35	0.371
Adults 9A preceded by 3 and 6	0.173	0.278	0.303	0.309	0.267

subgroup, but without the perceptual invariant. Next we can take the cases in which the colour is invariant but no subgroup is apparent and finally the cases where the shape is invariant and still no subgroup apparent. Finally we can take the residue. We then obtain the following picture (Table XII).

The figures given in Tables XI and XII can be interpreted as the probabilities of errors being committed in the categories given. Subtraction from unity will give the probabilities of making a correct prediction in these same categories.

It seems not altogether surprising that where the subgroup structure is brought out by a perceptual aid, either in the form of a colour being kept constant (yellow) or in the form of a shape being kept constant (circle), the error-probabilities should be lowest. It is noticed also that the adults' error probabilities are everywhere lower than those of the children, again not very surprising, although in many cases there is not a great deal of difference. On a purely perceptual hypothesis one might argue that the same error-probabilities should be expected every time the same colour is played as the colour of the symbol in the window, whatever the particular colour might be, particularly since the "shape scheme" is the same in all three cases. This, however, is not so. When yellow is played against yellow, the error-probabilities are everywhere consistently lower than if red is played against red, or if orange is played against orange. It seems unlikely that there is any other reasonable explanation of this, apart from the fact that,

(i) the yellow states and operators form a group,

(ii) that all subjects will have encountered this group, i.e., the 3-group, before, albeit in the framework of *different symbols*.

It seems likely that some kind of transfer is taking place which can reasonably be attributed to the recognition on the part of the subjects of the three-structure, already learned; or at least encountered by them on the previous day or on the day before that.

Let us now see what happens if, instead of withdrawing the subgroup structure, yet keeping the colour or shape constant, we withdraw the perceptual constancy of the symbols used, but keep the subgroup structure. In the 9A group there are two subgroups, namely

$$(y_2, 1, x_2) \quad \text{and} \quad (y_4, 1, x_4)$$

in which both the shapes and the colours are varied. We can see, by looking at the probability tables, that the error-probabilities increase in every case, but not by so much as if we remove the group structure and keep the perceptual constants. There is one exception, the adults' Group 3-9A, who for some unaccountable reason seem to find the combinations of "square" symbols very much easier than any other types of combinations, the error-probability on a square against square combination being as low as 20%.

One way of explaining this in the case of the adults' is that on this particular task they appear to be behaving more in conformity with the kinds of prediction that could be made on S-R assumptions than on structural assumptions, although the trend seen with the children is still observable.

It is also seen that, keeping the colour constant is easier than keeping the shape constant, assuming no structural "aid" is provided. This is in line with findings in other experiments, the trend is more clearly noticeable with the children, who are known to be better at colour than at shape discrimination. An adult has had so much practice at both these that the differences are minimal, if indeed they exist at all, with the simple and familiar shapes used in the present experiment.

The adults do not appear to find the residue any more difficult than the other parts of the matrix, with the exception of the part in which they are helped both perceptually and structurally. The children however, particularly those who had the 6-9A treatment, find the residue almost impossible to learn, as 89% error-probability would already be equal to what they would get if they knew nothing at all. The 3-9A children, however, only have a 60% error-probability in this area. It is possible as has already been suggested that the embeddedness of the immediately preceding task, i.e. of the 3-group, is a help in sorting out not only the structure embedded, but also the rest of it. With the 6-9A group, where there is overlapping instead of embeddedness, there is no such effect.

CONCLUSIONS

After analysing subjects performances on a variety of tasks of this kind we were led to the conclusion that when subjects are required to learn the properties of mathematical groups, when these are embodied in an electrical machine, the method of learning could be more elegantly handled by a model which assumes that structural learning is taking place in addition to simple S-R learning. In some of our experiments the complexity of some of the tasks was such that success was possible only if subjects were systematic in their behaviour. In this way the load on short term storage could be kept to a minimum thus leaving as much cognitive capacity as possible free for trying out rules or codes which would provide a shorthand way of structuring and handling a range of otherwise seemingly disconnected items of information. This emphasis on the intimate relationship between thought and memory has occurred in recent years in the work of other experimental psychologists interested in human information processing. For example, on Broadbent's (1958) view the human organism has a limited central capacity for processing information and if this capacity is taken up in one way at one time, then that amount of its capacity will not be available for other things. Applying this now to our own findings we can see that if a subject regards the learning of a mathematical group as the storage of many items of disconnected information, namely stimulus-response-outcome triads, then there will be no capacity remaining for the subject to try out possible means of coding this information, which, if successful, will appreciably reduce the load on memory. In fact the models we have explored are increasingly efficient in their potential for reducing information which has to be stored, and replacing such memory load by more and more widely applicable rules, which with a minimum of information in store, can then regenerate the whole mathematical group. Doubtless some simple rote learning is essential to hold enough items in short term storage for the subject to 'turn round on his own schemata' (Bartlett, 1932) and try out possible rules or codes which will then reduce such information to manageable proportions. Moreover on this view it may well be that by starting with the more complex task first, i.e., 6 followed by 3 rather than 3 followed by 6, or 5 followed by 3 rather than 3 followed by 5, has the effect of forcing subjects quickly to abandon the attempt to solve the problem on the basis of memory alone and may accelerate the tendency to try out various coding mechanisms. Certainly such a view fits well with our findings in our first experiments reported in *Thinking in Structures* where we observed that the overall performance on the four element group and the two element group was better when presented in the 4-2 order than the 2-4 order. At the same time we found a greater tendency to use higher order strategies amongst subjects thrown in at the deep end with the 4-2 order than those let in gently with the 2-4 order. This tendency was observed again in the later series of experiments where we compared the 3-5 order with the 5-3 order and the 3-6 order with the 6-3 order.

One final general observation concerning the methodology of studying the learning of structures and for that matter concept formation in general is worth recording. It is that we have seen repeatedly that if our measures of performance had been confined to total numbers of errors made in any phase of the task, we might have been tempted to conclude that an explanation of our results in terms of an S-R model was more adequate than it is. In fact it is only by analysing the component parts of subjects performances that we begin to understand the *process* of learning taking place rather than merely degree of success or failure on a particular task. It is only by such fine-grained analyses of performance that we can really critically evaluate the merits of competing theoretical explanations of the learning taking place in such experiments.

In this respect we would bracket our model amongst the information processing type models reviewed by Hunt (1962) and agree with him that whilst "... ... they do not provide us more parsimonious explanation of some data than previous models do" ... yet "... they do make more detailed predictions within a given setting". In the case of our experiments it was only by examining in detail what was happening in particular cells of an overall matrix (the structure to be learned) that we were able to differentiate between S-R and structural type models concerning what may have been occurring *between* input and output *in the course of* the learning of a structure. It would seem to us that it is knowledge of this kind which we must seek if we are to understand the *process* of learning complex tasks sufficiently well to facilitate and accelerate them in the classroom situation as well as the laboratory.

University of St. Andrews, St. Andrews, Scotland

REFERENCES

Bartlett, F. C., *Remembering*, Cambridge University Press, 1932.
Bartlett, F. C., Thinking, in Whittaker, J. G. (ed.), *Introduction to Psychology*, Saunders, Philadelphia, 1965, pp. 319–349.
Broadbent, D. E., *Perception and Communication*, Pergamon Press, London, 1958.
Diénès, Z. P. and Jeeves, M. A., *Thinking in Structures*, Hutchinson Educational, London, 1965.
Diénès, Z. P. and Jeeves, M. A., *The Effects of Structural Relations Upon Transfer*, Hutchinson Educational, London, 1970.
Hunt, E. B., *Concept Learning. An Information Processing Problem*, Wiley, New York, 1962.
Mandler, G., *From Association to Structure*, Psychol. Rev. **90** (1962), 415–427.

PAUL J. KELLY

TOPOLOGY AND TRANSFORMATIONS
IN HIGH SCHOOL GEOMETRY

There is a simple and natural way of regarding Euclidean geometry that has been largely ignored in the teaching of geometry. This is the view that Euclidean geometry is the mathematical study of the size and shape attributes of physical objects. The motivation of the study is that such attributes are intrinsically interesting and also that the information obtained is useful. It is my belief that this functional approach to the subject not only provides a natural dynamics to the development but leads to surprisingly sophisticated and vital mathematical concepts. Though I shall sketch the basis for only the innovative topics, topology and transformations, the view I am advocating leads with equal naturalness to other aspects of the subject, such as vectors or analytic geometry.

The natural frame for Euclidean geometry is three-space and not the plane. Even so, it is understandable that the initial study should concentrate on simple figures and be concerned with basic concepts. However, from the functional point of view, this natural beginning is just a beginning and not an adequate basis for continuation. One cannot seriously suppose that the simple figures of the traditional course are realistically representative of physical objects.

The natural question, "What is an object?" translates to the mathematical question, "What properties of a geometric set make it the analog of a physical object?". An obvious requirement is that the set be bounded. But the key property of the "oneness" of the set as an object is connectedness. This can be expressed by the condition that each two points of the set be the ends of an arc in the set. However, this intuition simply shifts the difficulty to "What is an arc?". Though an answer to this presents difficulties, one partial answer is quite clear, namely, that a segment is an arc. Thus convexity provides a very simple form of connectedness. Since neither boundedness nor convexity are dimensional in character, we can have one-, two-, or three-dimensional sets with these properties.

Further refinements of an "object-set" can be found in considering its inside, outside, and edge. The set interior to a sphere has a simple definition, and the sphere center is as "inside" this set as anything could be. Using this as a comparison, a point P is defined to be interior to set $\mathscr{S}$ if the sphere interior of some sphere at P is contained in $\mathscr{S}$. The set of all points interior to $\mathscr{S}$ then forms the interior of $\mathscr{S}$. Since the definition applies to any set, the interior of the complement to $\mathscr{S}$ is also well defined. The edge points, or boundary points, to $\mathscr{S}$ and its complement are now seen to be those points that are not in the interior of either $\mathscr{S}$ or its complement. An examination of different examples shows the usefulness of the concepts of open sets and closed sets. A natural requirement for an object-set is that it be closed.

One now has a reasonable definition for a large class of object-sets, namely, those that are closed, bounded, and convex. With the extra condition that the set be linear, or planar and non-linear, or non-planar, one obtains examples of different dimensions. Moreover, the boundaries in the plane case now provide a class of simple closed curves that include the circle and convex polygons as special cases. The simple, closed convex surfaces include the convex polyhedra, sphere, cone, and cylinder as particular examples.

The development exhibits the creative aspects of definitions. It shows the system growing in an organic way with new perspectives encompassing former ones. It exhibits the role of analogy between dimensions, and there is a wealth of material to challenge the imagination and intuition. For example, it is clear that if two object-sets intersect, then their union has a new type of connectivity because a segment path of at most two steps joins each pair of points. Since the union is also closed and bounded, it follows that such unions provide an extended class of object-sets. The relation of the interior and boundary of the union to those of the initial sets is an interesting question for students to settle.

It is not necessary, and in a short paper it is not possible, to discuss all the pedagogical possibilities opened up by these simple topological notions. But it is worth observing that it is much easier to prove, for example, that the union of two closed and bounded sets is closed and bounded than it is to establish when and how two circles intersect. As for the concepts themselves, leaving proof aside, I believe these are wholly accessible to grammar school children and will some day be part of the elementary curriculum.

Returning to the extended class of geometric figures, it is natural to ask how we can investigate the properties of such general objects. There are many answers to this question, and they provide natural interconnections of elementary geometry with different parts of more advanced mathematics. But basic elementary geometry itself suggests one very fruitful clue. In the study of triangles, the concepts of congruence and similarity play a central role. Congruence simply means that two objects are identical in size and shape while similarity means they are identical in shape but not necessarily in size. Since both concepts make sense for any kind of object-set, not just triangles, this gives us a starting point. How can we define congruence and similarity for general geometric figures?

Let us consider object $\mathscr{S}$. If we move it and call it $\mathscr{S}'$ in its new position, then $\mathscr{S}$ and $\mathscr{S}'$ are surely congruent. We can think of each point X in $\mathscr{S}$ moving to a position X' in $\mathscr{S}'$. The correspondence of X and X', for all X in $\mathscr{S}$, establishes a 1-1 correspondence of $\mathscr{S}$ with $\mathscr{S}'$ and this correspondence is clearly distance-preserving. That is, $d(X, Y)=d(X', Y')$ for all X, Y in $\mathscr{S}$. What this suggests is that a necessary condition for congruence of two objects should be the existence between them of a 1-1 distance-preserving correspondence. But does the existence of such a correspondence imply that the objects are so alike in size and shape that it is reasonable to call them congruent?

It is not at all difficult to show that the last question has an affirmative answer. The heart of the matter lies in two familiar facts. The first is that linear betweenness is defined by the distance triangle equality, and the second is the side-side-side congruence theorem. Using these two facts, it becomes apparent that if a 1-1 correspondence preserves distance, then it also preserves betweenness, segments, angle measure, and every other attribute we associate with congruence. Dropping the physical notions, a reasonable and purely mathematical definition for the congruence of two figures is the existence between them of a 1-1 distance-preserving correspondence.

The physical ideas leading to the definition of congruence also suggest how one obtains congruence-correspondences. The physical ways of moving an object, by translation, reflection, or rotation, can easily be defined in purely mathematical terms, and doing so opens up the whole subject of geometric transformations. In the process one sees the correspondence between two objects as a sub-correspondence in a motion of space. It is a short and natural step from motions to similarities. Two objects are then defined to be similar if one is the image of the other in a similarity. What started as a search for object congruence and object similarity leads to something more general, since the mapping definitions apply to arbitrary sets.

It is again not possible in a short space to discuss all the pedagogical possibilities this treatment presents. But one of the taproots of modern mathematics, the invariants of transformations, has been exposed. Functions appear in a more general context than that of numerical functions. In the combinations of motions and of similarities there is material that is intriguing on physical grounds and which is rich in exploratory problems, even for mediocre students. The closure property for motions and for similarities arises in the context of its extreme usefulness and not as an item in an abstract list of group properties. However, from the closure property it is a natural step to the other properties of a group, and one can scarcely avoid finding the various subgroups of motions and similarities. Finally, the mappings which were introduced to extend the meaning of congruence and similarity are seen to provide a new and powerful tool for solving geometric problems.

There is a frustrating paradox about the viewpoint and the mathematics I have just sketched. The viewpoint is psychologically natural and the object oriented mathematics is, I think, appropriate for young students. The mathematics is unquestionably important in contemporary terms, and there is nothing in the program that is revolutionary in a technical sense. One doesn't have to change the axioms or the setting of the geometric system. The underlying philosophy is largely independent of methodology. That is, one could achieve the same objectives by synthetic means, or with analytic geometry, or with vectors. Yet, despite these facts, the program in its human implications is revolutionary because it calls for a revolt against a number of fixed ideas and attitudes that have long obstructed the effective teaching of geometry.

The most basic of the fixed and erroneous notions about geometry, and one widely held by high school teachers, is that the content of elementary Euclidean geometry consists of what is taught in our schools as elementary Euclidean geometry. This wrong and selfdefeating definition derives in part from the long tradition of a particular course, and it puts the subject in a straitjacket. The fact, of course, is that, as the state of knowledge about any subject changes, so does the state of elementary knowledge change. This has happened in Euclidean geometry as in other parts of mathematics. Yet, there is very little in most of our geometry textbooks that would be conceptually new to a 19th-century high school teacher.

The reason, in my opinion, that our secondary teachers are so ignorant about Euclidean geometry is that the subject is not taught as such in the mainstream of college mathematics. The "as such" is crucial here. It is virtually impossible for any college mathematics major to avoid Euclidean geometry as part of his education. But the "Euclidean space" and the "euclidean norm" that occur in his course in topology or linear algebra do not, in his mind, have any identification with Euclidean geometry. It is not only possible, but not uncommon, that a student who can manipulate the matrices of linear transformations may be unaware that the product of reflections in parallel planes is a translation. It is quite likely that in his college experience he will never use Euclidean motions or similarities to solve an interesting geometric problem.

Not only does our education of high school teachers fail to give them an intelligent understanding of Euclidean geometry, it gives them a bias against object-oriented mathematics and against the scientific aspects of geometry. They tend to believe that "genuine" mathematics is necessarily abstract, general, and analytic or algebraic. Such an attitude is harmful to the teaching of elementary algebra or function theory, but it is disastrous to the teaching of elementary geometry. Starting with two misconceptions, one about the nature of geometry and the other about the nature of mathematical thinking, it is not surprising that high school geometry teachers handle the subject in a narrow, dull, and uninspired way.

A high school course in Euclidean geometry could be, and should be, an introduction to a wide range of new and interesting ideas, a challenge to the imagination and intuition of the student, and an opening of doors to new mathematical perspectives. To construct a conceptual model of the world is audacious. To obtain from the model, by logic, new and unsuspected information about the world is the magic of the human intellect. Precisely because "the world is so full of a number of things" the range of questions to ask and objects to study is inexhaustible. That we manage to conceal nearly all of this from geometry students is an odd tribute to our skill as educators.

University of California, Santa Barbara, Calif., U.S.A.

VICTOR KLEE

THE USE OF RESEARCH PROBLEMS
IN HIGH SCHOOL GEOMETRY

Despite curricular changes associated with "the new math", many high school mathematics students do not realize that the body of mathematical knowledge is constantly growing. That is, they regard mathematics as an essentially "dead" science, while being at least vaguely aware of research progress in physics, chemistry, and biology. In order to help them appreciate the lively nature of mathematics, and thus to quicken their interest in the subject, they should be told about recent mathematical discoveries and about some of the as-yet-unsolved problems that occupy the attention of mathematical researchers. This is not feasible at the high school level in most areas of mathematics, but it is feasible in Euclidean and analytic geometry. By way of illustration, some suitable problems are presented below. The discussions are not complete, but should suffice to convey the flavor of each problem. References to more detailed discussions are also included. All but one of the problems appear in two films that I have recently completed [16], and additional references concerning them can be found in the viewer's manuals for those films.

CONFIGURATIONS OF POINTS AND LINES

If a finite set of points in the plane is such that the line through any two of the points passes also through a third point of the set, must the points all lie on one line?

Though proposed in 1893 by the famous British mathematician, J. J. Sylvester [21], this problem remained unsolved for approximately fifty years. See Kelly and Moser [13] for references to various (affirmative) solutions of the problem and for additional results related to it. The solution of Kelly reported by Coxeter [4, 5, p. 66] is quite suitable for inclusion in high school geometry courses. Suppose the points of the set are not all collinear. Consider all pairs consisting of a line L that passes through two or more points of the set and a point p that belongs to the set but is not on L. Among all such pairs, there is one for which the distance from p to L is a minimum. This L passes through precisely two points of the set, for otherwise two points of the set lie on L on the same side of the foot of the perpendicular from p, and the minimizing property of the pair (L, p) is contradicted (Figure 1).

The students may be given a glimpse of the beautiful duality of projective geometry by discussing the following dual form of Sylvester's question: If a finite family of lines in the plane is such that any point lying on two of the lines lies also on a third line of the family, must the lines all be parallel or all pass through a common point?

The above problems have been solved, but they have many unsolved relatives. Let a line be called *ordinary*, relative to a given finite set of points in the

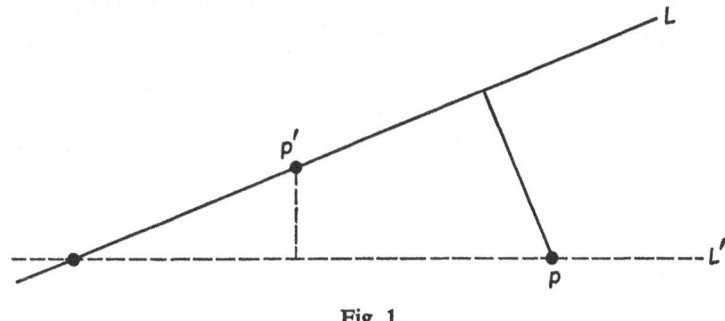

Fig. 1.

plane, provided that it passes through precisely two points of the set. The affirmative answer to Sylvester's question amounts to saying there is at least one ordinary line for any finite set whose points are not all collinear. But should there not be many ordinary lines when the set consists of many points? Dirac [8] conjectured there are at least $[n/2]$ ordinary lines whenever the set consists of n points, $[n/2]$ being the greatest integer that does not exceed $n/2$. Dirac's conjecture is unsettled, though Kelly and Moser [13] show there are at least $3n/7$ ordinary lines.

For additional information on Sylvester's problem and its relatives, see Crowe and McKee [7], Grünbaum [10], and some of their references.

PARTITIONING A RECTANGLE

Can a rectangle be partitioned into an odd number of triangles of equal area? This problem, which is due to F. Richman, was discussed by Thomas [22] and recently solved (in the negative) by Monsky [18]. Monsky's proof is not accessible at the high school level, though the students might be challenged to seek their own proof. (They would almost surely fail, but would probably learn some geometry in the process.) In any case, parts of Thomas's discussion can be made accessible. One might prove, for example, that ratios of areas of triangles are preserved by any plane transformation of the form $x' = ax$, $y' = by$ $(a > 0 < b)$. From this it follows that if one rectangle can be partitioned into a certain number n of triangles of equal area, then so can every other rectangle. It can then be shown that any rectangle partitionable into n triangles of equal area is also partitionable into $n+2$ triangles of equal area.

PARTITIONING A TRIANGLE

When can a triangle be partitioned into a given number n of triangles similar to it?

This problem is due to Freese, Miller, and Usiskin [9], whose discussion is readily accessible to high school students. A triangle T can be partitioned into two triangles that are similar to each other if and only if T is an isosceles triangle or a right triangle. If (and only if) T is a right triangle, the two triangles in the partition can be made similar to T itself. Further partitioning into triangles similar to T is then possible, and hence any right triangle can be partitioned into any given number of triangles similar to itself.

Any triangle whatever can be partitioned into 4 triangles similar to itself, and hence into 7, 10, 13, ... such triangles (Figure 2).

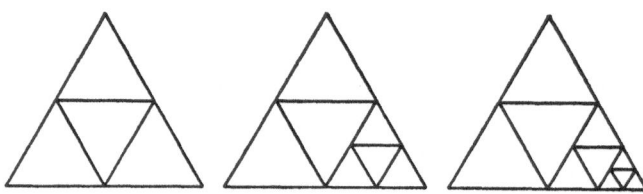

Fig. 2.

Similarly, any triangle can be partitioned into 6 (and hence into 9, 12, 15, ...) as well as into 8 (and hence into 11, 14, 17, ...) triangles similar to itself. However, only a right triangle can be partitioned into 3 triangles similar to itself. Finally, it was shown by Usiskin and Wayment [24] (and others) that a triangle T can be partitioned into 5 triangles similar to T if and only if T is either a right triangle or has angles of 30°, 30°, and 120°. It is not known exactly which triangles T can be partitioned into 5 triangles that are similar to each other but not necessarily to T.

COLORING THE PLANE

What is the smallest number c of colors that can be used for coloring all the points of the plane so that no two points at unit distance receive the same color?

This problem was discussed by Hadwiger [11]. It is known that $4 \leqslant c \leqslant 7$, and both of these inequalities can be proved for (or even by) high school students. To prove $4 \leqslant c$ we must show three colors do not suffice. Suppose there is a coloring in three colors – say red, white, and blue – and consider the following figure associated with an arbitrary red point r.

As all the edge-lengths are 1, and as two points at unit distance cannot receive the same color, the other two vertices of r's triangle must be white and blue, whence the point r' is red. Rotating the figure about r produces a circle of red points r'. As that circle plainly contains two points at unit distance, it follows that three colors do not suffice and $c \leqslant 4$.

The construction [11] showing $c \leqslant 7$ involves the tessellation of the plane by regular hexagons and another tessellation based on that one.

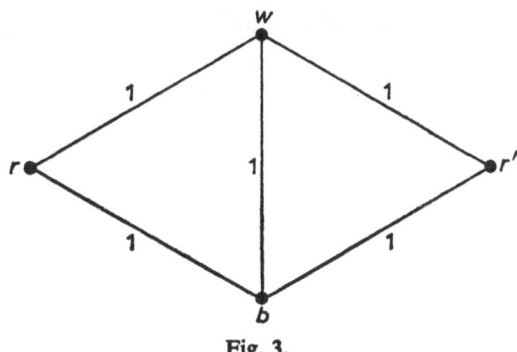

Fig. 3.

ILLUMINATING A TWO-DIMENSIONAL ROOM

In a two-dimensional room whose walls are mirrors forming a simple closed polygon, must the entire interior be illuminable from some point? The problem asks whether there must be a point in the room from which a point source of light could illuminate the entire interior, using reflected rays as well as direct rays. (Assume that a light ray terminates if it hits a vertex of the polygonal boundary.) This or related problems have been discussed by Penrose and Penrose [19] and Klee [15]. Though the problem is unsolved in the case of a polygonal boundary, there is a beautiful example (Figure 4) of a smoothly bounded region not illuminable from any point. (This is P. Ungar's modification of an example in [19].) The upper and lower parts of the boundary, above and below the broken lines, are semi-ellipses with foci $p, p', q,$ and q'. To prove the region is not illuminable from any point, one

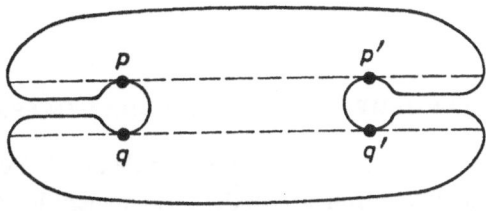

Fig. 4.

uses the fact that, in an ellipse, any light ray passing through one of the foci is reflected through the other focus.

CONFIGURATIONS OF POINTS AND CONICS

Does there exist a plane configuration consisting of six points and six congruent conics such that each conic passes through five of the points? The problem has been discussed by Kelly [12] and Blumenthal [3, p. 154]. It is easy to find, for each $n < 6$, a plane configuration consisting of n points and n congruent conics such that each conic passes through precisely $n-1$ of the points (see Figure 5 for $n=3$). And from the fact that five points determine a conic it follows that no such configuration can exist for $n \geqslant 7$. However, the case $n=6$ is unsettled.

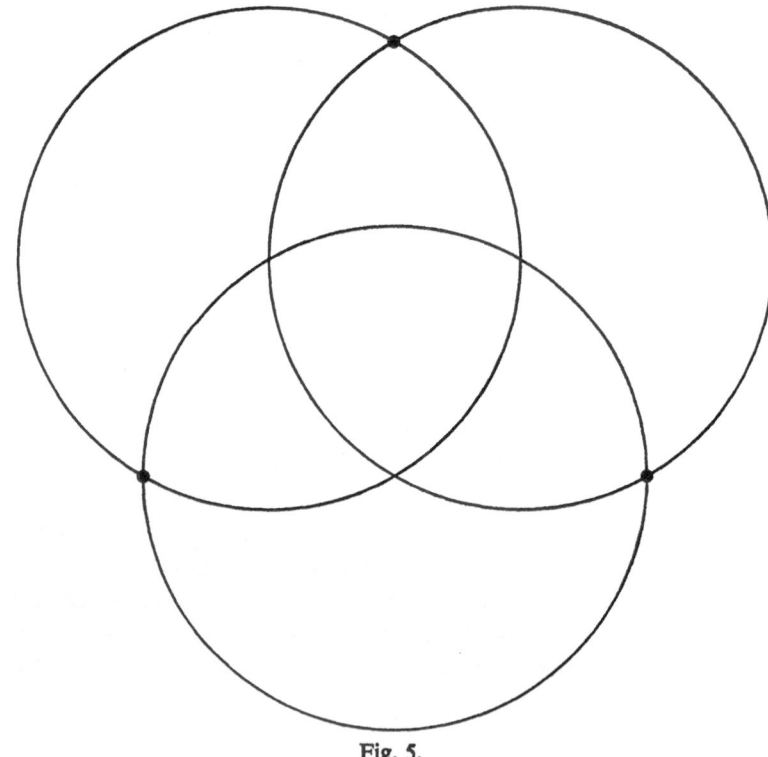

Fig. 5.

ARRANGEMENTS OF TETRAHEDRA IN SPACE

What is the maximum number t of solid tetrahedra that can be arranged in space so that any two tetrahedra in the arrangement have a two-dimensional intersection?

This problem is due to Bagemihl [1]. It is also the subject of an entire book by Baston [2] and a short exposition by Klee [14]. For the corresponding problem in the plane – arranging triangular regions so that any two of them have a one-dimensional intersection – the maximum is easily seen to be 4. To see that the maximum t for tetrahedra is at least 8, consider four tetrahedra with a common apex above the plane having as their bases the four triangles bounded by solid lines in Figure 6, and four more tetrahedra with a common apex below the plane having as their bases the four triangles bounded by broken lines in Figure 6. As any two of these eight tetrahedra have a two-

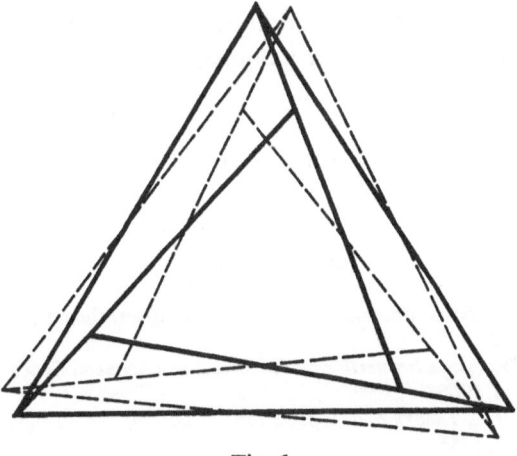

Fig. 6.

dimensional intersection, it follows that $8 \leqslant t$. Bagemihl's proof that $t \leqslant 17$ is accessible to high school students, but Baston's proof that $t \leqslant 9$ takes about two hundred pages! The unsolved problem is to decide whether the maximum t is 8 or 9.

A class discussion of the above problem might well include an example (Tietze) [23, pp. 77–78] showing that if the shapes of the solids are not restricted, a suitable infinite sequence of solids can be arranged in space so that any two solids in the arrangement have a two-dimensional intersection.

ARRANGEMENTS OF POINTS ON A SPHERE

How should a given number n of points be arranged on a sphere so as to maximize the minimum distance determined by the points?

One may regard the points as misanthropes, each of whom so hates the other misanthropes that he wants to get as far away from them as possible. In the corresponding plane problem – arranging n points on a circle so as to maximize the minimum distance – the points should be placed at the vertices of an inscribed regular n-gon. On the sphere, the problem has been solved only for $n \leqslant 12$ and for $n = 24$. For excellent expositions of this problem and its history, see Coxeter [6], Meschkowski [17], and Van der Waerden [25]. Coxeter's account is especially attractive. It is not all accessible at the high school level, but the historical and descriptive parts of it – Kepler's cubic close-packing, the Newton-Gregory controversy, the problem of the pollen grains, etc. – would surely be of interest to a high school geometry class. For references to some later work, see Robinson [20].

A class discussion of the above problem could include a proof of the best arrangements for $n \leqslant 6$, the observation that when $n = 6$ the best distance for $n - 1$ points is the same as that for n points, and Robinson's conjecture [20] that this happens precisely when n is 6, 12, 24, 48, 60, or 120.

University of Washington, Seattle, Wash., U.S.A.

REFERENCES

[1] Bagemihl, F., 'A conjecture concerning neighboring tetrahedra', *Am. Math. Monthly* 63 (1956) 328–329.
[2] Baston, V. J. D., *Some Properties of Polyhedra in Euclidean Space*, Pergamon Press, Oxford, England, 1965.
[3] Blumenthal, L. M., *Theory and Applications of Distance Geometry*, Clarendon Press, Oxford, England, 1953.
[4] Coxeter, H. S. M., 'A problem of collinear points', *Am. Math. Monthly* 55 (1948) 26–28.
[5] Coxeter, H. S. M., *Introduction to Geometry*, Wiley, New York, N.Y., 1961.
[6] Coxeter, H. S. M., 'The problem of packing a number of equal nonoverlapping circles on a sphere', *Trans. N.Y. Acad. Sci.* (2) 24 (1962), 320–331.
[7] Crowe, D. W. and T. A. McKee, 'Sylvester's problem on collinear points', *Math. Magazine* 41 (1968), 30–34.
[8] Dirac, G. A., 'Collinearity properties of sets of points', *Quart. J. Math. (Oxford Series)* 2 (1951), 221–227.
[9] Freese, R. W., A. K. Miller, and Z. Usiskin, 'Can every triangle be divided into n triangles similar to it?', *Am. Math. Monthly* 77 (1970), 867–869.
[10] Grünbaum, B., 'The importance of being straight', in *Proceedings of the Twelfth Biennial Seminar (Time Series and Stochastic Processes; Convexity and Combinatorics)*, Canadian Mathematical Congress, Montreal, Canada, 1970, 243–254.
[11] Hadwiger, H., 'Ungelöste Probleme, Nr. 40'', *Elemente Math.* 16 (1961), 103–104.
[12] Kelly, L. M., 'Covering problems', *Math. Magazine* 19 (1944–45), 123–130.
[13] Kelly L. M. and W. O. J. Moser, 'On the number of ordinary lines determined by n points', *Can. J. Math.* 10 (1958), 210–219.
[14] Klee, V., 'Can nine tetrahedra form a neighboring family?', *Am. Math. Monthly* 76 (1969), 178–179.
[15] Klee, V., 'Is every polygonal plane region illuminable from some point?', *Am. Math. Monthly* 76 (1969), 180.
[16] Klee, V., *Shapes of the Future, I and II.* Two films produced by the Educational Development Center, Newton, Mass., for the Individual Lecture Film Project of the Mathematical Association of America, 1970 and 1971. Available from Modern

Learning Aids, P.O. Box 302, Rochester, N.Y.

[17] Meschkowski, H., *Unsolved and Unsolvable Problems in Geometry* (trans. from German by J. A. C. Burlak), F. Ungar, New York, N.Y., 1966.

[18] Monsky, P., 'On dividing a square into triangles', *Am. Math. Monthly* 77 (1970), 161–164.

[19] Penrose, L. S. and R. Penrose, 'Puzzles for Christmas', *The New Scientist* 25 December 1958, 1580–1581, 1597.

[20] Robinson, R. M., 'Finite sets of points on a sphere with each nearest to five others', *Math. Ann.* 179 (1969), 296–318.

[21] Sylvester, J. J., 'Mathematical Question 11851', *Educational Times* 59 (1893), 98.

[22] Thomas, J., 'A dissection problem', *Math. Magazine* 41 (1968), 187–190.

[23] Tietze, H., *Famous Problems of Mathematics*, Graylock Press, Baltimore, Md., 1965.

[24] Usiskin, Z. and S. G. Wayment, 'Partitioning a triangle into 5 triangles similar to it', *Math. Magazine* 44 (1971), to appear.

[25] Van der Waerden, B. L., 'Pollenkörner, Punktverteilungen auf der Kugel und Informationstheorie', *Naturwissenschaften* 7 (1961), 189–192.

HOWARD LEVI

GEOMETRIC ALGEBRA
FOR THE HIGH SCHOOL PROGRAM

SUMMARY. Plane affine geometry is developed from four axioms, up to the construction of a field (in general non-commutative), introduction of coordinates and derivation of equations for lines. The development is intended to be intelligible and rewarding to better high school students.

INTRODUCTION

High school geometry in the U.S.A. today means either Euclidean or real affine geometry and whatever coordinates are used are based on distances. This is due partly to history, partly to the lack of a usable alternative. We present here an approach to general affine geometry. It is in competition with several items of mathematical mythology:

(a) that vector spaces are more basic, and affine geometry should be studied as a subdivision of linear algebra;

(b) that projective geometry is more basic and affine geometry should be studied as one of its subdivisions;

(c) that this subject is too special to concern most users of mathematics;

(d) that it is too difficult for the high school student.

The reader who holds such *a priori* objections to the program is asked to suspend them while he examines it.

We offer, in addition to the program itself, these observations in reply:

(a) An extensive counter to this point is given in [8].

(b) Friends of projective geometry know how smoothly things go in the presence of the "Fundamental Theorem" (triple transitivity), and regret that it is not so in the general case. Affine geometry is different in that the analog of this theorem (our Axiom IV on double transitivity) is available with no loss of generality [cf. 7].

(c) This development shows numbers emerging from activities quite different from counting and measuring, and illustrates one of the major achievements of modern mathematics, namely, the broadening of our concept of what numbers are and of where, and how, to find them. Moreover coordinates which are not real numbers are not exotic. The field of complex numbers and finite fields are in every day use in quite down to earth mathematics.

(d) Many existing treatments of this subject are beyond the high school level. The author believes that this one is not and invites the reader to see for himself.

AXIOMATIC BASIS

Our primitive terms are a set $\mathscr{S}$ called the *plane*, whose elements are called *points* and a set $\mathscr{L}$ of subsets of $\mathscr{S}$ whose elements are called *lines*.

AXIOM I. Distinct points A and B belong to exactly one line.

DEFINITION. Lines p and q are *parallel* if either $p = q$, or $p \cap q = \emptyset$.

AXIOM II (parallel postulate). For each point A and each line p there is exactly one line which contains A and is parallel to p.

NOTATION. AB means the line which contains the distinct points A and B. $p \parallel q$ means that lines p and q are parallel. $A*p$ means the line which contains A and is parallel to p.

THEOREM I. If p and q are lines, then either $p \cap q$ is a singleton, or $p \parallel q$.

PROOF. Suppose $p \cap q$ is not a singleton. If this intersection is empty, then $p \parallel q$. If it has more than one point, then, according to Axiom I, we have $p = q$ and again $p \parallel q$.

THEOREM II. If p, q, r are lines for which $p \parallel q$, $q \parallel r$, then $p \parallel r$.

PROOF. If $p \nparallel r$, then $p \cap r$ is a singleton $\{A\}$. The inference that distinct lines p and r contain A and are parallel to q contradicts Axiom II.

COROLLARY. The relation of parallelism is an equivalence relation on $\mathscr{L}$ each of whose equivalence classes is a partition of $\mathscr{S}$.

AXIOM III. There are three non-collinear points.

THEOREM III. Every line contains at least two points.

PROOF. Let p be a line and let A, B, C be non-collinear points. Then p is not parallel to at least one of AB, BC, CA, say AB. Then AB and $C*AB$ have no common point and each has a point in common with p.

DEFINITIONS. Let p and q be lines and let r be a line parallel to neither. The *parallel projection* from p to q in the direction of r is the function from p to q whose value at each point X of p is the point of intersection of q and $X*r$. An *affinity* is a parallel projection or a finite composition of parallel projections.

Notations. $p \xrightarrow{r} q$ means the parallel projection from p to q in the direction of r. A (p, q) affinity is an affinity from p to q.

THEOREM IV. Let p be a line. Relative to composition, the set of (p, p) affinities is a group.

PROOF. The definition of affinity implies that if f and g are (p, p) affinities, then so is $f \circ g$. Associativity is automatic since affinities are transformations. To show that the identity function on p is an affinity let A be any point on p and let B be any point not on p. Then $p \xrightarrow{AB} B*p \xrightarrow{AB} p$ is an affinity and is the identity function on p. To see that every (p, p) affinity has a (p, p) affinity for inverse, observe that if f is $p \xrightarrow{a} b \rightarrow \cdots \rightarrow r \xrightarrow{s} p$, then $p \xrightarrow{s} r \rightarrow \cdots \rightarrow b \xrightarrow{a} p$ is f^{-1}.

Notation. We denote this group by Aff(p,p) and its identity by i_p.

THEOREM V. If A and B are distinct points and if A' and B' are distinct points, then there is an affinity f such that $f(A)=A', f(B)=B'$.

PROOF. Case I. A, B, A', B' are collinear. If $A=A'$, $B=B'$, then i_{AB} is an affinity with the required property. If $A \neq A'$ let P be a point not on AB and let Q be the point of intersection of $B*AP$ and $B'*A'P$.

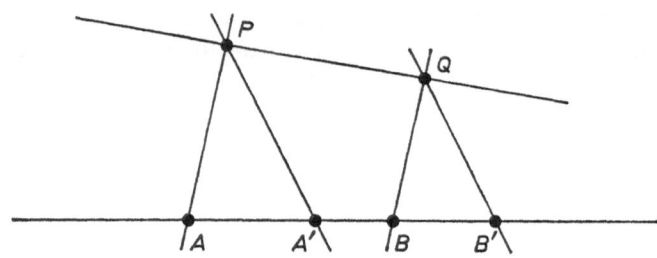

Since P is on AP and Q is not, we have $P \neq Q$, and the affinity $AB \xrightarrow{AP} PQ \xrightarrow{PA'} AB$ sends A to P to A' and B to Q to B'.

Case II. If the given points are not collinear, then one of A', B', say A' is not on AB. Let g be $AB \xrightarrow{AA'} A'B'$, and, referring to Case I, let h be an affinity for which $h(g(A))=A'$, $h(g(B))=B'$. Then, the affinity $h \circ g$ makes the required assignments.

AXIOM IV. If f and g are affinities and if A and B are distinct points for which $f(A)=g(A), f(B)=g(B)$, then $f=g$.

GROUPS OF AFFINITIES

We recall that a fixed point of a transformation is a point which is its own image under that transformation.

THEOREM VI. Let O and I be distinct points. Then the set of (OI, OI) affinities which have O as a fixed point is a subgroup of Aff(OI, OI). If X is any point of OI other than O, then there is exactly one element f of this group for which $f(I)=X$.

PROOF. It is clear that if $f(O)=O$ and $g(O)=O$, then $f(g(O))=O$, and $f^{-1}(O)=O$. This suffices to show that these affinities constitute a sub-group as stated. The second part of the theorem follows by applying Theorem V and Axiom IV.

THEOREM VII. Let A and A' be distinct points of the line p. Then there is exactly one fixed point free (p, p) affinity which transforms A to A'.

PROOF. Let B and B' be points of p distinct from A and A', respectively, and reproduce the construction described in the proof of Theorem V for the affinity which transforms A to A', B to B'. If PQ has a point on AB, then this point is a fixed point of the affinity. Thus our affinity is fixed point free if and only if $PQ \parallel AA'$ and it is completely determined by the pair A, A'.

COROLLARY. Let f and g be fixed point free (p, p) affinities. Then either $f \circ g$ is also fixed point free, or it is i_p.

PROOF. Suppose $f \circ g$ has a fixed point A, that is, $f(g(A)) = A$. Then f and g^{-1} both have the same value at $g(A)$, namely, A. Then $f = g^{-1}$ and $f \circ g = i_p$.

COROLLARY. Let p be a line. Then the fixed point free (p, p) affinities, together with i_p are a subgroup of Aff(p, p).

Notation. The subgroup of Aff(p, p) of our last corollary is denoted by $\mathscr{A}(p)$.

THEOREM VIII. $\mathscr{A}(p)$ is an invariant subgroup of Aff(p, p) (that is, if $f \in$ Aff(p, p) and if $g \in \mathscr{A}(p)$, then $f^{-1} \circ g \circ f \in \mathscr{A}(p)$).

PROOF. We are to show that $f^{-1} \circ g \circ f$ is either fixed point free or is i_p. Suppose it had a fixed point X. We deduce from $f^{-1}(g(f(X))) = X$ that $g(f(X)) = f(X)$. Then g has the fixed point $f(X)$. Therefore, $g = i_p$ whence, $f^{-1} \circ g \circ f = f^{-1} \circ i_p \circ f = f^{-1} \circ f = i_p$.

THEOREM IX. Let A, B, C be non-collinear points, let f_B be $AB \xrightarrow{AC} BC$, f_C be $BC \xrightarrow{BA} CA$, and f_A be $CA \xrightarrow{CB} AB$. Then, for every point X of AB, $f_A f_C f_B f_A f_C f_B(X) = X$.

PROOF. $f_A f_C f_B$ maps B to A and A to B. So does the inverse of this affinity. It follows from Axiom IV that this affinity is its own inverse and thus $f_A f_C f_B f_A f_C f_B = i_{AB}$

THEOREM X. $\mathscr{A}(p)$ is commutative.

PROOF. Let f and g belong to $\mathscr{A}(p)$. If $f = g$ or if $f^{-1} = g$ or if either f or g

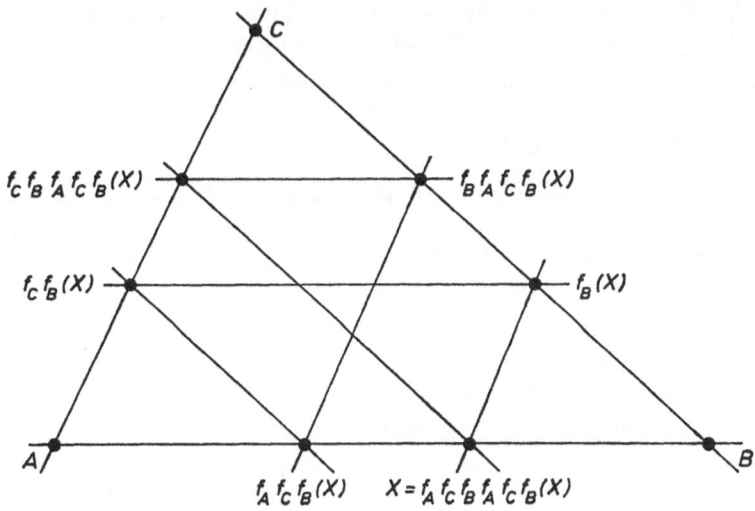

is i_p, the conclusion that $f \circ g = g \circ f$ is immediate. If not, A, $f(A)$, $g(A)$, $f(g(A))$ are distinct. Let C be a point not on p and reproduce the construction of Theorem IX, using $f(g(A))$ as B and $g(A)$ as X. Notice that f is $p \xrightarrow{AC} EE' \xrightarrow{CB} p$ and g is $p \xrightarrow{AC} FF' \xrightarrow{CB} p$. Therefore, under g, the point $f(A)$ goes to F' to B, whence $g(f(A)) = f(g(A))$. It follows from Theorem VII that $f \circ g = g \circ f$.

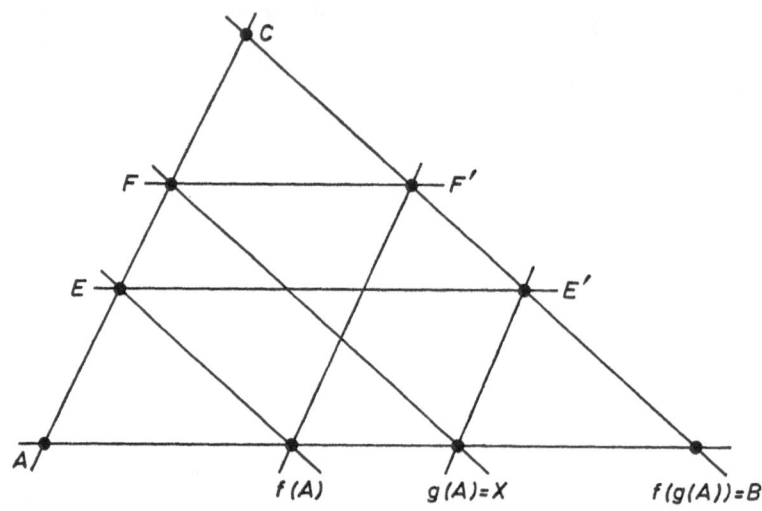

CONSTRUCTION OF THE FIELD

Let O and I be distinct points. We give line OI the structure of a field whose zero is O and whose one is I. If A and B are points of OI distinct from O then m_A^B is the affinity which has O as fixed point and transforms A to B. We denote by a_C^D the element of $\mathscr{A}(OI)$ which transforms C to D.

DEFINITIONS. If X and Y are points of OI, then

(i) $\qquad X + Y = a_O^X(Y)$

(ii) $\qquad X \cdot Y = m_I^X(Y)$ if $X \neq O$ and $X \cdot Y = O$ if $X = O$.

THEOREM XI. Line OI is a field, relative to the given operations.

PROOF. (1) $X + O = X$. This is so because $X + O = a_O^X(O) = X$

(2) $X + Y = Y + X$. Follows from $X + Y = X + a_O^Y(O) = (a_O^X \circ a_O^Y)(O) = (a_O^Y \circ a_O^X)(O) = Y + X$.

(3) $(X + Y) + Z = X + (Y + Z)$. We have $X + Y = a_O^{X+Y}(O)$ and $X + Y = (a_O^Y \circ a_O^X)(O)$. It follows from Theorem VII that $a_O^{X+Y} = a_O^X \circ a_O^Y$. Evaluate the expression on the left at Z, to obtain $(X + Y) + Z$ and the right one at Z to obtain $X + (Y + Z)$.

(4) $-X=a_X^O(O)$. From $i_{OI}=a_X^X=a_O^X \circ a_X^O$ we have $O=(a_O^X \circ a_X^O)(O)=$ $a_O^X(a_O^X(O))=X+a_X^O(O)$.

(5) $X \cdot O=O \cdot X=O$. We have $X \cdot O=m_I^X(O)=O$. The other fact is part of the definition of multiplication.

(6) $X \cdot I=X=I \cdot X$. We have $X \cdot I=m_I^X(I)=X$, if $X \neq O$, and we already know that $O \cdot X=O$. Since m_I^I is i_{OI}, we have $X=m_I^I(X)=I \cdot X$.

(7) $X \cdot (Y \cdot Z)=(X \cdot Y) \cdot Z$. We evaluate $m_I^{X \cdot Y}$ at O, I, Z, obtaining the respective values $O, X \cdot Y, X \cdot (Y \cdot Z)$. We evaluate $m_I^X \circ m_I^Y$ at O, I, Z obtaining O, $X \cdot Y, X \cdot (Y \cdot Z)$. Comparing the values at O and I we infer that $m_I^{X \cdot Y}=$ $m_I^X \circ m_I^Y$. Equating the values at Z gives our desired equation. If either X or Y is O these affinites are not defined but in this case the equation in question is a direct result of part 5.

(8) If $X \neq O$, then $X^{-1}=m_X^I(I)$. From $i_{OI}=m_I^X \circ m_X^I$ we have $I=(m_I^X \circ m_X^I)(I)=X \cdot m_X^I(I)$.

(9) $X \cdot (Y+Z)=(X \cdot Y)+(X \cdot Z)$, We know from Theorem VIII that there is a W such that $m_I^X \circ a_O^Y=a_O^W \circ m_I^X$. The left side evaluated at O is $X \cdot Y$ and the right side has the value W at O. We infer $W=X \cdot Y$. Now evaluate these expressions at Z. The left one has value $X \cdot (Y+Z)$ there and the right one (using $W=X \cdot Y$) has value $(X \cdot Y)+(X \cdot Z)$.

(10) $(Y+Z) \cdot X=(Y \cdot X)+(Z \cdot X)$. The proof of this statement is omitted because the only proofs the author has been able to devise are relatively long and unenlightening. The attached diagram indicates one proof. Dotted lines indicate applications of Desargues' Theorem and a preliminary reduction to $Y=I$ has already been carried out. In the absence of a better proof the author suggests telling the students "it can be shown that..."

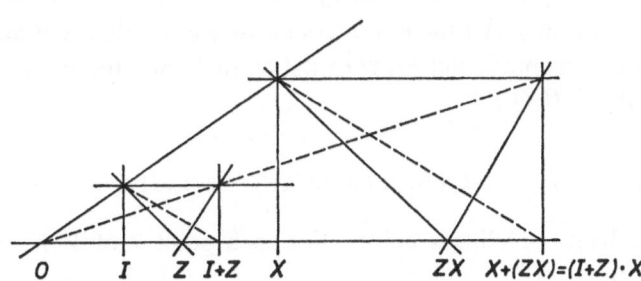

COORDINATES ON LINES

We have shown how to construct a field from a pair of distinct points. It is not difficult to show that if field F is constructed from the pair (O, I) and field F' from the pair (O', I'), then the affinity which maps O to O' and I to I' is an isomorphism of F to F'. We do not enter into the details nor do we have any need other than an esthetic one for the result. In what follows we imagine that a definite pair (O, I) has been chosen and refer to the resulting field as "the field". We denote it by $\mathscr{F}$.

THEOREM XII. Let f be an affinity from line OI to itself. Then for every X of OI we have $f(X) = (f(I) - f(O)) \cdot X + f(O)$.

PROOF. $a_O^{f(O)} \circ m_I^{f(I) - f(O)}$ is an affinity whose value at O is $f(O)$ and whose value at I is $f(I)$. It must therefore be the affinity f. Since the stated value for f at X is the value of this affinity at X our result follows.

COROLLARY. Let A and B be elements of $\mathscr{F}$, with $A \neq O$. Then the function f whose value at each point X of OI is $A \cdot X + B$ is an affinity from OI to itself.

PROOF. According to Theorem XII it is the affinity whose value of O is B and whose value at I is $A + B$.

DEFINITION. Let A and B be distinct points of line p. The (A, B) *coordinate system* for p is the (p, OI) affinity whose value at A is O and whose value at B is I.

Theorem XIII. Let f be the (A, B) coordinate system and g be another coordinate system for line AB. Then, for every X of AB we have $g(X) = (g(B) - g(A)) \cdot f(X) + g(A)$.

PROOF. Let h be the affinity from OI to itself for which $h(O) = g(A)$, $h(I) = g(A)$. Then $h \circ f$ and g have the same values at A and B and must therefore be equal for every X of AB. Combining this fact with the expression for h given by Theorem XII yields our result.

Corollary. Let f be a coordinate system for line p. Then a function g from p to OI is a coordinate system for p if and only if there are elements L and M of $\mathscr{F}$, with $L \neq O$, such that $g = L \cdot f + M$.

COROLLARY. Let f be a coordinate system for line p and let g be a coordinate system for line q. A function h from p to q is an affinity if and only if there are field elements L and M, with $L \neq O$, such that, for every X of p, we have $g(h(X)) = L \cdot f(X) + M$.

COORDINATES FOR THE PLANE AND EQUATIONS OF LINES

Let Q, X, Y be non-collinear points. For each point A of $\mathscr{S}$ let A_X be the intersection point of QX and $A*QY$ and let A_Y be the intersection point of QY and $A*QX$. We define $x(A)$ as the (Q, X) coordinate of A_X and $y(A)$ as the (Q, Y) coordinate of A_Y. Then x and y are functions from $\mathscr{S}$ onto $\mathscr{F}$. For each field element u the set $x^{-1}(u)$ is a line parallel to QY and the set $y^{-1}(u)$ is a line parallel to QX. The pairing $A \leftrightarrow (x(A), y(A))$ is a one-to-one correspondence from $\mathscr{S}$ to $\mathscr{F} \times \mathscr{F}$. It is called the QXY coordinate system for $\mathscr{S}$ and $(x(A), y(A))$ are the QXY coordinates of A.

THEOREM XIV. If line p is not parallel to QY, then x, restricted to p, is a

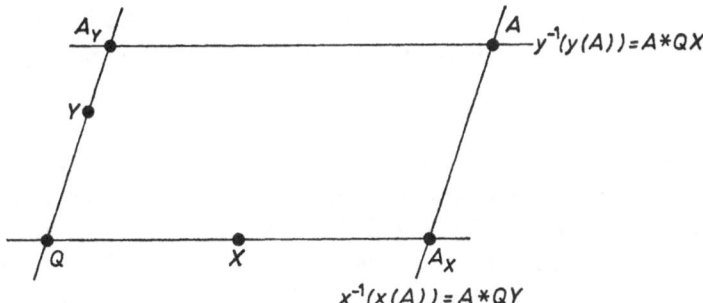

$$x^{-1}(x(A)) = A * QY$$

coordinate system for p, and if p is not parallel to QX, then y, restricted to p, is a coordinate system for p.

PROOF. Let f be the (Q, X) coordinate system. If p is not parallel to QY, then pairing each point A of p with A_x is the parallel projection $p \xrightarrow{QY} QX$ and the restrictions of x to p is the affininity obtained by following this parallel projection by f. Similarly if p is not parallel to QX then the restriction of y to p is an affinity from p to OI and hence is a coordinate system for p.

THEOREM XV (parametric equations for lines). Let A and B be distinct points and let t be the (A, B) coordinate system for AB. Then, for every point W of AB,
$$x(W) = (x(B) - x(A))t(W) + x(A)$$
$$y(W) = (y(B) - y(W)t(W) + y(A).$$

PROOF. If AB is not parallel to QY, then Theorem XIV tells us that x is a coordinate system on AB, and our formula for $x(W)$ follows from Theorem XIII. If AB is parallel to QY, then $x(W) = x(A)$, for every W of AB, which is precisely what the formula reduces to in this case. The formula for $y(W)$ is derived similarly.

COROLLARY. Let a, b, c, d be field elements with not both a and b zero. Then the set of points whose coordinate set is $\{(at+c, bt+d) : t \in \mathscr{F}\}$ is a line.

PROOF. Let A be the point whose coordinates are (c, d) and let B be the point whose coordinates are $(a+c, b+d)$. Then the set in question is the line AB.

COROLLARY. If line p is not parallel to QY, then there are elements d and e of $\mathscr{F}$ such that the point with coordinates (u, v) is on p if and only if $v = du + e$.

PROOF. Choose the coordinate system t of Theorem XV to be x restricted to p (this can be done by selecting A to be the intersection of QY with p and B to be the intersection of $X*QY$ with p), and then apply Theorem XV.

SOME SUPPLEMENTARY POINTS

A mathematics program at this level cannot consist exclusively of a chain of deductions to be followed. Here are some suggestions for activity on the part of the student, which could tend to enrich the program pedagogically.

Use of diagrams and constructions

Many of the statements of this course can be interpreted and tested physically. We can regard a dot on a flat piece of paper as a "point" and the tracing of a straight edge as a "line". One could then propose a test for Axion IV by drawing lines p, q, r, s, and following the fate of some points A, B, C, D under an affinity $p \rightarrow q \rightarrow r \rightarrow s$; carrying out the construction of Theorem V to obtain a new affinity which agrees with this one at A and B, and verifying that it also agrees at C and D.

Many of the other theorems admit similar checks, especially the verification of the field axioms. The author believes that such checks have a double benefit. In the first place it helps the student see what the statements mean and thus helps him learn subject. It also provides an antidote to the corrupting influence on communication of the mass media. These tests show him that language can function as a precise and responsible instrument.

Study of certain famous configurations

We owe to Hilbert the recognition of the connection between Desargues' and Pappus' theorems and the associativity and commutativity properties of the companion geometric algebra. These theorems are implicitly present here, in in a modernized form, in terms of mappings rather than in terms of configurations, and it might be useful to compare the two versions.

One version of Desargues' Theorem states that if lines p, q, r are concurrent at O, if A, A' are points of p, B, B' are points of q, C, C' are points of r, if $AB \parallel A'B'$, $BC \parallel B'C'$, then $AC \parallel A'C'$. Let us consider, under the same hypotheses, the affinities $p \xrightarrow{AB} q \xrightarrow{BC} r$ and $p \xrightarrow{AC} r$. They both send O to O and A to C. According to Axiom IV they must be the same. It follows that $p \xrightarrow{AC} r$ must also send A' to C', whence $AC \parallel A'C'$.

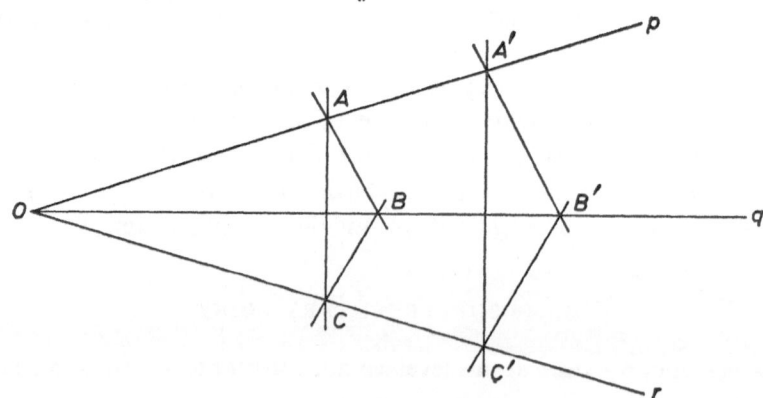

Another version of Desargues' Theorem is that is line p, q, r are distinct and parallel, and if A, A', B, B', C, C' are as before, then again $AC \parallel A'C'$.

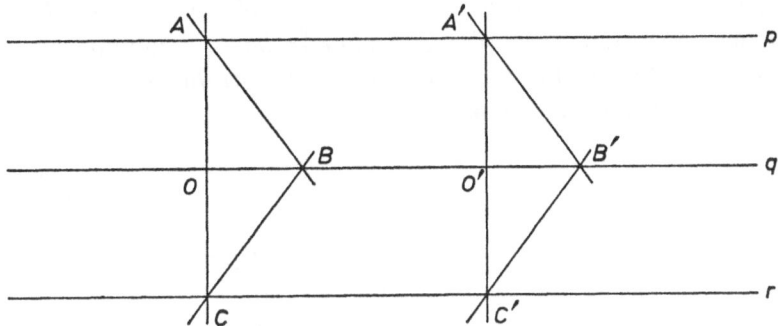

Let O and O' be the respective intersections of AC and $A'C'$ with q. According to Theorem VII the fixed point free (q, q) affinity which transforms O to B is unique. It can be effected either by $q\xrightarrow{AC}p\xrightarrow{AB}q$, or $q\xrightarrow{AC}r\xrightarrow{CB}q$. The fact that these affinities have the same value at O' expresses the conclusion to the theorem.

Pappus' Theorem also has two versions which arise here. One is that if p and q are distinct parallel lines, if A, B, C are points of p, if A', B', C' are points of q, if $AB' \parallel A'B$, $AC' \parallel A'C$, then $BC' \parallel B'C$. This fact is readily deducible from the fact that $\mathscr{A}(p)$ is commutative, and is equivalent to the commutativity of addition.

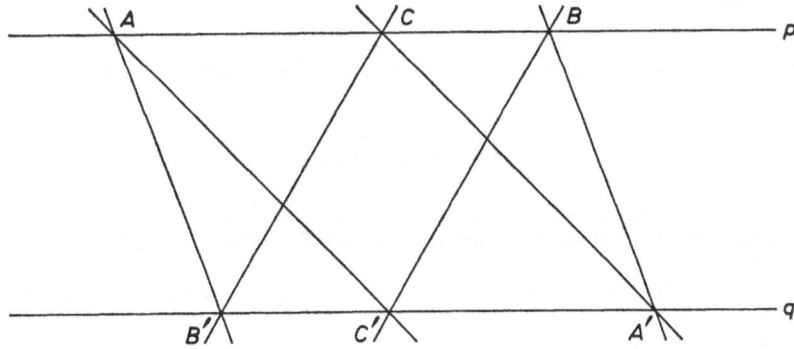

If p and q are not parallel, the corresponding fact is equivalent to the commutativity of multiplication, and is not deducible from our four axioms. Let us introduce a new axiom which is a disguised form of the second instance of Pappus' Theorem.

Axiom V. Let f, g, h be parallel projections from line p to line q. Then $fg^{-1}h=hg^{-1}f$.

THEOREM. $X\cdot Y=Y\cdot X$.

PROOF. Let R be any point not on OI, let f be $OR\xrightarrow{RX}OI$, let g be $OR\xrightarrow{RI}OI$, and let h be $OR\xrightarrow{RY}OI$. Then $m_I^X=fg^{-1}$ and $m_I^Y=hg^{-1}$. By multiplying on the right by g^{-1} the equation of our axioms becomes $fg^{-1}hg^{-1}=hg^{-1}fg^{-1}$, which is equivalent to $m_I^Y m_I^X=m_I^X m_I^Y$.

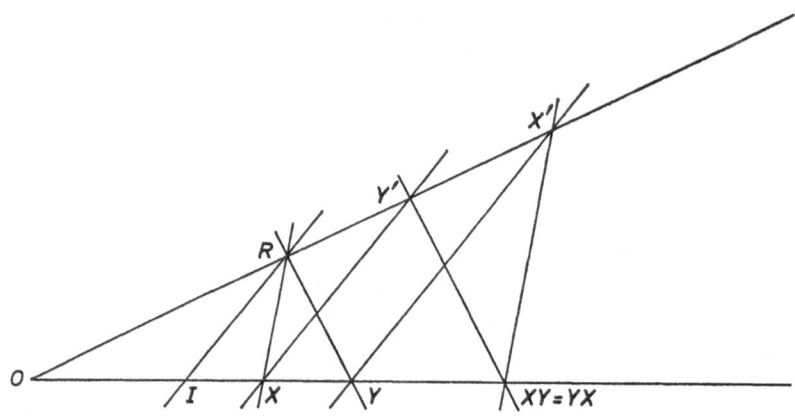

One fact which cannot be explored in this way but which has a certain educational value is a discussion of Fano's Theorem on the intersection of the diagonals of a parallelogram. Their non-intersection and its connection with the field having characteristic two is accessible and has a certain therapeutic shock value.

Lehman College, City University of New York, New York, N.Y., U.S.A.

BIBLIOGRAPHY

[1] Artin, E., *Geometric Algebra*, Interscience, New York, 1957.
[2] Bachmann, F., *Aufbau der Geometrie aus dem Spiegelungsbegriff*, Springer, Berlin, 1959.
[3] Baer, R., *Linear Algebra and Projective Geometry*, Academic Press, New York, 1952.
[4] Blumenthal, L. M., *A Modern View of Geometry*, Freeman, San Francisco, 1961.
[5] Hodge, W. V. D., and D. Pedoe, *Methods of Algebraic Geometry*, Cambridge, 1947.
[6] Levi, H., *Foundations of Geometry and Trigonometry*, Prentice-Hall, 1960.
[7] Levi, H., 'Plane Geometries in Terms of Projections', Proc. Am. Math. Soc. **16** (1965), 503–511.
[8] Levi, H., 'Foundations of Geometric Algebra', Rendiconti di Matematica 1969 (3–4), Vol. 2, Ser. VI., 639–670.
[9] Levi, H., *Topics in Geometry*, Prindle, Weber and Schmidt, Boston, 1968.
[10] Pickert, G., *Projektive Ebenen*, Springer, Berlin, 1955.

KARL MENGER

THE GEOMETRY RELEVANT TO MODERN EDUCATION

I. INTRODUCTION

For some time now, the shortcomings of Euclid's Elements have been recognized in precollege teaching; and the age that identified the study of geometry with reading Euclid's book is definitely past. But it has also become clear that a presentation of the whole of Euclidean geometry with modern rigor would require more time than secondary schools can apportion to this task. As a result of this situation, many educators feel that they are forced to adopt one of two courses: either to spend too much time on the teaching of geometry or to present the subject without meeting the requirements of modern rigor – a task in which they are supported by textbooks some of which actually have much greater shortcomings than Euclid's Elements. Two ways out of this dilemma have been proposed.

The first is to give up the teaching of geometry altogether – a step welcomed by those students who doubt that geometry is, to use a term in vogue, relevant. But young people who are not taught geometry miss the great experience of seeing a deductive theory developed before their eyes – a theory, moreover, that is accompanied by mental pictures and drawings on paper which motivate its assumptions and illustrate its conclusions. The study of rings and fields in algebra cannot compensate for this loss.

The second proposed way out of the dilemma, initiated by G. D. Birkhoff, is the deduction of geometry from postulates involving arithmetical ideas, in particular the profound concept of real number. This amounts to a synthetic presentation of something akin to analytic geometry. A related method is the embedding of geometry into linear algebra and its presentation as a theory of vector spaces over the real numbers. But although these ideas can be presented in the form of deductive systems they deprive the student of the experience of developing such a theory from assumptions about really simple and purely geometric concepts such as points, lines, and incidence. And just such a treatment is of particular importance because of certain applications.

The applicability of geometry to physics, chemistry, and biology, to engineering and economics and, of course, to other branches of mathematics has been stressed, as has the beauty of geometry, by all those who oppose the discontinuance of the teaching of geometry. It is true, that as far as some of these applications are concerned, a postulational theory of vector spaces is preferable to one in terms of points, lines, and incidence. What I wish to emphasize in this paper, however, are applications of a totally different kind, namely applications to reasoning in general – thus, as it were, applications of the *spirit* of synthetic (synthetic *synthetic* rather than synthetic *analytic*!) geometry to a wide spectrum of topics ranging from the most abstract to the most practical – from philosophy to politics. There is no better immunization against the pseudo-inferences and sham proofs that abound in extra-mathematical reasoning and in daily life than to be exposed in one's youth to a truly airtight deductive theory, however modest in scope. And, as the ancient Greeks discovered, there is no better material for a rigorous treatment than facts about space. *The study of a modicum of synthetic geometry is an experience no one should miss in the course of a modern secondary education: it has far-reaching social relevance.*

I propose a third way out of the dilemma. It is based on the conviction that geometry should be included in the curriculum, that it ought to be taught with perfect rigor, and that it should be presented within a reasonable span of time. The way to achieve these aims is *to abandon the idea of teaching euclidean geometry in its entirety and to present only a part or an aspect of euclidean geometry – but that part or aspect with absolute rigor – as well as some simple related theories which, from the points of view of various students, are "relevant".*

As the material to be presented I have selected the following chapters:

I. The elements of affine plane geometry.

II. The theory of a self-dual fragment of the affine plane [1] with applications to kinematics [2].

III. Certain principles of noneuclidean geometry.

If the quality of the students permits it and more time is available, then these chapters should be supplemented by expositions, strictly separated from the deductive theories, of stimulating intuitive material on linear and planar patterns and polyhedra, on sphere, torus, and Möbius strip, and on some interesting simple curves as well as the most general curve concept [3].

This choice of material should make it clear that no attempt is being made to convert the course in geometry into a course in logic. Nor is it my purpose to make the students either over- or under-estimate deductive reasoning, but rather simply to let them see how this mental process works. I actually believe that such a course will, when competently taught, implant in students more deductive methodology than they could get from a first course in logic (on the algebra of classes and propositions). In addition, the course offers some (to the students) unexpected and exciting revelations about space.

The course outlined in this paper is designed for everyone receiving a modern secondary education. Those who plan to specialize in mathematics, science or engineering will need more geometry – additional chapters and additional methods. But the course contains elements that they, like everyone else, should know and understand and yet rarely learn. Hence, if the course were confined to those who do *not* plan to specialize in science or engineering, then some of these young people might well develop a better insight into certain aspects of geometry and general deductive methodology than would some future scientists and engineers and perhaps some future teachers of mathematics, who then might perpetuate the underrating of the social relevance of the study of geometry. Moreover, it is quite likely that, when an inspiring teacher teaches such a course some non-scientifically minded students might change their minds.

Before outlining the material I wish to discuss from educational, psychological, and sociological points of view what this approach achieves.

(1) The student learns that the deductive geometric system must start from unproven assumptions in terms of undefined concepts and that, within that theory, these initial (so-called *primitive*) elements *remain* unproven and undefined. He can easily be made to realize that this is a fundamental feature of any deductive theory.

(2) He learns that the original model of the primitive elements of the geometric theory are objects and facts of the physical world, which the undefined concepts and unproven assumptions are designed to reflect. Perennial occupation with these objects has created in man something called geometric intuition, which is a second, if less tangible, motivation for the development of deductive theories about space.

(3) He is shown that, within the totality of the propositions of a theory, the postulates are not distinguished as such; that they may be replaced by other assumptions from which they and their consequences can be proved; and that one may even choose different primitive concepts, in terms of which those originally chosen can be defined. Of course, in such an alternative development of the theory, *its* primitive elements remain unproven and undefined.

(4) He learns that definitions in terms of undefined terms can, when skillfully chosen, be of great practical importance as abbreviations that simplify theorems and proofs; but that, in principle, definitions can be dispensed with since the definiendum can always be replaced by the definiens.

(5) He learns what proving really means: combining and transforming the assumptions according to the rules of logic. The particular choice of the material makes this nature of the proofs fully transparent since not only are the assumptions free of references to numbers but their logical combinations and transformations do not utilize the theory or, except for pairs, triples, and

quadruples, even the concept of sets. For example, a line is not defined as a set of points. What rigor of geometric proofs means can be illustrated in the not infrequent situation where a student wants to use in a proof some plain and simple idea or some evident proposition but is told by the teacher: "That idea may be intuitively clear to you, and that proposition can indeed be verified in drawings on paper. But even so, since they have been not included among the basic concepts and the postulates of the theory, you must not use them unless you either derive them from that basis (by a definition or a proof) or formulate them explicitly as additions to that foundation." The fact that experience and intuition, while essential in inspiring and motivating the formulation of the foundations, are excluded from the strictly deductive process is very important for the student's understanding of deductive theories in general.

(6) He learns not only that, given a set of assumptions, certain propositions can be proved but also a complementary fact of paramount importance: that certain other propositions *cannot* be proved from those assumptions. He sees examples of systems (including planes containing altogether only a finite number of points) that satisfy certain postulates but not certain other propositions – a fact that *proves* that the latter cannot be proved from the former. The significance of this insight for the intellectual education and even the practical life of the student cannot be overestimated.

(7) He learns that the self-dual fragment of the affine plane is a sufficient basis for pre-limit calculus [4]. In that fragment, one can determine the area polygons of a step line, and the slope step line of a polygon, and establish their inverse character, which is at the core of calculus.

(8) He finds that one and the same theory may be capable of a variety of interpretations – an illustration of what Ernst Mach called the *Denkökonomie* of mathematics. By means of a simple dictionary, the geometry of the self-dual fragment of the affine plane is translated into Galileo's kinematics of a two-dimensional world (that is, a one-dimensional space in time).

(9) The study of the principles of noneuclidean geometry brings the student another revelation of fundamental importance: that one may vary the assumptions of a theory substantially and develop an incompatible sister theory, either one self-consistent if the other one is. Any mystery is dispelled by displaying to the students noneuclidean geometries in situations with which they are perfectly familiar: one such geometry on a sheet of paper, and another one on the streets of a modern metropolis governed by what I have called the *taxicab geometry* [5], whose circles are squares. In particular, various ways of measuring the distance between two points are expounded. It is undeniable that the taxi fare for a trip to a point three blocks East and

four blocks North of the starting point is equal to the fare for *seven* blocks, whereas as the crow flies as well as on the basis of Euclid's assumptions, the initial and the terminal point of this trip are the length of *five* blocks apart.

(10) The student will become familar with *various applications of geometry to nature*, in fact, with the possibility of applying even mutually incompatible theories. The teacher can at least report, if not explain, to him that just as the self-dual fragment of the affine plane can be applied to Galilean kinematics, a certain non Euclidean plane (more specifically, a fragment of the fragment just mentioned) can be applied to the kinematics of Einstein's special theory of relativity [6].

(11) It is perhaps not too much to hope that the preceding ten points, when competently presented by able teachers, may convince a good many students of the relevance of geometry. Without counting intrinsic interest and beauty of geometry at all they may realize its relevance in two directions: in direct applications to science and in the formation of a *Weltbild* that far transcends geometry and science. But one should also try to reach anti-intellectuals and diehard fighters for practicality, who might even say that they don't need relativity theory and that certainly the study of planes that contain only finite numbers of points, is outright ridiculous. "Indeed", the teacher should admit, "the discovery of such planes at the turn of the century began as a quite esoteric study. But 25 years later, statisticians found that the tool they needed in order to devise the most efficient tests to compare certain competing substances as to their relative merits were just those planes with only finite numbers of points. And today the results of the geometry of such planes are used by the United States Department of Agriculture in testing fertilizers." This is a use of ivory tower mathematics that even advocates of superpracticality should find earthy enough.

II. THE ELEMENTS OF AFFINE PLANE GEOMETRY

At the outset, the teacher should present examples of propositions about the plane from Euclid's Elements and Hilbert's Grundlagen, such as Euclid, Propositions 14, 15 and Hilbert I1, I2, I3, Theorem 1, III1, III2, III5, Theorems 11 and 20. Then he should say that his course will be confined to propositions of the type of I1–3, and that congruence of segments or angles and perpendicularity will not be considered. The restricted system of propositions that will be studied is called affine plane geometry.

Affine plane geometry deals with *points* and straight lines, briefly, *lines* (denoted by capital and lower case letters, respectively) and the relation of *incidence* (denoted by $\circ$, read: *on*). That P and *l* are incident is expressed by writing: $P \circ l$ or, synonymously, $l \circ P$. What are points and lines?

On a blackboard or a sheet of paper, points are chalk dots or pencil marks, and lines are rows of such dots drawn by means of a straightedge. (Student: "Isn't the definition of straight lines in terms of straightedges circular?" Teacher: "No. For the definition of a straightedge is not based on the concept of straightness. A straightedge is a rod which, when flipped over, yields the same line as before – in contrast to bent or dented rods. Hence one may use straightedges in defining straight lines.")

In affine geometry, points and lines are undefined objects, subject only to unproven assumptions expressed in three or four postulates; and they remain undefined, and the assumptions remain unproved throughout the course. The student must be made fully aware of the fact that, since neither proving nor defining can go on ad infinitum, each deductive theory must necessarily start from unproven assumptions in terms of undefined concepts (see Introduction, Point 1).

On the blackboard and on a sheet of paper one readily observes some relations between points and lines; for example that, by means of a straightedge, one can join any two points by one and only one line. (If one joins points by circles, drawn by means of a 25 cent piece, whose diameter is an inch, then one observes that points that are an inch apart can be joined by one and only one such circle; points that are closer to one another, by exactly two such circles; points that are farther than an inch apart, by none.) This observation concerning points and lines on a blackboard or on paper will be reflected in one of the assumptions (namely, in Postulate I) about the undefined objects called points and lines of affine geometry (see Introduction, Point 2).

On a sheet of paper one further observes many pairs of lines that do not intersect. In particular, if l is a line and P is a point not on l, then one can draw many lines through P that do not intersect l. On a larger sheet, the extensions of many of those lines do intersect the extension of l. If l is, say, horizontal, then some of those extensions meet l on the right side of P, others on the left side, still others not at all, although extensions of some of these last lines do meet an extension of l on an even larger sheet. Extrapolating one gets the idea that also the whole plane contains a whole line through P that fails to intersect the whole line l. Affine geometry is based on the assumption (expressed in Postulate II, which in this form is due to Proclus and Playfair) that there is exactly one such line on P. It is said to be parallel to l.

Before formulating the postulates the teacher should define the concept of parallelism in terms of the undefined concepts. The lines l and l' are said to be *parallel* (written $l \mid\mid l'$) if no point is on both l and l'.

The first three postulates then read as follows.

I. *If P and Q are distinct points, then there is one and only one line that is both on P and on Q.* (It is called the *join* of P and Q and denoted by PQ or, synonymously, by QP.)

II. *If l is a line and P is a point not on l, then there is one and only one line that is on P and parallel to l.* (That line will be denoted by $\pi(l, P)$.) The teacher should point out that Postulate II might be expressed in mere terms of points, lines, and incidence without the use of the definition of parallelism; but that it is convenient to use the definition because it shortens the theorems and their proofs (see Introduction, Point 4).

III. *There exist three points A, B, C that are noncollinear;* that is to say, such that no line is on A and on B and on C.

Student: "Why assume that there are such points? Isn't this a simple, evident fact?" Teacher: "Geometry tries to deduce as much as possible from as few and as simple assumptions as possible. Hence simplicity of a proposition does not disqualify it from serving as a postulate but, on the contrary, recommends it. Nor does the fact that the proposition is evident make its formulation unnecessary. For a rigorous deductive theory is based exclusively on the explicitly formulated postulates. In developing the theory one must not resort to other sources, in particular, one must not have recourse to intuition. Postulate III will actually be used in the deduction of theorems that could not be proved on the basis of Postulates I and II alone." (Later, in discussing Postulate III', the teacher may present a system consisting of exactly two points and three lines: the join of the two points, and two parallel lines, one on either point. This system satisfies Postulates I and II, whereas Postulate III and many of its consequences do not hold.)

Before drawing conclusions from these assumptions the teacher should point out that since the postulates are exceedingly simple one cannot expect spectacular consequences; that the purpose of this chapter of the course is rather to obtain modest conclusions by *absolutely impeccable reasoning*, in other words, to give real *proofs* of whatever theorems are asserted.

Theorems rigorously deducible (see Introduction, Point 5) from Postulates I and II alone include: If R is on PQ and $R \neq P$, then Q is on PR; and each point that is on PQ is on PR and vice versa. If two lines l and m are not parallel or, as one also says, if l and m *meet*, then there is one and only one point that is on both l and m; (it is called the *intersection* or the *meet* of the lines l and m and denoted by lm or, synonymously, by ml). Three distinct lines that are on one and the same point are said to be *concurrent*. If P, Q, R are noncollinear points, then the lines PQ, QR, RP are nonconcurrent, mutually nonparallel, and distinct. If l, m, n are nonconcurrent, mutually nonparallel lines then lm, mn, nl are noncollinear. (The teacher should

explain why the word "mutually" cannot here be omitted.) Since Postulate
III has not been used in obtaining this result, it follows that one might
replace that postulate by the following assumption (see Introduction,
Point 3).

III'. *There exist three nonconcurrent and mutually nonparallel lines.*

If l, m, n are nonconcurrent and nonparallel, then each line meets at least
two of these three lines.

From Postulates **I, II, III** it follows that *each line is on at least two points
and each point is on at least three lines.* *Proof:* The points A, B, C being
noncollinear, the lines AB, BC, CA are nonconcurrent. Hence each line
meets at least two of them. Say l meets AB and BC. If l is not on B, then l
has two distinct intersections with these two lines. If l is on B and is not
parallel to AC, then l has a point $\neq B$ in common with AC. Finally, if
$l = \pi(AC, B)$ then, by Postulate **II**, l is not parallel to $\pi(AB, C)$ and thus
has a meet $\neq B$ with that line. In any case, l is on at least two points. The
second half of the theorem is proved in a similar way.

Since there is a line $\pi(AC, B)$, there certainly exists a point $\neq A, B, C$,
namely the meet of $\pi(AC, B)$ and $\pi(AC, B)$, call it D. Student: "There
exists even a fifth point, namely the meet of $\pi(AC, B)$ and $\pi(BC, A)$ – call
it E, which is also distinct from A, B, C." Teacher: "That this point E is
distinct from A, B, C is an excellent remark. But one cannot prove that E is
distinct from D. In fact, on the basis of Postulates **I, II, III**, *one cannot prove
that there is any fifth point besides A, B, C, D.* For there is a system of
altogether four points and six lines satisfying the Postulates **I, II, III**, namely
the vertices and edges of a tetrahedron." Student: "But a tetrahedron is a
solid and we are studying plane geometry." Teacher: "We are studying the
consequences of Postulates **I, II, III** concerning undefined elements called
points and lines – assumptions which reflect properties of points and lines
in the ordinary plane as represented (or at least in part represented) by a
blackboard. That these assumptions are also satisfied by other objects and,
on the other hand, do not yield all properties of the points and lines in the
ordinary plane merely indicates that further assumptions are needed in
order to rule out some of these extraneous systems and to guarantee further
properties of the points and lines of the ordinary plane. For example, in the
latter there are more than two points on *each* line. The system of the vertices
and edges of a tetrahedron demonstrates that on the basis of Postulates
I, II, III one cannot even prove that there are at least three points on *any*
line. This suggests the introduction of the following postulate.

IV. *There are three collinear points.*

The student can easily deduce from **I, II, III, IV** that there exists a point besides *A, B, C, D*; and that, conversely, if there is a fifth point besides *A, B, C, D*, then three collinear points exist. Hence Postulate **IV** might be replaced (see Introduction, Point 3) by:

IV'. *There are more than four points.*

From **I, II, III, IV** one readily deduces that *each* line is on at least three points; that by means of parallel lines the points on any two lines can be brought into one-to-one correspondence; that if each line is on exactly n points, then each point is on exactly $n+1$ lines, and there are altogether n^2 points and $n(n+1)$ lines. While there are no systems with 5, 6, 7 or 8 points the teacher can exhibit a plane with 9 points. There are no systems with 10, 11, ..., 15 points. There is one with 16 points, each line on exactly 4 points. The student jumps to the conjecture that there are systems with 25, 36, 49, ... points. At this point the teacher can at least report facts about finite planes, e.g., that there is none with 36 points, that no one knows whether there are any with 100 or 144 points; that the decision of these questions by computing machines, while possible, would take even fast machines so much time that it is economically unfeasible to let them solve the problem; and that such finite planes are used in the design of experiments (see Introduction, Point 11).

At this point, the beginner has achieved the *general* insight (as described in the Introduction) that he can derive from affine geometry. How far beyond this point the teacher can go into *special* (that is to say, mathematically interesting) parts depends upon the quality of his students and the time at his disposal for this purpose. I should advise against teaching Desargues' Law, which is too technical for a general introductory course. But if the quality of the students and the available time permit it the teacher might spread over the affine plane a lattice of points with two integral coordinates (or possibly even a rational net) on the basis of a fifth assumption whose formulation is simpler than Desargues' Law; for example, *if in two quadrangles five sides are parallel, then so are the sixths sides* [7]. In this way, a connection with algebra would be established. But we must not let the beauty and mathematical depth of this theory keep us from questioning its relevance for those beginning students who are not particularly interested in mathematics.

III. A SELF-DUAL FRAGMENT OF THE AFFINE PLANE

In an affine plane, points and lines play different roles. Whereas there are pairs of lines that are not on a point, namely parallel lines, each pair of points is on a line. By retaining all points of an affine plane but disregarding one line and all lines parallel to it, however, one obtains a fragment of the affine plane in which points and lines play precisely equal roles. This fact is also expressed by saying that points and lines of the fragment are *dual* to one another; and the fragment is said to be governed by the *principle of duality* or to be *self-dual*.*

On a blackboard or on a sheet of paper, one may disregard, for example, all vertical lines. Since no points have been deleted from, or adjoined to, the affine plane nonparallel lines remain nonparallel in the fragment, and parallel lines remain parallel. But in the fragment there are pairs of points that are not on any line, namely those points which in the affine plane are on a vertical line. They may be called *vertical points* – the dual counterpart of parallel lines; and just as in affine geometry one speaks of a line being parallel to another line, in the geometry of the fragment one may speak dually of a point being vertical to another point.

While the points and lines of the fragment obviously do not satisfy Postulates **I** and **II** one can make the following assertions:

A. *If P and Q are distinct points, then there is at most one line that is on both P and Q* (if there actually is one it is called the *join* of P and Q). *If l and m are distinct lines, then there is at most one point that is on both l and m* (if there actually is one it is called the *meet* of l and m).

B. *If l is a line and P is a point not on l, then there exist*

(1)   *one and only one line that is on P and parallel to l;*
(2)   *one and only one point that is on l and vertical to P.*

Proposition **B** is a self-dual form of the traditional Parallel Postulate [8].

Among the undefined elements of the deductive affine geometry which are called lines, there are none that would warrant being called vertical. But one can choose any arbitrary line and all lines parallel to it and disregard *them*; and one may define *vertical points* as points which in the remaining fragment are not on any one and the same line. One wants, however, to describe the system of the points and lines of the fragment intrinsically, that is, without reference to an affine plane from which it has been obtained.

In the deductive theory of the self-dual fragment, the undefined concepts are *point*, *line*, and *incidence*; and, by definition, the lines l and l' are *parallel* if no point is on both l and l' – just as in affine geometry. But in this theory one also defines the points P and P' to be *vertical* if no line is on both P and

* Other self-dual fragments, for example, what remains of the affine plane when a point and all lines on that point are disregarded (see loc. cit. [1]), will not here be considered.

P' and, instead of Postulates **I** and **II**, uses Propositions **A** and **B** as postulates. Since in the entire geometry of the fragment, points and lines play equal roles one wants also to replace Postulates **III** and **IV** by self-dual assumptions. The following single postulate is sufficient for this purpose.

C. *There exist two nonparallel lines and two nonvertical points such that the meet of the former is on the join of the latter.*

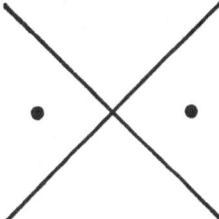

One can easily prove that a system satisfying Postulates **A**, **B**, **C** contains at least nine points and nine lines. The existence of more points and/or lines cannot be proved since there is a system consisting of exactly nine points and nine lines which satisfies the three postulates for a self-dual fragment. This system is nothing else but the classical Pappus configuration. On the torus, it can be represented by 9 points and 9 closed curves [9].

The principle of duality is often considered as a demonstration of the fact that mathematical theories may be capable of different interpretations. In a plane governed by this principle, points may be interpreted as lines and vice versa without affecting the validity of the theorems. Historically, the principle, which arose from the study of projective planes, was the first demonstration of this fact.

As an object of elementary studies, the self-dual fragment of the affine plane has great advantages over the projective plane. Both demonstrate the principle of duality equally well. But in the fragment it is unnecessary to define ideal points and a line at infinity, whose introduction in the projective plane often leads to misunderstandings and fruitless discussions among beginners. One can illustrate the fragment on a blackboard in the same sense in which one can illustrate the affine plane. Moreover, while a certain degree of sophistication is needed in order to understand the homogeneous point and line coordinates in a projective plane a student knows after one day's instruction in analytic geometry that vertical lines are exceptional since only the nonvertical lines have equations of the form $y=ax+b$. Such a student will understand that just as each point of the plane (the whole plane or the fragment!) can be described by a pair of numbers, namely abscissa and ordinate, each line of the fragment (that is, each nonvertical line) can be described by two numbers, namely slope and vertical intercept, and that points with equal abscissae are vertical just as lines with equal slopes are parallel. He might even grasp a purely arithmetical model of the fragment obtained by *defining* a point as a pair of numbers (x, y), a line as a pair of numbers $\langle a, b \rangle$, and incidence by a bilinear relation between the four numbers. In this self-dual model, the points (x, y_1) and (x, y_2) are vertical and the lines $\langle a, b_1 \rangle$ and $\langle a, b_2 \rangle$ are parallel.

An important feature of the self-dual fragment of the affine plane is the fact that, in a way to be described elsewhere, it can be used as the geometric basis of pre-limit calculus.

Another reason for teaching the fragment is the fact that, as I noticed more recently [10], the geometry of the fragment can be translated into the kinematics of a two-dimensional world (that is, a one-dimensional space in time). The geometric-kinematic dictionary reads as follows:

point	world point or event *
line	(uniform) motion
on	on
vertical points	simultaneous events (not accessible
(not joined by a line)	to one another by a motion)
parallel lines	equikinetic motions
(having the same slope)	(having the same velocity)

Simple constructions make it possible to determine relative motions; for example, one readily finds the motion m_4 which appears to an observer engaged in the motion m_3 just as a given motion m_2 appears to an observer engaged in the motion m_1.

The belief in an absolute space can be expressed in the selection of a line and all parallel lines as (absolute) *rest*. Galilean relativity theory recognizes only relative rest of two observers, expressed by the fact that their motions are isokinetic.

IV. NONEUCLIDEAN GEOMETRY

Becoming acquainted, however slightly, with the ideas of Bolyai and Lobachevsky after having studied some traditional geometry is one of the greatest intellectual experiences; and no one should go through a secondary school without hearing about noneuclidean geometry and its history.

Probably the simplest approach to the deductive theory of the (not self-dual) Bolyai-Lobachevsky plane is through the geometry of a sheet of paper

---

* Strictly speaking, a world point may be the scene of several events.

or of a blackboard. The noneuclidean propositions about points and lines (incidence, parallelism, order, and arrangement) can be verified by considering the points of a rectangle and the segments joining two points of the contour. It is, after all, an extrapolation of experiences about such bounded domains – and a rather questionable extrapolation, for that matter – that motivates Euclid's Parallel Postulate (that is, Postulate II, see p. 6). The actual fact is that on a blackboard, however large, a point is on *many* lines not meeting a given line. And this is what Bolyai and Lobachevsky postulated.

NONEUCLIDEAN PARALLEL POSTULATE. *If l is a line and P is a point not on l, then more than one line on P is parallel to l.* (In noneuclidean geometry, any two lines that do not meet are said to be parallel.)

One of the greatest of the more recent advances in the field of noneuclidean geometry – though it has been completely ignored in all books on the subject – is a development of the geometry of the Bolyai-Lobachevsky plane (including the theories of congruence and perpendicularity) from a few very simple postulates in terms of points, lines, and incidence by the Notre Dame school of geometry 1937–47 [11]. The importance of the possibility of such a development lies in the fact that an axiomatic of euclidean geometry in these terms is impossible.

Naturally in a course covering only the affine part of euclidean geometry and excluding euclidean congruence one will also exclude noneuclidean congruence, that is, the Klein-Hilbert theory of congruence in a convex domain (or, more specifically, in an ellipse) based on cross ratios. But the teacher can at least show how an *order* can be defined for the points on a line in a Bolyai-Lobachevsky plane and how, on the basis of a few utterly simple assumptions about points, lines, and incidence (including of course the Noneuclidean Parallel Postulate) the order so defined can be shown to possess the traditional properties of linear order. (In the affine plane, even this seems to be impossible since with regard to the only known affine definition of order in terms of points, lines, and incidence, given by M. Pieri, there do not exist simple postulates guaranteeing that order in this sense has the properties of order [12].)

A postulational theory of linear order is in itself one of the most appropriate topics for a course of the type envisaged in this paper. Its undefined concepts are *points* and a ternary *betweenness* relation, *Q is between P and R* being expressed by *PQR*. The theory may be based on the following postulates:

(1)   *If PQR then $P \neq Q \neq R \neq P$.*
(2)   *If PQR then RQP.*

(3)   *If PQR then* not *PRQ.*
(4)   *If PQR and PRS, then QRS.*
(5)   *If PQR and PRS, then PQS.*
(6)   *If PQR and QRS, then PQS.*

A final assumption expresses a property of the line rather than of between-ness.

(7)   *Of three points on a line at least one is between the other two.*

Examples of propositions that can be proved from these assumptions include the following three theorems. If *PQR* and *PQS*, then *QRS* or *QSR*. E. H. Moore's Theorem: Any four points of a line can be labelled *A, B, C, D* in such a way that

$$ABC, \quad ABD, \quad ACD, \quad BCD.$$

If there are more than four points on a line, then Properties (5) and (6) are consequences of (1), (2), (3), (4), and (7).

In noneuclidean geometry, the concept of betweenness (*undefined* in the postulational theory of order) can be *defined* in terms of points, lines, and incidence; and the properties of betweenness (*postulated* in the theory of order) can be *proved* on the basis of simple assumptions concerning these undefined noneuclidean concepts (see Introduction, Point 3).

In the noneuclidean plane, according to a definition of F. P. Jenks [13], the point *Q* is *between* the points *P* and *R* (written *PQR*) provided that the following condition is satisfied: If *p* on *P*, and *r* on *R*, are any two meeting lines, then each line on *Q* meets at least one of the lines *p*, *r*. Hence *P* is *not* between *P* and *R* if there exist lines *p*, *q*, *r* on *P*, *Q*, *R*, respectively, such that *p* and *r* meet while neither of them meets *q*.

The postulates from which the theory of betweenness can be developed are, with the single exception of the Noneuclidean Parallel Postulate (replacing Euclid's Parallel Postulate II and therefore referred to as Postulate II'), also valid in the affine plane. In fact, the noneuclidean Postulates I, III, and IV are identical with those of affine geometry. In noneuclidean geometry we furthermore assume two propositions which in affine geometry are consequences of Postulates I, II, III, IV whereas they do not follow from I, II', III, IV.

V. *Each line is on at least one point.*

VI. *If l and m are meeting lines, and P is a point on neither line, then P is on a line meeting l but not m (and on a line meeting m but not l).* If a pair of meeting lines is called a *cross*, then VI can also be expressed by saying that for each cross, each point not on the cross is on a parallel cross. For this reason, **VI** will also be referred to as the CROSS POSTULATE.

Student: "Why assume Postulate **V**? It has not been assumed in affine geometry. And how could there be a line without any point on it?" Teacher: "All our mental pictures of lines include of course many points. But I must once more ask you to remember that we do not base our theorems on intuition. Only propositions deduced from the postulates are admitted in the theory. The proposition **V** follows from **I**, **III**, and **IV** in conjunction with Euclid's assumption of unique parallels. The assumption that several parallels exist is in this respect weaker. In conjunction with **I**, **III**, and **IV** it does not rule out lines without points. For if to any system of points and lines satisfying these assumptions you adjoin a line on which there is no point, or several such ghost lines, then the new system still satisfies all the assumptions. Hence if in the noneuclidean plane we wish to rule out ghost lines we must include them by a special assumption. On the other hand, if we do make this assumption **V**, then it can be proved that each line is on more than four points (so that Properties **5** and **6** of betweenness are consequences of Properties **1–4** and **7**). Here is a context in which assumption **II'** is stronger than **II**."

Postulates **I–VI** yield the following theorem.

If PQR, then the three points P, Q, R are collinear. Proof: If *P*, *Q*, *R* are noncollinear, then let *q* be any line on *Q* not meeting the line, *p*, joining *P* and *R*. By the Noneuclidean Parallel Postulate **II'** here is a line on *R* which is distinct from *p* and parallel to *q*. The lines *p*, *q*, *r* demonstrate that *Q* is not between *P* and *R*.

This theorem does not hold in the affine plane. Under the assumption of Euclid's Parallel Postulate, if *P*, *Q*, *R* are any three points, and *p* and *r* are meeting lines on *P* and *R*, respectively, then each line *q* on *Q* meets at least one of the lines *p* and *r*, since *q* is parallel to at most one of two meeting lines. Hence of any three points in an affine plane each is between the other two in the sense of Jenks' definition. This goes to show that in affine geometry this definition is inappropriate. In the noneuclidean plane, on the other hand, the ternary relation so defined has the properties of betweenness.

Properties **1** and **2** are immediate consequences of the definition. We next prove Properties **3** and **4**.

If PQR, then not PRQ. Proof: Let *q* be a line on *Q*. By the Noneuclidean Parallel Postulate, there exist two lines, r_1 and r_2, on *R* that are parallel to *q*. By the Cross Postulate, there is a line *p* on *P* meeting r_1 but not r_2. If *PQR*, then *q* meets *p*. The meeting lines *p* and *q* and the line r_2, which is parallel to both of them, demonstrate that not *PRQ*.

If PQR and PRS, then QRS. Proof: Let *q* and *s* be any two meeting lines on

Q and S, respectively, and r a line on R parallel to s. It must be proved that r meets q. By the Cross Postulate, there exists a line p on P that meets s but not q. Because of PRS, the line r, which is parallel to s, meets p. Since PQR, the line q, which is parallel to p, meets r, as claimed.

A last assumption, Postulate **VII**, that guarantees Property 7 of betweenness leads to the completion of the theory of linear order in the affine plane.*

If instead of a bounded domain in the affine plane we study a vertical strip in the self-dual fragment, then we obtain a (not self-dual) geometry that is closely related to the Einstein-Minkowski kinematics of a 2-dimensional world (a 1-dimensional space in time). In fact, if instead of deleting from the self-dual fragment all points whose abscissae are outside of a certain interval and retaining all (nonvertical) lines we delete all lines whose slopes are outside of a certain interval and retain all points, then we have a geometry which by a geometric-kinematic dictionary can be translated in the very kinematics of the special theory of relativity, just as the geometry of the self-dual fragment can be translated into Galilean kinematics [14].

V. ANOTHER NONEUCLIDEAN GEOMETRY

There is a second approach to a noneuclidean geometry, though of a variety different from Bolyai's and Lobachevsky's. Yet that geometry is not unimportant because everyone inadvertently uses it and it reveals that there are essentially different practical ways of measuring distances. It is the geometry of a modern city built in rectangular blocks and with two-way streets [15].

One may define the distance between two points as the length of the shortest path joining the points. It is for the distance so defined that one pays in a taxicab. Unless two points are on one and the same street or avenue, they are joined by several shortest paths (that is, paths of equal length that are shorter than all other paths). For example, an intersection is joined with the intersection 4 blocks East and 3 blocks North by $(4+3)!/4!\, 3!$ shortest paths. The two intersections are 7 blocks apart, and this is the distance one has to pay for, even though in a right triangle with the sides 4 and 3 the hypothenuse is 5.

The other points that are 7 blocks from the first intersection are the

---

* Postulates I-VII also yield the entire theory of *planar* order including the famous Law of Pasch. This law, which can be proved from the seven noneuclidean postulates, is known as the Postulate (or Axiom) of Pasch since in affine and euclidean geometry it cannot be proved from the other assumptions while its validity is so essential for the development of the theory of order that Pasch and his followers have postulated it. By means of this law one can prove Jordan's famous theorem about simple closed polygons (stating that such a polygon divides the plane into exactly two parts).

intersections 7 blocks East, 6 blocks East and 1 block North, 5 blocks East and 2 blocks North, and so on; altogether 28 points on a square in diamond position so that *pour épater le bourgeois* or, more specifically, in order to flabbergast philosophers, one may say that in this geometry the circles are squares. This situation is of course well known from Minkowski's distance $|x_1 - x_2| + |y_1 - y_2|$ between the points (x_1, y_1) and (x_2, y_2). But while philosophers and high school teachers are not very familiar with Minkowski metrics they all use taxicabs.

Students are quite interested in the taxicab geometry – in fact, so much so that they are even eager to study the still stranger and more complicated taxicab geometry in a city with one-way streets [16].

Illinois Institute of Technology,
Chicago, Ill., U.S.A.

BIBLIOGRAPHY

[1] K. Menger, 'Self-dual fragments of the plane', *Am. Math. Monthly* **56** (1949) 545–546; and Chapter 5 of the book *Studies in Geometry* by L. M. Blumenthal and Karl Menger, San Francisco, 1970.

[2] *Studies in Geometry*, Section 5.4.

[3] *Studies in Geometry*, Part 4.

[4] K. Menger, *Calculus. A modern Approach*, Boston 1956, Chapters 1 and 2.

[5] K. Menger, *You will like geometry*, Museum of Science and Industry, Chicago, 1952.

[6] *Studies in Geometry*, Section 5.13.

[7] For related possibilities, see Pickert, *The Projective Plane*.

[8] *loc. cit.*[1].

[9] *Studies in Geometry*, Section 5, pag. 193.

[10] *loc. cit.* [2].

[11] See the papers by F. P. Jenks, J. C. Abbott, H. F. DeBaggis, B. J. Topel, and J. Landin in *Reports of a Math. Colloquium* (II), Issues 1–8; and K. Menger, 'On algebra of geometry and recent progress in noneuclidean geometry', *The Rice Pamphlets* **27** (1940) 41–78.

[12] Cf. K. Menger, *loc. cit.*[11]. See also H. S. M. Coxeter, *The Real Projective Plane*.

[13] F. P. Jenks, 'A New Set of Postulates for Bolyai-Lobachevsky Geometry', *Reports of a Math. Colloquium* (II) **1** (1938) 46 and *Proc. Nat. Ac. Sci. U.S.A.* **24** (1938) 486–490.

[14] *Studies in Geometry*, Section 5.13.

[15] *loc. cit.*[5].

[16] H. C. Curtis, *Am. Math. Monthly* **60** (1953) 416.

G. PAPY

A FIRST INTRODUCTION TO THE NOTION OF
TOPOLOGICAL SPACE

1. The linearization of theories, one of the most characteristic traits of the attractive face of the mathematics of today, appears best when the fundamental notions of differential and integral calculus are put in place – along with the rest of the calculus. Infinitesimal analysis reigns supreme today in nearly all the fields of applied mathematics. Linear algebra and infinitesimal calculus (including the first elements of differential and integral calculus) remain the two essential goals of all worthwhile teaching of mathematics at the secondary level.

2. In certain countries the teaching of infinitesimal analysis at the secondary level began more than half a century ago. In others, one is always involved in the discussion of the opportuneness of introducing such a course three centuries after the discovery of differential calculus by Newton and Leibniz.

Certain university professors regard infinitesimal analysis as a privileged pursuit to be jealously guarded against profanation at the secondary level. They imagine that they alone can teach the delicate notions of infinitesimals without ruining them. No doubt they believe, more or less conscientiously, that such notions should remain under the protection of the elite, and that throwing them to the whims of the uncultured would put in peril a whole precious heritage of intellectual and spiritual values. This typical caste reaction illustrates, by contrast, one of the golden rules of progress: all important discoveries are first reserved for a small elite and finish by appearing sufficiently simple and clear to be understood and used by humans who are not visited by any inspired ideas nor invested with special talents and virtuousity.

3. Analysis at the secondary level has not always been an uncontested success in convincing in a definitive manner a bewildered population. Such courses often remind one of cook-book recipes, designed to be applied with blind confidence and with no light on the inner function of this mysterious machine for solving problems. Such teaching contains nothing formative, nothing to provoke any profound reactions, and scarcely leaves any durable traces. The fundamental notion of limit is left smothered in metaphysical ideas too delicate to handle and too nebulous to be put in context with other mathematical notions that have been already clearly established.

4. Infinitesimal analysis – comprising differential and integral calculus – is one of the most important discoveries in the history of mathematics, of science, and of thought. Its gestation period was difficult, its liberation from foggy metaphysical ideas was a long and painful work. This progress did not consist in the solution of a single, isolated problem. It marked in fact the entrance into a new universe where a multitude of notions and concepts – often isolated for the first time – began converging, meeting, intersecting, and mutually influencing one another in subtle interaction. Mathematical analysis is at the crossroads. The simultaneous introduction of a large number of new notions has created a superposition of difficulties partly responsible for the failures often incurred in the teaching of analysis both at the secondary level and in the universities.

5. Teaching analysis is unthinkable without the explicit notion of function. It is possible to teach elementary geometry without making explicit use of this concept. This is what Euclid did and what traditional teaching continues to do. If it were possible today to teach elementary mathematics bypassing entirely the notion of function, then it would perhaps be justified to ignore it in geometry. Since the notion of function is inevitable, there is no reason to miss the eminent services it can render in geometry. All plane transformations – translations, central and axial reflections, homotheties, rotations, similitudes – are functions. The groups of geometry are groups of functions. The group law is composition, which classical analysis hides in the theory – so badly named – of functions of functions.

6. Ten or fifteen years ago, during a mathematics teachers convention, two excellent senior high school teachers crossed swords, with great courtesy, on the optimum age for the introduction of the notion of function, of real valued functions of a real variable, as a matter of fact.

- I tried in vain with pupils of sixteen. In spite of all precautions and patience, the notion does not enter. The brain is not developed enough. But at seventeen, no problem; the time is ready. What was impossible to inculcate – no matter how hard the effort – at sixteen, penetrates without difficulty at seventeen.
- I wonder, dear colleague, if it is possible to be as positive as you. Permit me to tell you that I also tried to introduce functions at sixteen. Of course, I don't claim that all the pupils completely assimilated the notion; but they didn't all die of it, and the answers to the exercises showed that they correctly manipulated the proposed functions....
- Allow me, dear colleague, to be somewhat skeptical about the import of the conclusion you seem to draw.... But perhaps our apparent opposition

comes from a misunderstanding about the meaning of the terms we are using. We must not confuse the assimilation of the notion of function with some exercises of manipulation which can be performed without the notion itself being assimilated. Let us not forget that for the pupils the notion of function is difficult, subtle, and delicate. In order to avoid an empty parrot-like verbalism, one must not skip over certain stages.

– I believe we almost agree. Exercises with some simple numerical functions may well serve to introduce progressively the notion at sixteen. But, in any case, before that age nothing can be done along that line for lack of necessary maturity.

At that moment, from the rear of the room came a loud question:

– From what age do you believe children understand the phrase "...has as mother..."?

The question sounded very strange in a discussion on the teaching of mathematics; it provoked some grumbling, from whence finally came out, badly articulated, the strange answer:

– From always!

A moving, naive, and magnificent answer from expert teachers of the 12-to-18 group, for whom – very clearly – the student begins at 12 (and, without doubt, ends at 18)!

Modern reform of the teaching of mathematics has abolished all those groupings.

Today, *Frédérique* teaches the notion of function to 8-year-olds who have been taught modern mathematic since the age of six. To do so she uses multicolored graphs and appropriate pedagogical techniques. [EE], [MM1], [EG], [EM1], [EM2], [EM3]. This initiation to modern mathematics at the age of six [EG], [EM1], seven [EM2], eight [EM3], twelve [MM1], fifteen [EE]... should interest any teacher who has charge of introducing the notions of modern mathematics at whatever age his pupils might be. The experience of modern reform repeatedly confirms that we are all equal before the unknown. The situations best able to initiate a new notion are almost independent of the age of the learner

7. ┌── At the secondary level ──────────────────────────────┐
   │   RELATION = set of couples*   │
   │   FUNCTION = relation not comprising two different couples with │
   │   same first component.   │
   └───┘

* Or ordered pairs.

The notion of relation – natural and without restriction – is simpler than that of function.

Functions are "relations with restriction".

The frame of relations imposes itself if one wishes to present functions in a context large enough to cover counterexamples which are close enough to the notion of function and which, at the secondary level, are nothing but "non-functional relations". Parallels, perpendiculars, divisibility, modular congruence, etc., attest to the unavoidable character of non-functional relations at the secondary level.

The set-theoretic presentation puts in evidence the conceptual link between functions and relations; the utilization of graphs – drawn or imagined – brings it out in a startling manner. A deep comprehension of that link facilitates intelligibility and helps the mastering of the concepts.

The proposed point of view does not represent an extremist set-theoretic tendency. It does not define the couple (x, y) as the set $\{\{x\}, \{x, y\}\}$ (after Wiener-Kuratowski). All one ought to know about couples at the elementary level is condensed in the *definition*

$$(x, y) = (a, b) \Leftrightarrow (x = a \wedge y = b).$$

This moderate set-theoretic-point of view, common language of the great majority of professional mathematicians all over the world, has great pedagogical virtues. It brings to the foreground, and presents as sets – that is, as objects of the theory – some notions which were formerly kept in the background. It simplifies the theory by suppressing the coexistence of notions which are separated only by subtle distinctions and which are often mutually definable. Thus, any relation is its own extension (unhappily named by Bourbaki, its graph).

That presentation – happily – restricts the abstract notion of relation. Thus membership is not a set. ($\{(x, y) \mid x \in y\}$ is a strict class and not a set.)

Later, much later, the children who will continue their mathematical studies could consider the membership from set to set as the class $E = \{(x, y) \mid x \in y\}$ – after Gödel.

Classes of couples – or *correspondences* – generalize relations without sacrificing anything of significance.

The membership correspondence E does not cover the membership from set to strict class, which leads to the notion of "binary predicate". At that level it is true that the "extension of the binary predicate", viz., "the class of all couples satisfying that predicate" does not always define the predicate itself. This distinction between "binary predicate" and "extension of that predicate" may be entirely avoided at the secondary level.

It suffices to think of the graphic representations of functions to realize

that the notion of "relations = sets of couples" is unavoidable at the secondary level. Therefore it is that notion of relation, and that one only, that one
must adopt.

A parallel presentation of other notions of relation is useless and harmful,
as long as one cannot settle the questions that would meaningfully suggest
and motivate such an effort.

All by-products of relations will be defined as attributes of "relations = sets
of couples". For more details see [F3].

8. It is with respect to problems of analysis that set theory started with
Cantor. It is via courses in differential and integral calculus that set theory
began to penetrate into the universities. It seems that the study of topological
properties of sets of points in the plane cannot be avoided in analysis. The
very nature of the subject imposes the adoption of a set-theoretic point of
view. Since analysis is one of the main goals of any worthwhile teaching of
mathematics at the elementary level, it is necessary to prepare the child to
adopt that point of view; above all, one must avoid any teaching which would
continue to reject it.

9. A strange small school hidden in an old castle surrounded by woods
gathered together the children that the occupying armies of that country at
war intended to exclude from the human community. For obvious reasons,
that underground institution did not have at its disposal specialized per-
sonnel for the different subjects nor for the various ages of the children. One
of its mathematics teachers, freshly graduated from the university, found
himself – without knowing it – in an exceptional pedagogical situation.
Sometimes, in the same week, he had to teach students ranging from four to
nineteen years of age; a lesson to children of eight or nine might be followed
by a course to a student of university age.

The situation was exceptional indeed, for it is not at all usual that the
same person teaches simultaneously at the primary level and at the university. Generally, the primary teacher does not have the opportunity to test at
the university, ten years later, the good or bad results of his teaching. Our
young teacher was in that situation, and for him time was shrunk, as it were,
because he saw one hour later the results of his teaching in grade 4 upon
students of age 19, because he taught the 9-year-olds in the same traditional
way he himself was taught, and in the way his 19-year-old students had also
been taught.

One day that young teacher was speaking about the square to a class of
9-year-olds. He asked the children to draw a square and put to them the
classical question, "Partition that square into four equal squares". In that

story the choice of the word "partition" was perhaps to have important consequences. He was evidently expecting the answer:

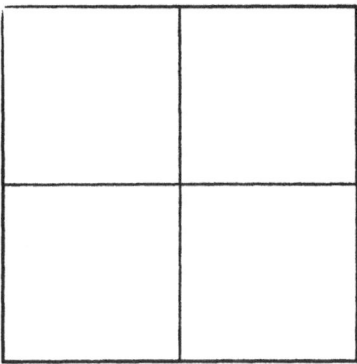

Expecting children to answer in accordance with the teacher's wishes gives rise to many surprises. To impose the teacher's point of view may be a grave pedagogical error.

While most of the children quickly drew the desired answer, one of the best pupils did not draw anything at all, and seemed lost in deep meditation. "How is it possible that Isaac has not yet got the right answer?", the teacher asked himself. He could not imagine that it was because of tiredness, lack of attention, or because he hadn't understood the question. The teacher did not jump to conclusions. He wanted to learn why the child did not answer and refused to decide for himself. He accepted the fact that the children could teach him how to teach them to learn.

– So, Isaac, you are not the first to give us the right answer today!

The questioned child gave, from under his beautiful black eyelashes, one of those sidelong looks whose secret is shared by beautiful women and by children. When his glance had rested on the copybooks of his classmates he imperceptibly shrugged his shoulders and said:

– You know very well that they don't have the right answer!

This was the second surprise for the teacher on that memorable day.... But his reaction was good. Without raising his voice, without condemning answer or child, he simply asked:

– Will you come to the blackboard and explain to us why?

After having drawn the "wrong answer", Isaac fattened out one point on a median

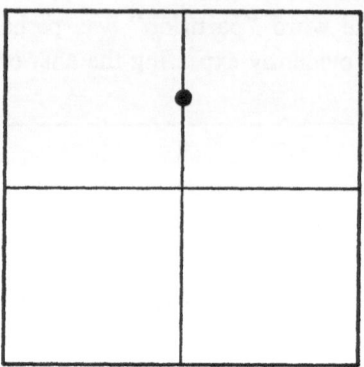

and added

– If in the partition this point goes with the small square on the left, then
this small square has one point more and that is no longer a partition into
equal squares. If it goes with the small square to the right, etc.

The young teacher was taken aback by the third surprise. Realizing that he
was in a very bad position and badly prepared for the answer, he panicked
and, in a very loud voice – as loud as his bad conscience – he shouted

– THE POINTS ON THOSE LINES

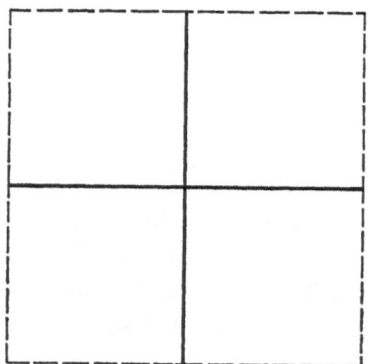

HAVE NO IMPORTANCE!

Confronted with that dictatorial, arbitrary assertion, Isaac answered
teasingly:

– IF THOSE POINTS HAVE NO IMPORTANCE, then OK.

A quick, logical look at that answer reveals that Isaac agrees without
approving. A pedagogically bad situation!

Perhaps in order to ease himself out of it, the teacher had tried what seemed to be a very good means to achieve a very good end. Here, alas, the goal was bad. He tried to give a motivation, an intuitive support, and an apparently plausible justification of his dogmatic assertion, "Those points have no importance".

"When one cuts with scissors, one doesn't ask oneself on what side the points along the cut are going". He took a sheet of paper, folded it in two, and then in two again, unfolded it, showed the cross,

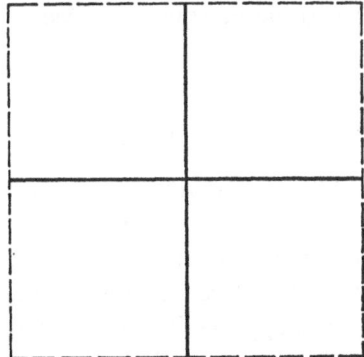

cut in into two pieces, and then in four pieces, emphasizing in a very peremptory manner that "the points on those lines have no importance". Thus, he forced the children to adopt the tailor's point of view, which unfortunately does not fit the mathematics created by Newton and Leibniz.

Some minutes later a new lesson would bring our teacher out of his dream and illusions. He would realize that he had just committed a pedagogical crime.

In the next class our hero had to teach a more or less modern course in analysis. With good sense, he considered as known some non-infinitesimal notions and took them as a basis of departure. He did not intend to redefine the square and the circle but to limit himself to refining those concepts, while making more precise the notions of open square, closed square, and perhaps some intermediate cases neither open nor closed. An enthusiastic lover of exact references and of technical labels, the teacher clearly announced:

— DEFINITION 7.12.32:

One calls *open square* the set of points interior to the square with the exclusion of the boundary points.

— DEFINITION 7.12.33:

One calls *closed square* the set of all the points of the square including those of the boundary.

Marvellous! Make judicious use of known notions to give precise definitions by means of a language which lacks precision. Marvellous, *in abstracto*! In that class there is only one such student, but he exists, and manifests himself with humor, and causes a new surprise in this memorable day in the life of our young teacher by stating:

- PROPOSITION 7.12.34: An open square is closed.

- PROPOSITION 7.12.35: Any closed square is open.

- Certainly my student has not heard well, or he was not attentive. Let us repeat the definition.
  Definition 7.12.32....   Definition 7.12.33....

That pedagogy of repetition is as common as it is inefficient. One can ask oneself by what miracle a person who could not understand a definition, a proposition, or a proof, could suddenly perceive its intelligibility thirty seconds later during a word-for-word repetition of the previous formulation. It is true that the procedure appears foolproof the second time around! In order to get peace, the questioned pupil shortens his torture by stating he has now understood. And there is nothing more like a pupil who has understood than one who has not, but gives the impression that he has. Full of mischief, our "unique" student caused surprise number 5 by repeating again, word-for-word, ...

- Proposition 7.12.34....   Proposition 7.12.35....

This time the teacher reacted properly:

- Very well, you enunciate two propositions... Prove them!

Samuel reflected for some moments, organized his plan of attack, and before the final thrust decided to establish clearly his ground:

- You agree that every closed square comes from an open one and that every open square comes from a closed one?
- Of course, answered the teacher who was beginning to be interested in that attempt at a proof of a proposition he knew to be false. Every open square can be obtained by removing the boundary of a closed one.
- OK, then. Here is a closed square. But as the folks on the boundary are removed, *those that were just behind them become the first* and constitute the new boundary. The open square one gets is thus closed, and the case is won.

Sixth surprise for the young teacher!

This student was very intelligent. He evidently still is, because he has become a university professor, in chemistry, as a matter of fact. At that time he was extremely brilliant: he solved in a very astute way the riddles of plane geometry. Yet it became suddenly evident that his microscopic vision of the Euclidean plane was unbelievably poor and strewn with errors. The teacher had the good taste not to develop a frontal attack on the proof. It is true that its logical organization was satisfactory, but the fallacy rested upon the erroneous sentence underlined in the above answer.

The young teacher drew a square on the blackboard

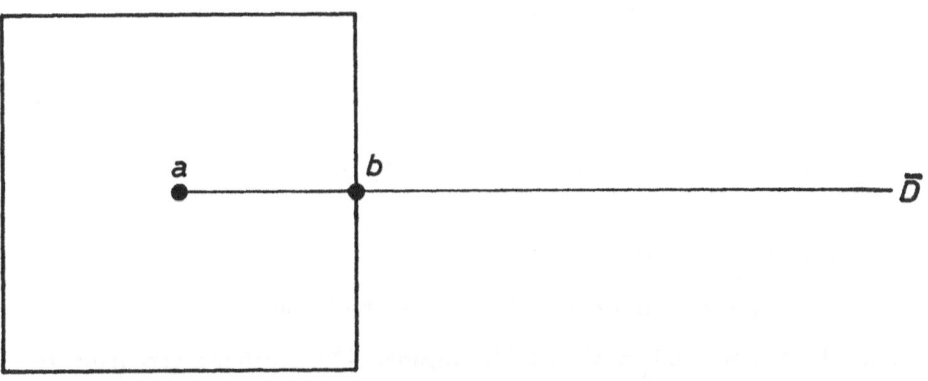

– Distance *ab*?
– 12 cm.
– *b* is at 12 cm. from *a* on the half-line $\bar{D}$.
– Draw a point on the half-line at 11 cm from *a*.
– ...

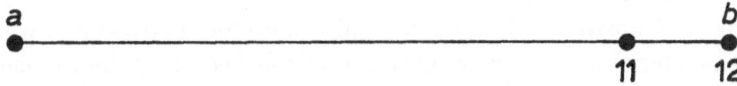

– Is that the last point strictly before b?
– No, there is one at 11.1.
– Is that the last one?
– No, there is one at 11.2.
– Is that the last one?

11.3, 11.4, ..., 11.9 were produced.
And then 11.91, 11.99, 11.999, ...

And one arrives at the conclusion that between a and b there exists no last point strictly before b. There does not exist a point just behind the point on the boundary! Samuel was crestfallen. He had nothing to say, but he was NOT CONVINCED!

The seventh salutary surprise for the teacher was to learn that

To be convinced by a proof	$\neq$	Not to be able to object to that proof

Nothing is more humiliating than to have to capitulate before a logical argument when one believes one has the truth. Only masochists could prefer punishment by logic to flagellation. Still a bachelor, our young teacher did not know that it is useless – and extremely lacking in gallantry – to try to convince a woman by logical reasoning. One must never beat a woman "not even with a flower" says the proverb. It could add also: "Above all, not with a syllogism".

– Your premises and your reasoning are surely impeccable, because you are so intelligent – said a handsome woman.

But she added immediately,

– It doesn't prevent your conclusion from being stupid!

Samuel was not at all convinced. He considered his teacher a very naughty sophist. He admitted he did not have the necessary perspicacity to object to him, but HE WAS NOT CONVINCED! Happily, he suddenly noticed on the desk the concrete evidence of the crime committed during the previous hour.

Taking between his fingers one of the cut squares, he asked candidly this embarrassing question: (Surprise #8!)

– Is this paper square OPEN or CLOSED?

A very enlightening surprise answer!

There is no answer to that Machiavellian question. Papers and cards constitute a material too coarse to illustrate the set-theoretic point of view. The pedagogical error committed during the first lesson appeared now in a glaring manner.

Little Isaac had spontaneously adopted the set-theoretic point of view. The treasures of pedagogy used by our young colleague had led him into the opposite direction and had prevented him from taking the same point of view.

RESULT: 10 years later – and for our young teacher, one hour later – it was necessary to use new treasures of pedagogy to counteract in the pupils the effects of the slow poison to which they had been exposed.

Naturally, it was necessary to invent a new intuitive support finer than the one of the postcards. The green-red traffic light convention plays that role.

With such means it is possible to introduce topological notions in a natural way, using the children's creativity.

I told elsewhere [F3] the story of the discovery of open sets and of continuous functions.

Université de Bruxelles, Belgium

WORKS OF FRÉDÉRIQUE AND PAPY

House of Mathematics for Children

Books for children

[GG]   Frédérique and Papy, *Graphs Games*, Thomas y Crowell, New York.

Books for teachers

[EG]   Frédérique and Papy, *L'Enfant et les Graphes*, Didier, Brussels.
         English translation, Algonquin, Montreal.
         Futch translation (in preparation), Didier, Brussels.
         Meulenhoff, Amsterdam.
         Frédérique, *Les Enfants et la Mathématique*, Didier, Brussels.
[EM1]   Vol. 1 (6 years-grade 1)
[EM2]   Vol. 2 (7 years-grade 2), being printed
[EM3]   Vol. 3 (8 years-grade 3), in preparation
         English translation, Algonquin, Montréal.
[M]   Papy, *Minicomputer*, Ivac, Brussels.
         Dutch translation, Ivac, Brussels.
         English translation (in preparation), Ivac, Brussels.

House of Mathematics for Teen-agers

Books for teen-agers

         Papy, *Mathématique Moderne*, Didier, Brussels, Vol. 1: *Ensembles–Relations, Les débuts de la géométrie. Le groupe des entiers rationnels.*
         English translation, Collier-Macmillan, London.
         Spanish translation, Eudeba, Buenos-Aires.
         Japanese translation, Nippon Hyoron-Sha, Tokyo.
         Dutch translation, Didier, Brussels.
         Rumanian translation, Tineretului, Bucuresti.
[MM2]   Vol 2: *Nombres réels et vectoriel plan.*
         English translation, Collier-Macmillan, London.
         Dutch translation, Didier, Brussels.
         Rumanian translation, Tineretului, Bucuresti.
[MM3]   Vol. 3: *Voici Euclide.*
         Dutch translation, Didier, Brussels; Meulenhoff, Amsterdam.
[MM5]   Vol. 5: *Arithmétique.*
[MM6]   Vol. 6: *Géométrie plane.*
[EE]   Papy, *Erste Elemente der Modernen Mathematik*, O. Salle, Frankfurt.
[EMG]   Papy, *Elemente der Modernen Geometrie*, 1. Heft., Klett, Stuttgart.

Books for teachers

[G]   Papy, *Groupes*, Presses Universitaires de Bruxelles.

English translation, Macmillan, London.
Italian translation, Feltrinelli, Milano.
Dutch translation, Plantyn, Antwerpen.

[F1]   Papy, *Géométrie affine et nombres réels*, P.U.B.
German translation, Vandenhoeck & Ruprecht, Göttingen.

[F2]   Papy, *Initiation aux espaces vectoriels*, P.U.B.
German translation, Vandenhoeck & Ruprecht, Göttingen.
Dutch translation, Plantyn, Antwerpen.

[F3]   PAPY, *Le premier enseignement de l'analyse*, P.U.B.
English translation, Algonquin, Montreal.

[G]   Papy, *Groupoïdes*, Labor, Bruxelles.
German translation, Vandenhoeck & Ruprecht, Göttingen.

GÜNTER PICKERT

THE INTRODUCTION OF METRIC BY THE
USE OF CONICS

1. It has been suggested in several publications [3, 4, 5, 7] that the teaching of systematic deductive geometry should begin with plane affine geometry. Reports of experiments in this direction are already available. I will not discuss the advantages or the disadvantages. Instead, I shall consider a consequence of this approach to geometry. How can the metric concepts (distance, orthogonality) be introduced in a well-motivated way if the plane affine geometry is already developed to such an extent that it can be described by a two dimensional vector space over the field of the reals. Of course the students already have intuitive and experimental knowledge of the simplest properties of the circle. So it might be appropriate to investigate certain curves in the affine plane with some of these properties and then to choose one of these curves as unit circle, defining distance and orthogonality by it. The lines which are described by linear forms obviously don't have the necessary properties of such curves so one is motivated to try the quadratic forms derived from the symmetric bilinear forms. In this way one gets to the curves of second degree with a center. In the title of this paper we refer to them as "conics". Thus creation of the metric notions as supplementary concepts not readily available from the outset opens to the student the possibility of metrics other than the usual Euclidean one. In particular, an indefinite quadratic form gives the Minkowski metric which plays an important role in the theory of relativity. Using conics in connection with the bilinear forms and related to the introduction of metric, in my opinion gives these objects the right place in comtemporary mathematics and saves them from being junked, a fate they might receive in retaliation for the overweight they have had in traditional teaching.

I used this introduction of metric in a course for prospective high school teachers and several times during in-service training courses for teachers. As I have been told, some teachers were influenced by these courses to try out this method in their classes (11th, 12th, and 13th grade of the Gymnasium, students of ages 16–19) with appropriate modifications of course.

2. Let us assume that the students already know the following representation of plane affine geometry by a two-dimensional vector space V over the field $\mathbf{R}$ of real numbers: Relative to a point of origin O, the points of the plane are in 1-1-correspondence with the elements of V by means of the mapping $X \rightarrow \overrightarrow{OX} (\in V)$. This mapping brings the lines of the planes into the cosets $\vec{a} + \mathbf{R}\vec{c}\,(\vec{c} \neq \vec{o} = \overrightarrow{OO})$ of the one-dimensional subspaces of V. In order to simplify the formulation, we identify every point X in the following with the vector $\overrightarrow{OX}$ and therefore the lines with the above mentioned cosets. In this sense lines are parallel if and only if they are cosets of the same subspace. To every one-dimensional subspace there is a *linear form* L of V, i.e., a mapping $L : V \rightarrow \mathbf{R}$ with

$$[\forall \vec{x}, \vec{y} \in V;\ a, b \in \mathbf{R}]\ L(a\vec{x} + b\vec{y}) = aL(\vec{x}) + bL(\vec{y}), \qquad (1)$$

so that the subspace is the kernel of L, i.e., the set $\{\vec{x} \mid L(\vec{x}) = 0\}$. The parallel lines, which are the cosets of this subspace then are the inverse images

$$L^{-1}\{c\} = \{\vec{x} \mid L(\vec{x}) = c\}$$

for each $c \in \mathbf{R}$. Here L cannot be the null form (with $L(\vec{x}) = 0$ for all $\vec{x} \in V$). Vice versa, every linear form L different from the null form gives, by its inverse images of the real numbers, a pencil of parallel lines. We remark that the set $L^{-1}\{1\}$ already determines L completely: Two different elements $\vec{a}, \vec{a}_1$ of the set by putting $\vec{a}_2 = \vec{a} - \vec{a}_1$ give a basis $(\vec{a}_1, \vec{a}_2)$ and one has

$$[\forall x_1, x_2 \in \mathbf{R}]\ L(x_1\vec{a}_1 + x_2\vec{a}_2) = x_1. \qquad (2)$$

This result may be phrased more geometrically: One gets the function value $L(\vec{x})$ by projecting $\vec{x}$ in the direction of $L^{-1}\{1\}$ onto $\mathbf{R}\vec{a}_1$ with $\vec{a}_1 \in L^{-1}\{1\}$ and taking the coordinate of this projection relative to $\vec{a}_1$ (see Figure 1).

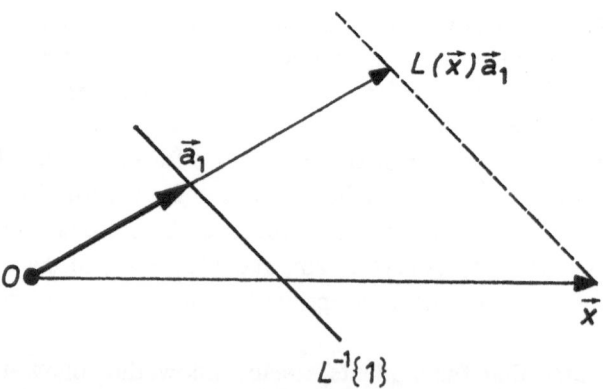

With the means of affine geometry we can compare the distances of pairs of points lying on parallel lines. By choosing a vector $\vec{e} \neq \vec{o}$ in a one-dimensional subspace, we may define for every $\vec{x} \in \mathbf{R}\hat{e}$, the length $|\vec{x}|$ of the vector $\vec{x}$ as the uniquely determined number $|x|$ with $\vec{x} = x\hat{e}$. The vector $\hat{e}$, having then the length 1, is called the unit vector of the subspace relative to the distance function $\vec{x} \to |\vec{x}|$. It is uniquely determined by this up to a multiplication by -1.

To introduce a metric in the plane we must choose a unit vector in every one-dimensional subspace. These unit vectors, considered as points, give a certain point set U of the plane with the property:

$$(3)$$

to every $\vec{x} \in V - \{0\}$ there exists a positive real number x with

$$x^{-1}\vec{x} \in U, \tag{3_1}$$

$$[\forall y \in \mathbf{R}] \, (y^{-1}\vec{x} \in U \Rightarrow y = x \quad \text{or} \quad y = -x). \tag{3_2}$$

This number, which is uniquely determined because of (3_2), is called the *length* $|\vec{x}|$ of the vector $\vec{x}$.

The set U will be called the *gauge curve* (in German: Eichkurve). Apparently it can't be a line, so a representation of U by a linear form as the inverse image of a real number is out of the question. Let us invent, therefore, another sort of mapping from V to $\mathbf{R}$ such that the inverse image of a certain real number, say 1, can be used as a gauge curve. By restricting our considerations to a line and therefore to a one-dimensional subspace $\mathbf{R}\hat{e}$ of V, such a mapping Q is easily constructed, e.g.,

$$[\forall x \in \mathbf{R}] \, Q(x\hat{e}) = x^2 \quad \text{and} \quad \{\vec{x} \mid Q(\vec{x}) = 1\} = \{\hat{e}, -\hat{e}\}.$$

In order to generalize this idea of a quadratic function to the two-dimensional case, we observe that the operation of squaring is the restriction of multiplication to the diagonal of $\mathbf{R} \times \mathbf{R}$ and that $\mathbf{R}$ is a (one-dimensional) vector space over $\mathbf{R}$. Therefore, we first try to formulate the concept of multiplication in V with values in $\mathbf{R}$. We can consider the multiplication in $\mathbf{R}$ as a mapping F, writing $F(x, y) = xy$, with the following properties:

$$[\forall x, y, z \in \mathbf{R}] \, F(x + y, z) = F(x, z) + F(y, z) \quad \text{(distributivity)}$$
$$[\forall a, x, y \in \mathbf{R}] \, F(ax, y) = aF(x, y) \quad \text{(associativity)}$$
$$[\forall x, y \in \mathbf{R}] \, F(x, y) = F(y, x) \quad \text{(commutativity)}.$$

These reformulations of well-known properties of a field have the advantage that they can be thought of as conditions for a mapping F of $V \times V$ into $\mathbf{R}$:

$$[\forall \vec{x}, \vec{y}, z \in V] \, F(\vec{x} + \vec{y}, z) = F(\vec{x}, \vec{z}) + F(\vec{y}, \vec{z}), \tag{4}$$

$$[\forall \vec{x}, \vec{y} \in V, a \in \mathbf{R}] \, F(a\vec{x}, \vec{y}) = aF(\vec{x}, \vec{y}), \tag{5}$$

$$[\forall \vec{x}, \vec{y} \in V] \, F(\vec{x}, \vec{y}) = F(\vec{y}, \vec{x}). \tag{6}$$

Now (4), (5) compared with (1) state that the "partial mappings" $\vec{x} \to F(\vec{x}, \vec{y})$

are linear forms (for every $\vec{y} \in V$) and also, according to (6), are the $\vec{y} \rightarrow F(\vec{x}, \vec{y})$ (for every $\vec{x} \in V$). Hence we call the mappings $F: V \times V \rightarrow \mathbf{R}$ with (4)–(6) the *symmetric bilinear forms* of V and we take these as the desired multiplications.

In order to help the students remember the way we arrived at this concept, it might be good to replace the "functional" notation $F(\vec{x}, \vec{y})$ with one usually used for operations, say $\vec{x} * \vec{y}$, but we will stick to the functional notation here.

Quadratic forms are now defined as those mappings $Q_F: V \rightarrow \mathbf{R}$, which can be derived from the symmetric bilinear forms F according to

$$[\forall \vec{x} \in V] \, Q_F(\vec{x}) = F(\vec{x}, \vec{x}). \tag{7}$$

Now we have to investigate whether

$$U_F = \{\vec{x} \mid Q_F(\vec{x}) = 1\}, \tag{8}$$

which is the inverse image of 1 under such a quadratic form Q_F, fulfills the condition (3) and could therefore be used as a gauge curve. From (5)–(7) we get

$$Q_F(x^{-1}\vec{x}) = x^{-2}Q_F(\vec{x}),$$

so that (3_1) for $U = U_F$, taking into account (8), is equivalent to

$$x^2 = Q_F(\vec{x})$$

and (3_2) is equivalent to

$$y^2 = x^2 \Rightarrow y = x \quad \text{or} \quad y = -x.$$

Since this implication is valid in every field, condition (3) for U_F simply becomes

$$[\forall \vec{x} \in V] \, (\vec{x} \neq \vec{o} \Rightarrow Q_F(\vec{x}) > 0). \tag{9}$$

A quadratic form O_F and also the symmetric bilinear form F with this property is called *positive definite*. By choosing such a quadratic form Q_F, we get a metric with gauge curve (8) and

$$|\vec{x}| = \sqrt{Q_F(\vec{x})}. \tag{10}$$

For (10) first we have the restriction $\vec{x} \neq o$, but the bilinearity of F gives $F(\vec{o}, \vec{x}) = 0$ and therefore $Q_F(\vec{o}) = 0$, and, since we naturally define $|\vec{o}| = 0$, we have (10) without any restriction on $\vec{x}$. From (9) with $Q_F(\vec{o}) = 0$ follows

$$[\forall \vec{x} \in V] \, Q_F(\vec{x}) \geqslant 0. \tag{9'}$$

From the intuitive point of view, the *unit disc*,

$$\{\vec{x} \mid |\vec{x}| \leqslant 1\} \tag{11}$$

has the property of being *convex*, i.e., with any two points it contains all the points between them. Now the question arises if this also holds with respect to our metric, introduced by (10). Since the points between the points $\vec{a}$, $\vec{b}$ are described by

$$(1 - t)\,\vec{a} + t\vec{b}, \quad 0 < t < 1 \tag{12}$$

we need a formula by which to express $Q_F(\vec{x} + \vec{y})$ by $Q_F(\vec{x})$, $Q_F(\vec{y})$ and eventually other terms. With the help of (4)–(7), we get

$$Q_F(\vec{x} + \vec{y}) = F(\vec{x} + \vec{y}, \vec{x} + \vec{y}) = F(\vec{x}, \vec{x} + \vec{y}) + F(\vec{y}, \vec{x} + \vec{y})$$
$$= F(\vec{x}, \vec{x}) + F(\vec{x}, \vec{y}) + F(\vec{y}, \vec{x}) + F(\vec{y}, \vec{y})$$

and therefore,

$$Q_F(\vec{x} + \vec{y}) = Q_F(\vec{x}) + Q_F(\vec{y}) + 2F(\vec{x}, \vec{y}). \tag{13}$$

This equation shows that the quadratic form Q_F completely determines the symmetric bilinear form F.

With $\vec{x} = (1-t)\,\vec{a}$, $\vec{y} = t\vec{b}$, $Q_F(\vec{a}) \leq 1$, and $Q_F(\vec{b}) \leq 1$, (13) gives

$$Q_F((1 - t)\,\vec{a} + t\vec{b}) \leqslant 1 + 2t(1 - t)\,(F(\vec{a}, \vec{b}) - 1).$$

Since $t(1-t)$ is positive for the points (12), the set (11) is therefore convex if

$$Q_F(\vec{a}), Q_F(\vec{b}) \leqslant 1 \Rightarrow F(\vec{a}, \vec{b}) \leqslant 1. \tag{14}$$

We now derive (14) from (9′) with the help of (13) and by putting $\vec{x} = \vec{a}$, $\vec{y} = t\vec{b}$:

$$Q_F(\vec{b})\,t^2 + 2F(\vec{a}, \vec{b})\,t + Q_F(\vec{a}) \geqslant 0.$$

This being true for all $t \in \mathbf{R}$ we must have

$$Q_F(\vec{b}) = 0 \Rightarrow F(\vec{a}, \vec{b}) = 0.$$

Therefore we can restrict the proof of (14) to the case $Q_F(\vec{b}) > 0$:

$$Q_F(\vec{b})\,(t + F(\vec{a}, \vec{b})\,Q_F(\vec{b})^{-1})^2 + Q_F(\vec{b})^{-1}\,(Q_F(\vec{a})\,Q_F(\vec{b}) - F(\vec{a}, \vec{b})^2) \geqslant 0$$

Putting $t = -F(\vec{a}, \vec{b})\,Q_F(\vec{b})^{-1}$ gives

$$F(\vec{a}, \vec{b})^2 \leqslant Q_F(\vec{a})\,Q_F(\vec{b}) \tag{15}$$

and therefore (14).

In proving (3) we used the existence (in $\mathbf{R}$) of the square root for every positive real number. Analysis of the proof shows that it remains valid if we substitute any ordered field K for $\mathbf{R}$ with K having the *Euclidean property*:

$$[\forall x \in K]\,(0 < x \Rightarrow [\exists y \in K]\, x = y^2). \tag{16}$$

Indeed, all the following results remain true for such a Euclidean field K taken instead of $\mathbf{R}$, but, since other geometric investigations need the Archimedian property and therefore at least (if not $\mathbf{R}$ itself) a subfield of $\mathbf{R}$ as field of scalars, we stick to $\mathbf{R}$.

The metrics introduced by (10) are not the only ones fulfilling condition (3), even if we want the convexity of (11). So we might choose a basis $(\check{e}_1, \check{e}_2)$ of V and define (see Figure 2)

$$U = \{x_1\check{e}_1 + x_2\check{e}_2 \mid |x_1| + |x_2| = 1\}, \tag{17}$$

which yields

$$|x_1\check{e}_1 + x_2\check{e}_2| = |x_1| + |x_2|. \tag{18}$$

This distance function is appropriate in situations where moving is allowed only in two directions (the directions of the basis vectors) as in a city with

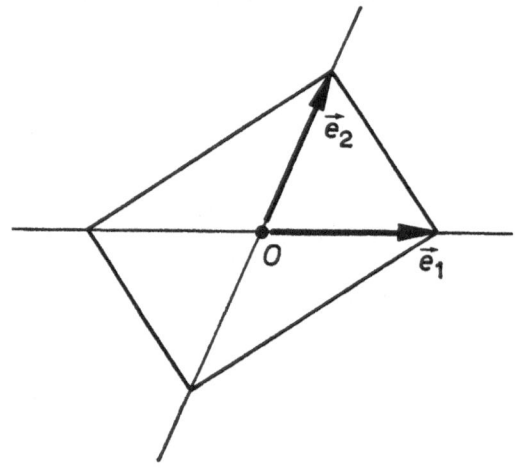

Fig. 2.

a grid of avenues and streets rectangular to each other. That this metric cannot be given by (10) with a quadratic form Q_F follows from the fact that $|\check{e}_1 - \check{e}_2| = 2$ and

$$0 \leqslant t \leqslant 1 \Rightarrow |\check{e}_1 + t(\check{e}_2 - \check{e}_1)| = 1,$$

whereas, according to (10), (13):

$$Q_F(\check{e}_1 + t(\check{e}_2 - \check{e}_1)) = 4t^2 + 2F(\check{e}_1, \check{e}_2 - \check{e}_1)t + 1 \neq 1$$

for all $t \neq 0$, $\tfrac{1}{2}F(\check{e}_1, \check{e}_1 - \check{e}_2)$.

In Section 4 it will be proved that through every point of U_F, with F being a positive definite symmetric bilinear form, there is exactly one line with only one point in common with U_F. From this fact we can also see that (18) is not compatible with (10) because every line $\check{e}_2 + \mathbf{R}(\check{e}_1 + t\check{e}_2)$ with $|t| < 1$ has only $\check{e}_2$ in common with U in (17).

The reason why the metric (18) is not considered as a "good" one if one starts from a vector space, lies in the fact that there exist too few automorphisms, i.e., linear mappings of V onto V which preserve the gauge curve and therefore the distance. Only the eight linear mappings which transform the basis $(\check{e}_1, \check{e}_2)$ into the eight different linearly independent ordered pairs out of the vector set $\{\check{e}_1, \check{e}_2, -\check{e}_1, -\check{e}_2\}$ are such automorphisms. Indeed, if we postulate the existence of a group G of linear mappings (of V onto V) sharply transitive on the so-called "flags", i.e., the pairs $(\mathbf{R}_+\check{e}_1, \mathbf{R}\check{e}_1 + \mathbf{R}_+\check{e}_2)$ for all bases $(\check{e}_1, \check{e}_2)$, (with $\mathbf{R}_+$ as the set of the positive real numbers), then G is exactly the automorphism group relative to a metric given by a positive definite quadratic form (see [1], p. 450, prop. 2). From a purely mathematical point of view, I consider this to be the best way to motivate introduction of a metric by a positive definite quadratic form, but didactically, this procedure seems to be too difficult.

3. We have now the task of determining all positive definite quadratic forms. Obviously a symmetric bilinear form F is completely determined by the values

$$Q_F(\bar{a}_1) = a_{11}, \quad Q_F(\bar{a}_2) = a_{22}, \quad F(\bar{a}_1, \bar{a}_2) = a_{12} \, (= a_{21}) \quad (19)$$

if $(\bar{a}_1, \bar{a}_2)$ is a basis. We have

$$F(x_1\bar{a}_1 + x_2\bar{a}_2, y_1\bar{a}_1 + y_2\bar{a}_2) =$$
$$a_{11}x_1y_1 + a_{12}(x_1y_2 + x_2y_1) + a_{22}x_2 y_2. \quad (20)$$

On the other hand, according to (20), every real number triple (a_{11}, a_{12}, a_{22}) or equivalently, every real symmetric matrix

$$\begin{bmatrix} a_{11} & a_{12} \\ a_{21} & a_{22} \end{bmatrix}$$

with respect to a basis $(\bar{a}_1, \bar{a}_2)$, determines a symmetric bilinear form with (19). Under what conditions for (a_{11}, a_{12}, a_{22}) is this bilinear form positive definite? To answer this question we first exclude the case in which the range of Q_F consists only of 0. In the latter case, according to (13), we would have $a_{11} = a_{22} = 0$ and also $a_{12} = 0$. If $a_{11} = a_{22} = 0$, we can therefore assume that $a_{12} \neq 0$. Formula (20), then gives

$$Q_F(x_1\bar{a}_1 + x_2\bar{a}_2) = 2a_{12}x_1x_2. \quad (21)$$

Obviously not even (9') is valid here, but (21) is interesting insofar as (the basis and the number a_{12} given) the curve U_F can be easily plotted by the students who will thereby meet again the graph of a function which is well known to them. In order to exclude also the deficiencies of (21), we may suppose $a_{11} \neq 0$ for the sequel. Before the students go on it might be good for them to gain some plotting experience here too by calculating first a lot of values for Q_F and then drawing not only U_F but also some other inverse images of numbers in the neighborhood of 1.

To facilitate the investigation, we try to change the basis for another basis $(\vec{b}_1, \vec{b}_2)$ with $F(\vec{b}_1, \vec{b}_2) = 0$, keeping the basis vector $\vec{a}_1$ unchanged: $\vec{b}_1 = \vec{a}_1$. This can be done by putting

$$\vec{b}_2 = a_{12}\vec{a}_1 - a_{11}\vec{a}_2.$$

Then $(\vec{a}_1, \vec{b}_2)$ is a basis because $a_{11} \neq 0$, and

$$F(\vec{a}_1, \vec{b}_2) = a_{12}a_{11} - a_{11}a_{12} = 0.$$

From (20) one gets

$$Q_F(\vec{b}_2) = a_{11}(a_{11}a_{22} - a_{12}^2)$$

and for the new basis $(\vec{b}_1, \vec{b}_2)$ we have

$$Q_F(x_1\vec{b}_1 + x_2\vec{b}_2) = a_{11}x_1^2 + a_{11}(a_{11}a_{22} - a_{12}^2)x_2^2. \tag{22}$$

From this equation we see that F is positive definite if and only if

$$a_{11} > 0, \qquad a_{11}a_{22} > a_{12}^2. \tag{23}$$

Regarding (19), part of (23) is already contained in (15), which, on the other hand, follows easily from (23) for linearly independent $\vec{a}, \vec{b}$ and is trivially valid (with the equality sign), if $\vec{a}, \vec{b}$ are linearly dependent.

If in (23) one has $a_{11} < 0$, instead of $a_{11} > 0$, then we have $Q_F(\vec{x}) < 0$ for all $\vec{x} \neq \vec{o}$, and the quadratic form Q_F (as well as the bilinear form F) is called *negative definite*. In this case U_F is the empty set. If we change the other condition in (23) to $a_{11}a_{22} = a_{12}^2$, then we still have (9') but not (9), and U_F is the union of the two lines

$$\pm a_{11}^{-1}\vec{b}_1 + \mathbf{R}\vec{b}_2,$$

a *pair of parallels*. The quadratic form Q_F (as well as the bilinear form F) is then called *positive semidefinite* and *negative semidefinite* provided $a_{11} < 0$ (instead of $a_{11} > 0$), in which case U_F is again empty. In the remaining case

$$a_{11}a_{22} < a_{12}^2, \tag{23'}$$

which also includes the case $a_{11} = a_{22} = 0$, $a_{12} \neq 0$, already investigated, the range of Q_F contains negative numbers as well as positive ones (and is in fact equal to $\mathbf{R}$). The quadratic form Q_F (as well as the bilinear form F) is then called *indefinite*. In the special case $a_{11} = a_{22} = 0$, we reach an analogue to (22), if we take $\vec{b}_1 = \vec{a}_1 + \vec{a}_2$, $\vec{b}_2 = \vec{a}_1 - \vec{a}_2$. We get $Q_F(\vec{b}_1) = 2a_{12}$, $Q_F(\vec{b}_2) = -2a_{12}$, $F(\vec{b}_1, \vec{b}_2) = 0$ according to (20) and therefore

$$Q_F(x_1\vec{b}_1 + x_2\vec{b}_2) = 2a_{12}(x_1^2 - x_2^2). \tag{22'}$$

The curves U_F in case (23) are now named *ellipses*, those in case (23')
hyperbolas. These names are easily explained according to their meaning in
Greek, if we write (eventually interchanging $\vec{b}_1$, $\vec{b}_2$) the right hand sides of
(22), (22') in the form $a^2 x_1^2 - b x_2^2$ with $ab > 0$ and choose a new point of
origin O' with $\overrightarrow{OO'} = a^{-1}\vec{b}_1$. Relative to O' and the basis $(\vec{b}_1, \vec{b}_2)$ the curve
U_F is then described by the equation

$$x_2^2 = 2ab^{-1}x_1 + a^2b^{-1}x_1^2. \tag{24}$$

If now $b < 0$ (ellipse), x_2^2 has a *defect* compared with $2ab^{-1}x_1$, but it has an
excess if $b > 0$ (hyperbola).* The "intermediate" case of the *parabola** with
an equation

$$x_2^2 = 2px_1$$

does not come up here, since we only consider curves with a *center*. The
latter is evident from

$$Q_F(-\vec{x}) = F(-\vec{x}, -\vec{x}) = -F(\vec{x}, -\vec{x})$$
$$= -(-F(\vec{x}, \vec{x})) = F(\vec{x}, \vec{x}) = Q_F(\vec{x})$$

which means that

$$\vec{x} \in U_F \Rightarrow -\vec{x} \in U_F,$$

i.e., the reflection in the origin transforms U_F into itself.

As we have shown, the decision whether U_F is an ellipse, a pair of parallels,
or a hyperbola, can be made by purely affine, and therefore (algebraically
speaking) rational methods, we only need an ordered field (not even a
Euclidean one) as a field of scalars. This well-known fact is often forgotten
in the teaching of geometry, due to the dominance the metric concepts
usually, but not legitimately, have over the affine concepts.**

4. As we have shown, among all those curves such as ellipses, pairs of
parallels, and hyperbolas, the ellipses only can serve as gauge curves with
condition (3). They alone have a non-empty intersection with every line
through the origin. Thus every ellipse with center O determines the length $|\vec{x}|$
for each vector $\vec{x}$ as described in (3). By (10) it also determines the positive
definite quadratic form Q_F and, according to (13), the symmetric bilinear
form F as well. The ellipse is then just U_F.

If the basis $(\vec{b}_1, \vec{b}_2)$ is chosen as in (22), we can go over to another basis
consisting of *unit vectors* (i.e., vectors $\in U_F$), namely, $\vec{e}_i = |\vec{b}_i|^{-1}\vec{b}_i$ $(i = 1, 2)$,

---

* The source of the names, introduced by Apollonius, were the Greek verbs: ἐλλείπειν =
to be deficient, ὑπερβάλλειν = to exceed, παραβάλλειν = to be equal.
** The affine point of view regarding the conics is described in [2, 6].

and we have then, according to (19), (20):

$$Q_F(x_1 \vec{e}_1 + x_2 \vec{e}_2) = x_1^2 + x_2^2. \tag{25}$$

This result, for which the Euclidean property (16) is essential, enables us to discover the connection between two different metrics. For any other positive definite symmetric bilinear form F' there also exists a basis $(\vec{e}_1', \vec{e}_2')$ with

$$Q_{F'}(x_1 \vec{e}_1' + x_2 \vec{e}_2') = x_1^2 + x_2^2. \tag{25'}$$

Now the two bases determine a linear mapping f of V onto V with

$$f(x_1 \vec{e}_1 + x_2 \vec{e}_2) = x_1 \vec{e}_1' + x_2 \vec{e}_2'. \tag{26}$$

According to (25), (25'), (26) we have

$$[\forall x \in V]\, Q_{F'}(f(\vec{x})) = Q_F(\vec{x}),$$

i.e., f is an *isomorphis m*of the metric plane V, F onto the metric plane V, F'. Due to this isomorphy the different ellipses (with center O) give essentially the same metric. This means that in order to introduce the metric concepts into affine geometry, we can take any ellipse as the gauge curve. This curve is then called the *unit circle*.

Having thus defined the distance by means of an ellipse as the gauge curve, we must also define the second fundamental metric concept, *orthogonality*. From elementary geometry the student knows that orthogonality can be characterized in many ways using the concept of distance. It can be done, for example, by means of a pythagorean triangle, or by a rhombus. Considering our method of defining distance, it would be adequate to also define orthogonality by a method which makes direct use of the unit circle. In this connection the students may remember the fact that orthogonal reflections with axes through the center O preserve this circle. So we have to investigate those affine reflections which leave the ellipse U_F invariant.

Let $\mathbf{R}\vec{b}$ with $\vec{b} \in U_F$ be the axis of such an affine reflection, the reflection having the direction $\mathbf{R}\vec{c}(\vec{c} \notin \mathbf{R}\vec{b})$. Then for every $\vec{x}$ the image $\vec{y}$ is determined by

$$\vec{x} + \vec{y} \in \mathbf{R}\vec{b}, \qquad \vec{x} - \vec{y} \in \mathbf{R}c. \tag{27}$$

Thus, if we represent

$$\vec{x} = \vec{b} + tc, \tag{28}$$

we get

$$\vec{y} = \vec{b} - tc. \tag{28'}$$

Under the assumption (28) (without any restriction on $\vec{c}$) and using (13) and $Q_F(\vec{b}) = 1$, $Q_F(\vec{x}) = 1$ proves to be equivalent to

$$(2F(\vec{b}, \vec{c}) + Q_F(\vec{c})\, t)\, t = 0. \tag{29}$$

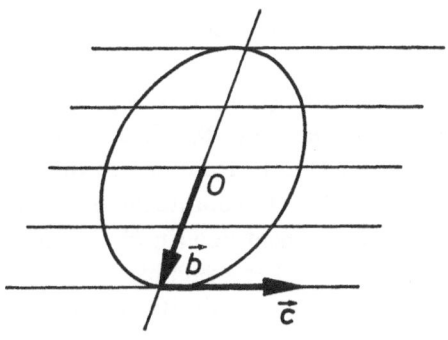

Fig. 3.

Because of $Q_F(\vec{c}) > 0$, (29) shows that the set

$$U_F \cap (\vec{b} + \mathbf{R}\vec{c}) \tag{30}$$

consists of $\vec{b}$ only, provided $F(\vec{b}, \vec{c}) = 0$, otherwise it contains exactly two elements. But, regarding (28′), (28), and the postulate that the reflections preserve U_F (30) must contain either exactly one element (namely $\vec{b}$) or at least three: $\vec{b}, \vec{x}, \vec{y}$. Thus according to our previous result, the set (30) is $\{\vec{b}\}$, and therefore

$$F(\vec{b}, \vec{c}) = 0 \tag{31}$$

This equation completely determines $\mathbf{R}\vec{c}$ as the kernel of the linear form $\vec{x} \to F(\vec{b}, \vec{x})$ which is not the null form (since $F(\vec{b}, \vec{b}) = 1$). So for every point $\vec{b}$ of the ellipse U_F there is exactly one line through $\vec{b}$ having only one point in common with U_F. This is called the *tangent line* in $\vec{b}$ to U_F. According to our motivation we should call it "orthogonal" to the diameter $\mathbf{R}\vec{b}$. Regarding (5) we therefore use (31) as the defining condition for orthogonality of vectors $\vec{b}, \vec{c}$ (without any restrictions) (see Figure 3).

From (28) and (31) joined with $Q_F(\vec{b}) = 1$, we get $F(\vec{b}, \vec{x}) = 1$. This leads to

$$\vec{b} + \mathbf{R}\vec{c} \subseteq \{\vec{x} \mid F(\vec{b}, \vec{x}) = 1\}.$$

Since the set on the right is a line, the equality sign is valid if $\vec{c} \neq \vec{o}$. Therefore

$$F(\vec{b}, \vec{c}) = 0, \quad Q_F(\vec{b}) = 1, \quad \vec{c} \neq \vec{o} \Rightarrow \vec{b} + \mathbf{R}\vec{c} = \{\vec{x} \mid F(\vec{b}, \vec{x}) = 1\}. \tag{32}$$

The tangent line in $\vec{b} \in U_F$ is thus the inverse image of 1 under the linear form $\vec{x} \to F(\vec{b}, \vec{x})$ and therefore (see the remark connected with (2)) determines this linear form. This prove again that U_F already determines the symmetric bilinear form F. It also provides a geometric construction of the value $F(\vec{a}, \vec{x})$

for any pair of vectors $\vec{a}$, $\vec{x}$, except for the trivial case $\vec{a} = \vec{o}$. The construction can be described as follows:

Determining $|\vec{a}|$ according to (3) by means of $U = U_F$, we put $\vec{b} = |\vec{a}|^{-1}\vec{a}$ and determine the tangent line in $\vec{b}$ to U_F. This line is the inverse image of 1 under the linear form $\vec{x} \rightarrow F(\vec{b}, \vec{x})$. According to the formulation following (2) we get the value of $F(\vec{b}, \vec{x})$ as the coordinate relative to $\vec{b}$ of the projection of $\vec{x}$ in the direction of the tangent line (in $\vec{b}$) on $\mathbf{R}\vec{b}(=\mathbf{R}\vec{a})$. According to (5) $F(\vec{a}, \vec{x})$ is then the product of this number and $|\vec{a}|$. Taking into account our definition of orthogonality we can shorten this description to the following way which is exactly the usual definition of the *inner* (or *scalar*) *product* of the vectors $\vec{a}$, $\vec{x}$: Project $\vec{x}$ orthogonally on $\mathbf{R}\vec{a}$, take the coordinate of this projection relative to the unit vector $|\vec{a}|^{-1}a$ and multiply by $|\vec{a}|$.

If we exclude the case $\vec{c} = \vec{o}$, we can as well postulate $Q_F(\vec{c}) = 1$. Then $(\vec{b}, \vec{c})$, (with (31) and $Q_F(\vec{c}) = 1$), can be taken as basis $(\vec{e}_1, \vec{e}_2)$ in (25) and vice versa every basis $(\vec{e}_1, \vec{e}_2)$ with (25) consists of orthogonal unit vectors. The reflection referred to in (27), now with $\vec{b} = \vec{e}_1$, $\vec{c} = \vec{e}_2$, maps $x_1\vec{e}_1 + x_2\vec{e}_2$ onto $x_1\vec{e}_1 - x_2\vec{e}_2$ and therefore, according to (25), U_F into itself. Together with the fact that the tangent line in $\vec{e}_1$ is parallel to $\mathbf{R}\vec{e}_2$, this is the content of the *theorem on the conjugate diameters* for an ellipse. In the next section we will prove this also for a hyperbola.

5. In the case of an indefinite quadratic form Q_F we can find a basis $(\vec{e}_1, \vec{e}_2)$ so that in analogy to (25):

$$Q_F(x_1\vec{e}_1 + x_2\vec{e}_2) = x_1^2 - x_2^2. \tag{33}$$

As in (25) we can take any point of U_F for $\vec{e}_1$. From (33) follows:

$$\{x \mid Q_F(\vec{x}) = 0\} = \mathbf{R}(\vec{e}_1 + \vec{e}_2) \cup \mathbf{R}(\vec{e}_1 - \vec{e}_2). \tag{34}$$

Regarding (29), the set (30) is therefore equal to $\{\vec{b}\}$ not only in case (31) but also if $\vec{c}$ belongs to the set (34). Thus there are three lines through $\vec{b}$ having only $\vec{b}$ in common with U_F, two of them parallel to the lines $\mathbf{R}(\vec{e}_1 + \vec{e}_2)$, $\mathbf{R}(\vec{e}_1 - \vec{e}_2)$, and the third given by the equation $F(\vec{b}, \vec{x}) = 1$ as in (32). As we see from (20) (with $a_{11} = 1$, $a_{12} = 0$, $a_{22} = -1$), this reads relative to the basis in (25) with $\vec{b} = b_1\vec{e}_1 + b_2\vec{e}_2$ as follows:

$$b_1 x_1 - b_2 x_2 = 1. \tag{35}$$

If we construct the third line for another point of U_F, different from $\vec{b}$ and $-\vec{b}$, it is not parallel to the line with Equation (35). So the third line can be discerned from the other two, using only the point set U_F. The line given by (35) is called the *tangent line* in $\vec{b}$ to the hyperbola U_F. It can be distinguished from the lines $\vec{b}+\mathbf{R}(\vec{e}_1+\vec{e}_2)$, $\vec{b}+\mathbf{R}(\vec{e}_1-\vec{e}_2)$ also by the fact that its points not equal to $\vec{b}$ belong to one connected subset of the complement of U_F. We take $(\vec{b}, \vec{c})$ for the basis $(\vec{a}_1, \vec{a}_2)$ in (22) and have $Q_F(\vec{c})<0$ according to (23'). Therefore

$$t \neq 0 \Rightarrow Q_F(\vec{b} + t\vec{c}) = 1 + t^2 Q_F(\vec{c}) < 1.$$

Each of the two terms

$$Q_F(\vec{b} + t(\vec{e}_1 \pm \vec{e}_2)) = 1 + 2t(\vec{b}_1 \pm \vec{b}_2)$$

(one with the upper sign, the other with the lower sign) has (for different values of t) values <1 as well as >1, since $(b_1-b_2)(b_1+b_2)=b_1^2-b_2^2=1$. The role of the *asymptotes* $\mathbf{R}(\vec{e}_1 \pm \vec{e}_2)$ is best seen if we use $\vec{a}_1=\vec{e}_1+\vec{e}_2$, $\vec{a}_2=\vec{e}_1-\vec{e}_2$ as basis vectors, which leads us to (21) with $a_{12}=2$. A line parallel to an asymptote, but different from it, has therefore exactly one point in common with the hyperbola; but to every tangent line there are parallel lines cutting the hyperbola in two points. This is a third way of distinguishing the tangent (in a point of the hyperbola) from the two lines (through this point) parallel to the asymptotes.

As in Section 4, for a basis $(\vec{e}_1, \vec{e}_2)$ with (33), the reflection with axis $\mathbf{R}\vec{e}_1$ and direction $\mathbf{R}\vec{e}_2$ (transforming $x_1\vec{e}_1+x_2\vec{e}_2$ into $x_1\vec{e}_1-x_2\vec{e}_2$), leaves U_F invariant. The tangent line in $\vec{e}_1$ is parallel to $\mathbf{R}\vec{e}_2$. Thus we have the theorem on the conjugate diameters for hyperbolas also.

The tangent line in $\vec{b}$ to the hyperbola U_F determines the linear form $\vec{x} \rightarrow F(\vec{b}, \vec{x})$ as was the case in Section 4. Since there are linearly independent elements $\vec{b}_1, \vec{b}_2$ in U_F (e.g., $\vec{e}_1$ and $\frac{5}{3}\vec{e}_1+\frac{4}{3}\vec{e}_2$) the mapping F is already determined:

$$F(y_1\vec{b}_1 + y_2\vec{b}_2, \vec{x}) = y_1 F(\vec{b}_1, \vec{x}) + y_2 F(\vec{b}_2, \vec{x}).$$

Thus an indefinite symmetric bilinear form F is also determined by the point set U_F. But then $U=U_F$ does not fulfill condition (3). If we nevertheless take it as gauge curve of a metric, the determination of a length $|\vec{x}|$ is possible only for vectors $\vec{x}$ with $Q_F(\vec{x})\geqslant 0$. This kind of metric is named after *Minkowski* who applied it to the theory of relativity. In this application the plane describes the kinematics of one-dimensional movements, the one spatial coordinate x and the time t being the two coordinates in a coordinate system of the plane. Allowed coordinate transformations must keep invariant the expression

$$x^2 - c^2 t^2,$$

where c (>0) is the velocity of light. This physical situation can be described by a bilinear form F with

$$Q_F(x\vec{e}_1 + t\vec{e}_1) = x^2 - c^2 t^2, \tag{36}$$

and therefore

$$Q_F(\check{e}_1) = 1, \quad Q_F(\check{e}_2) = -c^2, \quad F(\check{e}_1, \check{e}_2) = 0 \qquad (37)$$

In order to derive (37) we use (13) or compare (36) with (19), (20).

For another allowed coordinate system, if we want to get the same expression as in (26), we need a basis $(\check{e}_1', \check{e}_2')$ such that

$$Q_F(\check{e}_1') = 1, \quad Q_F(\check{e}_2') = -c^2, \quad F(\check{e}_1', \check{e}_2') = 0. \qquad (37')$$

The connection between the new (x', t') and the old (x, t) coordinates is given by

$$x\check{e}_1 + t\check{e}_2 = x'\check{e}_1' + t'\check{e}_2'. \qquad (38)$$

If the new coordinate system describes a "frame" moving with velocity v relative to the "frame" given by the first system, then $x'=0$ must be equivalent to $x=vt$ which, according to (38), gives

$$\check{e}_2' = k(v\check{e}_1 + \check{e}_2) \qquad (39)$$

with a certain number $k \neq 0$. The second equation in (37') with (36) gives

$$-c^2 = k^2(v^2 - c^2),$$

that is

$$k = \pm\sqrt{1 - v^2 c^{-2^{-1}}}, \qquad (40)$$

which implies $|v| < c$. For

$$\check{e}_1' = a\check{e}_1 + b\check{e}_2$$

the third equation in (37') with (39), (37), (20) yields

$$av - bc^2 = 0$$

whereas the first equation in (37') together with (36) leads to

$$a^2 - c^2 b^2 = 1.$$

So, using (40), we get

$$\check{e}_1' = \pm k(\check{e}_1 + vc^{-2}\check{e}_2). \qquad (41)$$

Substituting (39), (41) in (38) and comparing the coefficients of $\check{e}_1$, $\check{e}_2$ on both sides gives

$$x = k(\pm x' + vt') \qquad (42)$$
$$t = k(\pm vc^{-2}x' + t').$$

By eventually reversing the orientation of the new spatial coordinate axis we can omit here the $\pm$, and because of the "irreversibility of time" we have to omit the $\pm$ in (40). Thus (42) goes over into the *Lorentz transformations*, which usually are described with interchanged new and old coordinates and therefore with $-v$ instead of v.

6. In order to see that ellipses, hyperbolas, and parabola are planar sections of a cone and so deserve the name "conics", we go over to the three-dimensional affine space with basis $(\vec{e}_1, \vec{e}_2, \vec{e}_3)$. In the plane $\mathbf{R}\vec{e}_2 + \mathbf{R}\vec{e}_3$ we introduce a metric such that $\vec{e}_2, \vec{e}_3$ become orthogonal unit vectors. The unit circle with centre $\vec{e}_3$ then is

$$\{y_2\vec{e}_2 + y_3\vec{e}_3 \mid y_3^2 - 2y_3 + y_2^2 = 0\}. \tag{43}$$

We use it as base of a cone with vertex

$$\vec{s} = s\vec{e}_1 + s'\vec{e}_3 \quad (s, s' \neq 0)$$

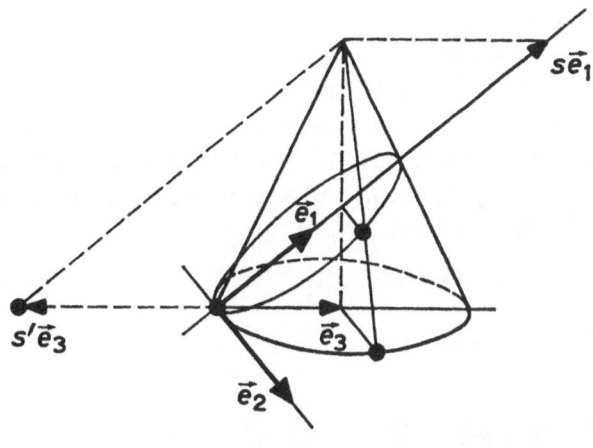

Fig. 4.

and determine the intersection of this cone with the plane $\mathbf{R}\vec{e}_1 + \mathbf{R}\vec{e}_2$ (see Figure 4). Now every point of the line joining $\vec{x}(=x_1\vec{e}_1 + x_2\vec{e}_2)$ with $\vec{s}$ can be written as

$$t\vec{x} + (1 - t)\,\vec{s} = (tx_1 + s(1 - t))\vec{e}_1 + tx_2\vec{e}_2 + s'(1 - t)\,\vec{e}_3 \tag{44}$$

for some $t \in \mathbf{R}$. The condition that (44) be on the plane $\mathbf{R}\vec{e}_2 + \mathbf{R}\vec{e}_3$ is

$$tx_1 + s(1 - t) = 0,$$

and this gives

$$t = -s(x_1 - s)^{-1}, \quad 1 - t = x_1(x_1 - s)^{-1}.$$

With this condition fulfilled, (44) is the point $\vec{y} = y_2\vec{e}_2 + y_3\vec{e}_3$ with

$$y_2 = -sx_2(x_1 - s)^{-1}, \quad y_3 = s'x_1(x_1 - s)^{-1}.$$

Point $\bar{x}$ belongs to the planar section of the cone if and only if point $\bar{y}$ belongs to (43). Therefore

$$s'^2 x_1^2 (x_1 - s)^{-2} - 2s'x_1 (x_1 - s)^{-1} + s^2 x_2^2 (x_1 - s)^{-2} = 0$$

is the equation of the planar section which, under the condition $x_1 \neq s$, is equivalent to

$$(s'^2 - 2s') x_1^2 + 2s'sx_1 + s^2 x_2^2 = 0. \tag{45}$$

Since (45) for $x_1 = s$ becomes $s^2(s'^2 + x_2^2) = 0$, which is incompatible with $s, s' \neq 0$, (45) is the equation of the planar section. Rewriting (45) with $p = -s's^{-1}$ as

$$x_2^2 = 2px_1 - (2s^{-1} + p) px_1^2, \tag{45'}$$

we see, comparing (45') with (24), that every ellipse, hyperbola, or parabola can be represented as a planar section of a cone.

If we want the cone not only to have a circle as base but also to be orthogonal (i.e., the axis orthogonal to the base), we must introduce a metric, not only in the plane $R\tilde{e}_2 + R\tilde{e}_3$, but also in the whole three-space. It should be consistent with the plane metric. We proceed as follows: With $\tilde{e}$ as unit vector orthogonal to $\tilde{e}_2$, $\tilde{e}_3$ and $h(>0)$ as the distance of the vertex from the base plane $R\tilde{e}_2 + R\tilde{e}_3$, we have

$$s\tilde{e}_1 + s'\tilde{e}_3 = \tilde{e}_3 + h\tilde{e},$$

that is

$$s\tilde{e}_1 = (1 - s') \tilde{e}_3 + h\tilde{e} \tag{46}$$

and therefore

$$s^2 = (1 + ps)^2 + h^2. \tag{46'}$$

The only restriction for s is

$$|s| \geqslant h$$

because exactly under this condition there exists $p \neq 0$ with (46') and thus a unit vector $\tilde{e}_1$ with (46). According to (46') the coefficient of x_1^2 in (45') becomes

$$- (2s^{-1} + p) p = (1 - (1 + ps)^2) s^{-2} = (h^2 + 1) s^{-2} - 1.$$

For every fixed h this term takes all values > -1 and $\leqslant h^{-2}$ because $|s| \geqslant h$ is the only restriction on s. On the other hand, the coefficient $2p$ of x_1 is of no importance for the shape of the planar section. It can be arbitrarily varied (remaining $\neq 0$) by a similarity mapping $\bar{x} \to k\bar{x}(k \neq 0)$, which gives the same result as going over to a new intersecting plane parallel to $R\tilde{e}_1 + R\tilde{e}_2$.

With our suppositions $\tilde{e}_1$, $\tilde{e}_2$ being orthogonal, we now need the theorem*

---

* For a proof along the lines followed in this paper see [6].

saying that every ellipse, or hyperbola, has a pair of orthogonal conjugate diameters, and that the analogue holds for the parabolas. We can now assume that in (24) the basis vectors are orthogonal as in (45'). Furthermore we use the fact that in the "vertex equation" (45), the coefficient of x_1^2, augmented by 1 is the square of the eccentricity. Then our calculations above show that an orthogonal cone with distance h between vertex and the plane of the circular base (of radius 1) gives as planar sections all the ellipses and parabolas, and those hyperbolas with eccentricity at most $\sqrt{1+h^{-2}}$.

Justus Liebig-Universität
Giessen, F.R.G.

BIBLIOGRAPHY

[1] R. Baer, 'Free mobility and orthogonality', *Trans. Am. Math. Soc.* **68** (1950), 439–460.
[2] M. Barner and F. Flohr, 'Die Kegelschnitte in vektorieller Behandlung', *Der Mathematikunterricht* **9** (1963), Heft 3, 89–111.
[3] G. Papy, *Géométrie affine plane et nombres réels*, Bruxelles, 1962.
[4] G. Papy, *Mathématique moderne* 1, 2. Bruxelles, 1963, 1965.
[5] G. Pickert, 'Deduktive Geometrie im Gymnasialunterricht', *Math. Phys. Semesterberichte* **10** (1964), 202–223.
[6] G. Pickert, 'Bilinearformen und Kegelschnitte', Les Répercussions de la Recherche Mathématique sur l'Enseignement, Séminaire C.I.E.M. Echternach 1965, pp. 177-191.
[7] H. Prade, 'Affine Geometrie im Mittelstufenunterricht des Gymnasiums', *Der Mathematikunterricht* **12** (1966), Heft 5, 5–36.

ANDRÉ REVUZ

THE POSITION OF GEOMETRY IN
MATHEMATICAL EDUCATION

What I am aiming at is to define a tentative general strategy for the teaching of geometry at the elementary and secondary levels, and in doing so I shall put problems more than I shall give solutions, but I think that the first task is to know where the problems are and what they are.

History has given geometry a particular position. Being the first mathematical theory that has been axiomatized, geometry became in the past for the mathematicians a model that they wanted to imitate in other fields of mathematics without being very often able to do so. To explore a question "more geometrico" was an ideal in mathematics and even outside of mathematics. The mentality of the men of the past centuries, up to the 18th, for whom geometry and perhaps arithmetics too, was the only domain where mathematical reasoning apparently took a perfect form is still reigning in the mind of many, if not almost all teachers of mathematics of the secondary school (at least, in continental Europe!).

Another belief which is still alive is that geometry does not study a model, and perhaps a rather good one indeed of the "real space" in which we live and act, but is the study of this very "real space" itself. This second belief which is of a more philosophical than mathematical nature, and the first one, the essential root of which is the ignorance of the state of mathematics in our century, are the main obstacles that meets everyone who wants to work for a better mathematical education at the secondary level.

If we want to discuss really the problem of the teaching of geometry, one can't avoid the question: Is it sensible to consider today that there exists a relatively independent part of mathematics which is called geometry?

As a mathematician, I will certainly answer: No! But, if I am asked if geometry must be taught, I feel compelled to say: Yes!

This apparent contradiction comes from the fact that it is not at all clear what must be understood under the name geometry. The narrowest meaning is that geometry is the theory of euclidean spaces, that is finite dimensional linear spaces over the field or real numbers, on which a scalar product is defined. There is no question about the importance of this structure, which has many applications in physics and mechanics, and there is no question, too, about the necessity of teaching at the secondary level geometry in this narrow sense.

But, is not the general idea of what geometry is or may be much richer and much wider?

To go on, I want to give what seems to me the general framework in which mathematics comes to existence and develops itself, and the explanation of why mathematics which is not at all a description of reality may be used to work in real life. This general framework is made of the triples, "situation, model, theory".

It is difficult to give a good definition of what "real life" or "reality" is, but there is no problem about the existence of a world in which we live, act and struggle, which we try to understand, but which remains ever richer than our whole science, and which will never be completely transparent neither for our eyes nor for our mind.

A situation is a portion of reality which is separated from its environment and which we want to consider in itself. It may be roughly described as a particular domain in which we want to act, and at the same time, as the nature of the action that we have in mind.

To consider a situation is a first process of abstraction: considering a situation, we keep only from reality the features that we think relevant to our action. By nature, a situation can't be strictly determined, and rather than of abstracting, one should speak of isolating, for a certain time, the situation.

The real abstraction process begins with the building of models. A model is a schematization of the situation to its true essential features, which must be described with mathematical terms, in order to make possible the use of mathematical tools to study it. Many models can be imagined for one situation, and many different situations may be represented by the same model. A difficult task is to choose, if possible, the best model. The qualities that are to be claimed from a model are:

(i) The easiness of use. For instance, linear spaces give a much easier model of space than the clumsy one which is to be found in Euclid's work.

(ii) The multivalence, i.e., the possibility for the model of bearing account for many different situations, and in that respect, too, linear spaces are multivalent models.

(iii) The adequacy of the model to the situation .The quality of this adequacy is not a mathematical question, but it is a vital question for many sciences, and it is not always sufficiently stressed and studied. If one uses a somewhat inadequate model, because of its conveniency and in spite of its inadequacy, one must be aware of the danger of drawing definitive conclusions about reality from the study of such a model.

The theory appears when one goes further in the process of abstraction and forgets all references to the situations, to study only the structure of the models in itself.

It is quite clear that the work of the mind goes in two opposite directions: from given situations to the creation of models and theories, but also from known theories to models for given new situations. The knowledge of theories may lead to the choice of models (if too much knowledge has not choked all creativity!), or give ideas to build new ones.

One must also remark that a man can be confronted with a situation of which he sees only certain aspects, and that the building of a model may compel him to throw a more acute look on the situation and discover features of which, at the beginning, he was not aware. To know more may help to see more, and for instance, the teachers of the kindergarten are not always aware of all the mathematics that may be found in the spontaneous work of their pupils.

A sound mathematical education should give a very important place to the passage in both opposite directions from situation to model and theory and from theory to situation and model.

Two dangers constantly threaten the teaching of mathematics:

(1) to make confusion between situations and models, and to study a model as if it were the situation itself, which prevents to consider any other one.

(2) to separate too strongly situation and model, and to disregard completely one of them. One can study models without ever making any reference to the situations which they schematize, one can also remain in the situation without seriously seeking to build a model, and without going out of the situation to think in a mathematical fashion. Mathematization is certainly a very important activity of human thought, and as such, must be stressed in mathematical education. But mathematization is a transition process from situation to model. Not to start from the situation, or to dwell in the situation without emerging of it, are two heavy mistakes, and unfortunately too, not very rare mistakes.

From the above point of view, one must ask, speaking of geometry, what are the situations, the models and the theories that belong to this domain which is called geometry.

As in every case, the most difficult is to know what the situations are, and in particular, what is this intuitive space that we perceive and in which we live. This problem is not to be solved by the mathematician alone, but with the aid of the psychologist, the sociologist (because it is clear that our ideas about space are not so "natural" as we think, but are deeply influenced by the culture in which we live), and of people who work with space as architects, mechanicians and also sculptors and painters. The main thing I should like to stress is the richness and the variety of situations that come under the name of space, and of which many are ignored by the classical geometry.

When speaking of intuitive space, we often think that it is a very good study for elementary schools and kindergarten, and that's right, but many think too that their study being done at that level, secondary teachers have got rid of it and have nothing to do with it. This view seems to me to be wrong and twofold wrong:

(1) intuitive space must not be neglected at any stage of education;

(2) models must intervene and practically do so quite automatically at the earliest stages of education.

At the other end, one can draw a list of mathematical theories that may be abstracted from geometrical models. Here is such a list and surely not an exhaustive one: (1) Linear Algebra; (2) Hilbert Spaces; (3) Topology (both general and algebraic topology); (4) Measure Theory; (5) Group Theory, including topological groups; (6) Lattice Theory; and, (7) naturally all kinds of "geometries": differential geometry, algebraic geometry.

A very important point is that each of these theories has something to de with "geometry" but that none of these has to do only with geometry. This means that geometry can't any longer be taught as an independent part of mathematics, and that the adjective "geometrical" must be applied to situations and models rather than to mathematical theories.

Concerning the above list, one must ask: How far must each of these theories penetrate in secondary school teaching? In my opinion, no definite answer should be given. Nevertheless a tentative answer is that all of these should be prepared and that some of these should be studied in an elementary form (at least, finite dimensional linear spaces and Hilbert spaces, measure theory, group theory). And now, I meet the hardest problem: How must we go, in teaching "geometry", from the situations to the theories. There are certainly many ways, and more than a good way, but I should like to emphasize some principles that can serve as landmarks to guide the teaching.

(1) Geometry is not to be taught separately, but in narrow connection with all parts of mathematics taught at the primary and secondary level.

(2) One must use as many approaches of real space as possible and for each approach build models: topological models, metric models, affine models, numerical models. The building of a theory that gives account of all these models will be the crowning of the work, and the motivation of a totally deductive theory.

(3) One must begin the teaching of geometry in the kindergarten. Every living creature has to place himself in space, to appreciate distances, direc-

tions, shapes, motions, deformations.... One must try to help the children to get the richest concrete experience of space. In fact, we may say that some pupils of the kindergarten have a richer geometrical experience than older pupils, but in general this experience is lost during the primary school years where nothing or nonsense is taught about geometry. And yet, at the very early stage, it is sure that children build very easily and spontaneously models of spatial relations. One should follow them and let them consider what part of reality is represented in the models and of what use the models can be.

(4) In the conditions of teaching that we have in almost every country, a totally deductive treatment of geometry can't take place before the age of 15–16. I don't know if this is a definitive situation, but I think that it must not mean that every kind of deduction is forbidden before this age. I believe that every opportunity of having the children make some deduction is to be used. To compel to do unmotivated deduction is as bad as to refuse deduction if the children want it, but the problem is that official syllabi very often make the choice once and for all.

I should like to make a final remark about the use of geometrical terms in many fields of mathematics, and the presence of geometrical images in our minds, even in fields where they seemed to be irrelevant. Such facts are often recorded to plead for the teaching of geometry in its most classical way. It should be very interesting to discuss the nature of these "geometrical images" which are present in the mind of the mathematician who is seeking for the solution of a problem, and even of the geometrical images that we cultivated when we were solving problems of geometry, and which were surely deeply different from the figures we drew on a paper.

It seems indeed impossible for many mathematicians to work without creating and contemplating in their mind images, which are often felt as being of a spatial nature, and which, yet, have little to do with the real space or with the mathematical space of euclidean geometry. The images that flash in the mind of the working mathematician are perhaps nearer to Picasso's paintings than to the draught of a mechanical engine. These images, even when they help to solve a geometrical problem, represent in the mind of the mathematician possibilities of action rather than concrete objects, and geometry is not for them the only source. In mathematical work, and in mathematical teaching, what is important is the activity of the mind, and these so-called geometrical images are the witnesses of this activity and are not bound to the study of concrete space.

Université de Paris
Paris, France

THOMAS G. ROOM

THE GEOMETRY AND ALGEBRA OF REFLECTIONS
(AND OF 2×2 MATRICES)

This outline of a segment of the course in mathematics prescribed in New South Wales (Australia) for the mathematically best 4 per cent of students in the final two years of high school* is based on the "Notes for Teachers" published by the State Education Department.

The aim of this segment of the course is to introduce students to matrix algebra in a way which presents this algebra as a piece of mathematics that grew out of attempts to provide solutions to a class of mathematical problems in a form in which the true nature of a problem was not overwhelmed by a mass of complicated and repetitive algebraic manipulation.

For such a course, in which facility in matrix manipulation has to be acquired as well as an understanding of the significance of matrix operations, it seems best to restrict the geometrical problems to which the matrix methods are to be applied to the rectangular Cartesian plane with a fixed origin. The matrices can then be restricted to having no more than two rows and columns. The properties of these small matrices do not differ in kind from those of larger (square) matrices, while the proofs of these properties can often be simpler and more direct than those required to establish them in the general case. It is possible therefore to strike quite deep into matrix territory and to provide, for example, an account of eigenvalues and similarity against a tangible background of realizable geometry.

Almost all the students taking the course may be expected to continue their mathematical studies at a university, and they should have acquired a firm basis for a first year course in linear algebra.

I. TRANSFORMATIONS OF THE PLANE

In relation to a fixed origin O we define a set of geometrical *operations* each of which associates with each point P in the plane another point P', i.e., we *transform* the plane into itself or *map* the plane onto itself.

(i) *Rotations ("Turns")*

For a given origin O and an angle of given magnitude α, the point P' asso-

---

* The organization of high school courses in Australia is described briefly in Appendix A.

ciated with each point P is defined by:

$$\angle \,{}^{*}POP' = \alpha \quad \text{(measure of an angle)}$$

$${}^{*}OP' = {}^{*}OP \quad \text{(signed length)}.$$

Write this relation between P and P' as

$$P' = \mathscr{J}_{\alpha}P.$$

(Initially we may write $\mathscr{J}_{\alpha}(P)$, since $\mathscr{J}_{\alpha}$ is the symbol of a geometric function of the points P, but the parentheses can be omitted without introducing ambiguity.)

$\mathscr{J}_{\alpha}$ is the symbol representing the operation of rotation which transforms P into P'. We refer to $\mathscr{J}_{\alpha}$ as an operator. From geometry we have

$$\mathscr{J}_{\beta}(\mathscr{J}_{\alpha}P) = \mathscr{J}_{\alpha+\beta}P = \mathscr{J}_{\alpha}(\mathscr{J}_{\beta}P).$$

Note the order of operations and operators: "$\mathscr{J}_{\alpha}$" followed by "$\mathscr{J}_{\beta}$" has to be written $\mathscr{J}_{\beta}(\mathscr{J}_{\alpha}P)$, and again we may omit the parentheses and write $\mathscr{J}_{\beta}\mathscr{J}_{\alpha}P$. In this case the operator $\mathscr{J}_{\beta}\mathscr{J}_{\alpha}$ which is the "resultant of $\mathscr{J}_{\alpha}$ followed by $\mathscr{J}_{\beta}$" (or the "result of compounding $\mathscr{J}_{\beta}$ with $\mathscr{J}_{\alpha}$") is the same as $\mathscr{J}_{\alpha}\mathscr{J}_{\beta}$.

Since the relation

$$\mathscr{J}_{\alpha}\mathscr{J}_{\beta}P = \mathscr{J}_{\alpha+\beta}P$$

is valid for all points P in the plane, we could write it simply as a relation among the operators,

$$\mathscr{J}_{\beta}\mathscr{J}_{\alpha} = \mathscr{J}_{\alpha+\beta} = \mathscr{J}_{\alpha}\mathscr{J}_{\beta}.$$

Now introduce the *identity operator* $\mathscr{J}$, with the property

$$\mathscr{J}P = P$$

for all P. Then from geometry

$$\mathscr{J}_{-\alpha}\mathscr{J}_{\alpha} = \mathscr{J}$$

and we can introduce the inverse operator $\mathscr{J}_{\alpha}^{-1}$ with the definition $\mathscr{J}P_{\alpha} = P' \Rightarrow P = \mathscr{J}_{\alpha}^{-1}P'$, so that

$$\mathscr{J}_{\alpha}^{-1} = \mathscr{J}_{-\alpha}.$$

$\mathscr{J}_{\alpha}^{-1}$ is the unique rotation which compounded with $\mathscr{J}_{\alpha}$ produces $\mathscr{J}$.

Combination of rotations

$$\mathscr{J}_{2\pi} \;\; = \mathscr{J}_0 = \mathscr{J},$$
$$\mathscr{J}_{2\pi+\alpha} = \mathscr{J}_{\alpha},$$
$$(\mathscr{J}_{\alpha})^{n} \;\; = \mathscr{J}_{n\alpha},$$
etc.

Groups of rotations.

(ii) *Reflections ("Symmetries")*

In relation to a given line a through O we define the operation $\mathscr{S}_a$ of reflection in such a way that, for any point P,

$$P' = \mathscr{S}_a P \overset{\text{def}}{\Leftrightarrow} (PP' \perp a \text{ and } a \text{ bisects } \ulcorner PP' \urcorner).$$

P' is the *reflection* of P in a.

Properties:

(1) $\mathscr{S}_a(\mathscr{S}_a P) = P$ for all P,

i.e., $\mathscr{S}_a^2 = \mathscr{I}$ or $\mathscr{S}_a^{-1} = \mathscr{S}_a$.

(2) $\mathscr{S}_b \mathscr{S}_a = \mathscr{I}_{2 \angle \,^*ab}$,

were $\angle \,^*ab$ is the measure of the angle from a to b, and $0 \leq \angle \,^*ab < \pi$. $\angle \,^*ab + \angle \,^*ba = 2\pi$.

$$\mathscr{S}_a \mathscr{S}_b = \mathscr{I}_{2 \angle \,^*ba} = \mathscr{I}_{2(\pi - \angle \,^*ab)} = \mathscr{I}_{-2 \angle \,^*ab}$$
$$= (\mathscr{I}_{2 \angle \,^*ab})^{-1} = (\mathscr{S}_b \mathscr{S}_a)^{-1}.$$

In fact, since

$$\mathscr{S}_a^2 = \mathscr{I},$$
$$\mathscr{S}_a \mathscr{S}_b \mathscr{S}_b \mathscr{S}_a = \mathscr{I}$$

and therefore

$$\mathscr{S}_a \mathscr{S}_b = (\mathscr{S}_b \mathscr{S}_a)^{-1}.$$

(3) Given a, b, c through O there is a single line d through O such that

$$\mathscr{S}_c \mathscr{S}_b \mathscr{S}_a = \mathscr{S}_d$$

(d is defined by $\angle \,^*ab = \angle \,^*dc$)

THEOREM. $\mathscr{S}_a \mathscr{S}_b \mathscr{S}_c = \mathscr{S}_c \mathscr{S}_b \mathscr{S}_a$.

(4) The fixed points under the operation $\mathscr{S}_a$ are the points of the line a, i.e.,

$$\mathscr{S}_a K = K \Leftrightarrow K \in a.$$

The line is a *point-by-point (pointwise) invariant*.

If $h \perp a$, and $H \in h$, then

$$H' = \mathscr{S}_a H \Rightarrow H' \in h$$

and conversely, so that h, $h \neq a$, is overall invariant under $\mathscr{S}_a$ if and only if $h \perp a$.

(iii) *Translations ("Displacements")*

If A, B are any two distinct points they determine an interval $\ulcorner AB \urcorner$ and a transformation $\mathscr{D}_{AB}$ of the plane in which, for any point $P \notin AB$

$$P' = \mathscr{D}_{AB}P \overset{\text{def}}{\Leftrightarrow} PP' \,\|\, AB \text{ and } PA \,\|\, PB';$$

if $P \in AB$, $\mathscr{D}_{AB}P$ is defined in two stages.

Properties:

(1) $\mathscr{D}_{AB}P \;\;= Q \Rightarrow \mathscr{D}_{PQ} = \mathscr{D}_{AB}$.

(2) $\mathscr{D}_{BC}\mathscr{D}_{AB} = \mathscr{D}_{AC} = \mathscr{D}_{AB}\mathscr{D}_{BC}$.

(3) $\mathscr{D}_{AB}\mathscr{D}_{BA} = \mathscr{I}$; $\mathscr{D}_{AB}^{-1} = \mathscr{D}_{BA}$.

(4) $\mathscr{D}_{AB}\mathscr{S}_a \;\;= \mathscr{S}_a\mathscr{D}_{A'B'}$ where $A' = \mathscr{S}_a A$, $B' = \mathscr{S}_a B$.

(iv) *Congruence Transformations*

The effect of an operator $\mathscr{I}$, $\mathscr{S}$, or $\mathscr{D}$ on any geometrical figure is to produce another geometrical figure, which, from geometrical considerations, is congruent to the original figure. The operators $\mathscr{I}$, $\mathscr{S}$, and $\mathscr{D}$ can therefore be described as generating transformations of the plane into itself or *maps* of the plane onto itself, and in particular as generating *congruence transformations*. The explicit properties which remain invariant under these transformations are:

	transforms into	
a point		a point
a line		a line
parallel lines		parallel lines
perpendicular lines		perpendicular lines
an interval of length d		an interval of length d
an angle of magnitude α		an angle of magnitude α but sense is reversed by reflection.

II. CONGRUENCE AND REFLECTIONS

There is very little logically developed geometry in the syllabus for the earlier years in the high school in New South Wales; the opportunity is taken here to provide practice in geometrical reasoning in the context of congruence transformations.

(a) Groups associated with the regular polygons in terms of $\mathscr{S}$ and $\mathscr{I}$.

(b) Reduction of a congruence transformation to the resultant of a sequence of three or fewer reflections.

From simple geometrical arguments, we can prove the following three relations:

(i) $a \cap b = 0 \Rightarrow \mathscr{S}_b\mathscr{S}_a = \mathscr{I}_{\angle \, *ab}$

(ii) $(l \perp k, \, m \perp k, \, l \cap k = L, \, m \cap k = M, \, N = \mathscr{D}_{LM}M)$

$$\Rightarrow \mathscr{S}_m\mathscr{S}_l = \mathscr{D}_{LN}.$$

That is, the resultants of reflections in two distinct lines is a
displacement if the lines are parallel and a rotation if they are not.

(iii)   If l, m, n are three distinct concurrent lines, then there is a unique
line k concurrent with them such that $\mathscr{S}_m\mathscr{S}_l=\mathscr{S}_n\mathscr{S}_k$.

Using these results we can reduce any sequence of transformations $\mathscr{S}$, $\mathscr{J}$, $\mathscr{D}$
to a sequence of three or fewer reflections. (The canonical reduction to a
glide-reflection or a rotation or a displacement is not included in the syllabus.)
   Appendix B contains two questions set in recent examinations.

III. CARTESIAN FORM OF ROTATIONS, REFLECTIONS: 2×2 MATRICES

(i) *Rotation* (Figure 1)

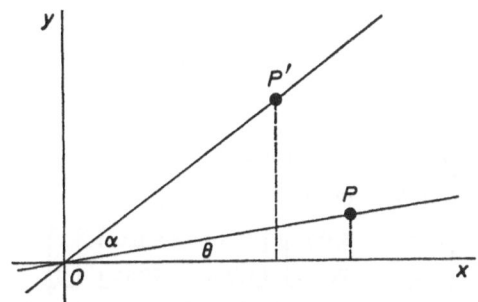

Fig. 1.

P is $(r \cos \theta, r \sin \theta)$
P' is $(r \cos (\theta + \alpha), r \sin (\theta + \alpha))$

so that

$$x' = r(\cos \theta \cos \alpha - \sin \theta \sin \alpha)$$
$$y' = r(\cos \theta \sin \alpha + \sin \theta \cos \alpha)$$

i.e.,

$$x' = \cos \alpha \cdot x - \sin \alpha \cdot y$$
$$y' = \sin \alpha \cdot x + \cos \alpha \cdot y.$$ (1)

Let us invent an algebraic operator $\mathbf{T}_\alpha$, and write these equations in the form

$$\begin{bmatrix} x' \\ y' \end{bmatrix} = \begin{bmatrix} \cos \alpha & -\sin \alpha \\ \sin \alpha & \cos \alpha \end{bmatrix} \begin{bmatrix} x \\ y \end{bmatrix}$$ (2)

or

$$\mathbf{r}' = \mathbf{T}_\alpha \mathbf{r},$$ (3)

where

$$\mathbf{r} = \begin{bmatrix} x \\ y \end{bmatrix}, \quad \mathbf{r}' = \begin{bmatrix} x' \\ y' \end{bmatrix}$$

are *vectors*, *column-vectors*, or *column-matrices*, and

$$T_\alpha = \begin{bmatrix} \cos\alpha & -\sin\alpha \\ \sin\alpha & \cos\alpha \end{bmatrix}$$

is the matrix of two *rows* $[\cos\alpha, \ -\sin\alpha]$ $[\sin\alpha, \ \cos\alpha]$ and two *columns*

$$\begin{bmatrix} \cos\alpha \\ \sin\alpha \end{bmatrix} \begin{bmatrix} -\sin\alpha \\ \cos\alpha \end{bmatrix} .$$

The symbolic relation (3) is a condensed form of (2) which in turn corresponds exactly to the two explicit Equations (1).

The matrix T_α is an *algebraic operator* transforming the vector **r** into the vector **r'**.

Let us follow the next step in the account of the geometric operators to discover how to combine the algebraic operators. I.e., consider the algebraic equivalent of

$$P'' = \mathcal{J}_\beta P' = \mathcal{J}_\beta \mathcal{J}_\alpha P .$$

$$\begin{bmatrix} x'' \\ y'' \end{bmatrix} = \begin{bmatrix} \cos\beta & -\sin\beta \\ \sin\beta & \cos\beta \end{bmatrix} \begin{bmatrix} x' \\ y' \end{bmatrix}$$

$$= \begin{bmatrix} \cos\beta & -\sin\beta \\ \sin\beta & \cos\beta \end{bmatrix} \begin{bmatrix} \cos\alpha & -\sin\alpha \\ \sin\alpha & \cos\alpha \end{bmatrix} \begin{bmatrix} x \\ y \end{bmatrix}$$

$$= \begin{bmatrix} \cos\beta\cos\alpha - \sin\beta\sin\alpha & \cos\beta(-\sin\alpha) - \sin\beta\cos\alpha \\ \sin\beta\cos\alpha + \cos\beta\sin\alpha & \sin\beta(-\sin\alpha) + \cos\beta\cos\alpha \end{bmatrix}$$

$$\times \begin{bmatrix} x \\ y \end{bmatrix}$$

$$= \begin{bmatrix} \cos(\alpha+\beta) & -\sin(\alpha+\beta) \\ \sin(\alpha+\beta) & \cos(\alpha+\beta) \end{bmatrix} \begin{bmatrix} x \\ y \end{bmatrix}$$

This is exactly $T_\beta T_\alpha \mathbf{r} = T_{\alpha+\beta}\mathbf{r}$ and gives the rules which have to be followed if the "matrices" T_α are to correspond exactly to the geometric operators $\mathcal{J}_\alpha$. This is the line of thought which led Cayley to the invention of matrices to represent the operators and matrix multiplication to represent the compounding of two operators.

$$T_0 = \begin{bmatrix} 1 & 0 \\ 0 & 1 \end{bmatrix} = 1, \quad \text{the unit matrix}.$$

$$T_{\frac{1}{2}\pi} = \begin{bmatrix} 0 & -1 \\ 1 & 0 \end{bmatrix}, \quad T_\pi = -1.$$

(ii) *General* 2×2 *Matrices*

The rule for multiplication shows up most clearly if we use double subscripts.

Take, in a matrix $\mathbf{A}$, a_{ij} to be the *element* in the *row* numbered i and *column* numbered j, so that $\mathbf{A}$ can be written as

$$\mathbf{A} = \begin{bmatrix} a_{11} & a_{12} \\ a_{21} & a_{22} \end{bmatrix},$$

then,

$$\begin{array}{cc} \text{column 1} & \text{column 2} \end{array}$$

$$\mathbf{AB} = \begin{bmatrix} a_{11}b_{11} + a_{12}b_{21} & a_{11}b_{12} + a_{12}b_{22} \\ a_{21}b_{11} + a_{22}b_{21} & a_{21}b_{12} + a_{22}b_{22} \end{bmatrix} \begin{array}{l} \text{row 1} \\ \text{row 2} \end{array}$$

or, if $\mathbf{AB} = \mathbf{C}$, then

$$c_{rs} = a_{r1}b_{1s} + a_{r2}b_{2s}.$$

Clearly in general

$$\mathbf{AB} \neq \mathbf{BA}.$$

THEOREM. Matrix multiplication is *associative*.

(iii) *Reflections*

If $y = x \tan \theta$ is the line l, write $\mathscr{S}_\theta$ for $\mathscr{S}_l$, and $\mathbf{S}_\theta$ for the matrix corresponding to $\mathscr{S}_\theta$. Then

$$\mathbf{S}_0 = \begin{bmatrix} 1 & 0 \\ 0 & -1 \end{bmatrix}, \qquad \mathbf{S}_{\frac{1}{4}\pi} = \begin{bmatrix} 0 & 1 \\ 1 & 0 \end{bmatrix}, \qquad \mathbf{S}_{\frac{1}{2}\pi} = \begin{bmatrix} -1 & 0 \\ 0 & 1 \end{bmatrix}.$$

$\mathbf{S}_\theta$ can be computed directly from the Cartesian diagram, but it is simpler to use the relation

$$\mathscr{S}_\theta \mathscr{S}_0 = \mathscr{J}_{2\theta},$$

that is

$$\mathscr{S}_\theta = \mathscr{J}_{2\theta} \mathscr{S}_0,$$

so that

$$\mathbf{S}_\theta = \mathbf{T}_{2\theta} \mathbf{S}_0 = \begin{bmatrix} \cos 2\theta & -\sin 2\theta \\ \sin 2\theta & \cos 2\theta \end{bmatrix} \begin{bmatrix} 1 & 0 \\ 0 & -1 \end{bmatrix}$$

$$= \begin{bmatrix} \cos 2\theta & \sin 2\theta \\ \sin 2\theta & -\cos 2\theta \end{bmatrix}.$$

(iv) *The Affine Transformation*

This transformation we define algebraically as a generalization of $\mathbf{S}$ and $\mathbf{T}$, and then investigate its geometric properties. The discussion depends on the section formulae:

$$x_3 = \frac{k_1 x_1 + k_2 x_2}{k_1 + k_2}, \qquad y_3 = \frac{k_1 y_1 + k_2 y_2}{k_1 + k_2},$$

which we can write as

$$(k_1 + k_2)\,\mathbf{r}_3 = k_1\mathbf{r}_1 + k_2\mathbf{r}_2,$$

or

$$k_1\,(\mathbf{r}_3 - \mathbf{r}_1) = k_2\,(\mathbf{r}_2 - \mathbf{r}_3).$$

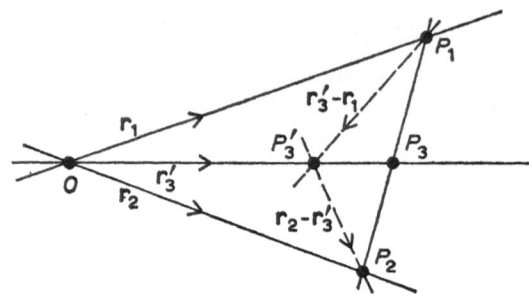

Fig. 2.

$$\boxed{k_1\,(\mathbf{r}_3' - \mathbf{r}_1) = k_2\,(\mathbf{r}_2 - \mathbf{r}_3')} \;\Rightarrow\; \boxed{k_1 * P_1P_3' = k_2 * P_3'P_2}$$

$$\boxed{P_1P_3' \parallel P_3'P_2} \qquad \left[\begin{array}{c}\text{for some value} \\ \text{of } \dfrac{k_1}{k_2}\end{array}\right]$$

$$\boxed{P_1, P_2, P_3' \text{ are collinear.}}$$

In matrix form the section formula can be written as:

$$\begin{bmatrix}(k_1 + k_2)\,x_3 \\ (k_1 + k_2)\,y_3\end{bmatrix} = \begin{bmatrix}x_1 & x_2 \\ y_1 & y_2\end{bmatrix}\begin{bmatrix}k_1 \\ k_2\end{bmatrix}$$

or

$$(k_1 + k_2)\,\mathbf{r}_3 = [\mathbf{r}_1 \quad \mathbf{r}_2]\,\mathbf{k}.$$

The transformation we are to consider is

$$\mathbf{r}' = \mathbf{Mr},$$

where

$$\mathbf{M} = \begin{bmatrix}m_{11} & m_{12} \\ m_{21} & m_{22}\end{bmatrix}.$$

When this is applied to the points of a line

$$(k_1 + k_2)\,\mathbf{r} = [\mathbf{r}_1\,\mathbf{r}_2]\,\mathbf{k},$$

$(k_1, k_2$ vary along the line) we find

$$\mathbf{M}\{(k_1 + k_2)\,\mathbf{r}\} = \mathbf{M}[\mathbf{r}_1\,\mathbf{r}_2]\,\mathbf{k},$$

which we can rewrite as

$$(k_1 + k_2)\, \mathbf{Mr} = [\mathbf{Mr}_1\ \mathbf{Mr}_2]\, \mathbf{k},$$

i.e., the transformed points are the points of the line joining the points $\mathbf{Mr}_1$, $\mathbf{Mr}_2$.

Write $\mathcal{M}$ for the corresponding geometric transformation. Then, if $\mathcal{M}P_i = P_i$ and $P_3 \in P_1 P_2$, we have $P_3' \in P_1' P_2'$ and

$$\frac{{}^*P_1 P_3}{{}^*P_1 P_2} = \frac{{}^*P_1' P_2'}{{}^*P_1' P_2'}.$$

Thus, under the transformation $\mathcal{M}$:

points	become	points
lines	become	lines
ratios of lengths of intervals on a line	become	equal ratios on the transformed line

so that

parallels	become	parallels.

But *distances* are altered and *angles* are altered.

IV. SOME PROPERTIES OF MATRICES UNDER MULTIPLICATION

(i) *The Inverse Matrix*

$\mathbf{M}^{-1}$ is defined by $\mathbf{M}^{-1}\mathbf{M} = 1$.

If

$$\mathbf{M} = \begin{bmatrix} a & b \\ a & d \end{bmatrix} \quad \text{and} \quad \mathbf{M}^{-1} = \begin{bmatrix} x & y \\ z & t \end{bmatrix}$$

we have

$$\begin{bmatrix} x & y \\ z & t \end{bmatrix} \begin{bmatrix} a & b \\ c & d \end{bmatrix} = \begin{bmatrix} 1 & 0 \\ 0 & 1 \end{bmatrix}$$

$$ax + cy = 1 \qquad az + ct = 0$$
$$bx + dy = 0 \qquad bz + dt = 1,$$

so that

$$\mathbf{M}^{-1} = \begin{bmatrix} \dfrac{d}{ad-bc} & \dfrac{-b}{ad-bc} \\[2ex] \dfrac{-c}{ad-bc} & \dfrac{a}{ad-bc} \end{bmatrix}$$

$$\mathbf{M}\mathbf{M}^{-1} = \mathbf{M}^{-1}\mathbf{M} = 1.$$

Thus $\mathbf{M}^{-1}$ exists if and only if $ad - bc \neq 0$.

THEOREM. If $\mathbf{A}$ and $\mathbf{B}$ are any two (non-singular) matrices,

$$(\mathbf{AB})^{-1} = \mathbf{B}^{-1}\mathbf{A}^{-1}.$$

(ii) *The Determinant of a Matrix*

Notation:

$$ad - bc = \det \mathbf{M} = \begin{vmatrix} a & b \\ c & d \end{vmatrix}.$$

If $\det \mathbf{M} = 0$, the matrix is singular (i.e., it has no inverse). Note that $\det \mathbf{T} = 1$, $\det \mathbf{S} = -1$, so that these matrices are always non-singular.

If the matrix $\mathbf{M} = \begin{bmatrix} a & b \\ c & d \end{bmatrix}$ is singular, then the affine transformation

$$\mathbf{M} \begin{bmatrix} x_0 \\ y_0 \end{bmatrix} = \begin{bmatrix} ax_0 + by_0 \\ cx_0 + dy_0 \end{bmatrix} = \begin{bmatrix} ax_0 + by_0 \\ (ax_0 + by_0)\,c/a \end{bmatrix}$$

(provided $a \neq 0$), is such that for all points $\mathbf{r}$ the transformed point $\mathbf{Mr}$ lies on the line $cx - ay = 0$, i.e., the *transformation is singular*, the whole plane being *mapped onto* the line.

By direct multiplication we find for any two matrices $\mathbf{M}$, $\mathbf{M}'$

$$\det(\mathbf{MM}') = \det \mathbf{M} \det \mathbf{M}' = \det(\mathbf{M}'\mathbf{M}),$$
$$\det(\mathbf{M}^{-1}) = (\det \mathbf{M})^{-1}.$$

(iii) *The Zero Vector and Zero Matrix*

$$\mathbf{o} = \begin{bmatrix} 0 \\ 0 \end{bmatrix} \qquad \mathbf{0} = \begin{bmatrix} 0 & 0 \\ 0 & 0 \end{bmatrix}.$$

For all $\mathbf{M}$, $\mathbf{Mo} = \mathbf{o}$, and $\mathbf{M0} = \mathbf{0M} = \mathbf{0}$.

The most general singular matrix may be written as

$$\begin{bmatrix} rr' & rs' \\ r's & ss' \end{bmatrix}.$$

For this matrix we have

$$\begin{bmatrix} rr' & rs' \\ r's & ss' \end{bmatrix} \begin{bmatrix} us' \\ -ur' \end{bmatrix} = \mathbf{o} \text{ for any } u.$$

THEOREM. The following conditions are equivalent:
  (i) $\mathbf{M}$ is singular,
  (ii) there exists a non-zero $\mathbf{u}$ such that $\mathbf{Mu} = \mathbf{o}$.

$$\begin{bmatrix} rr' & rs' \\ r's & ss' \end{bmatrix} \begin{bmatrix} us' & vs' \\ -ur' & -vr' \end{bmatrix} = \mathbf{0} \text{ for any } u, v,$$

$$\begin{bmatrix} hs & -hr \\ ks & -kr \end{bmatrix} \begin{bmatrix} rr' & rs' \\ r's & ss' \end{bmatrix} = \mathbf{0} \text{ for any } h, k.$$

Thus in matrix algebra there exist pairs of matrices $\mathbf{M}$ and $\mathbf{N}$, both non-zero but such that $\mathbf{MN}=0$. If we are given $\mathbf{AB}=0$ we cannot deduce that either $\mathbf{A}$ or $\mathbf{B}$ is the zero matrix, but only that, if neither $\mathbf{A}$ nor $\mathbf{B}$ is the zero matrix, then both are singular.

(iv) *The Transposed Vector and Matrix*

If

$$\mathbf{a} = \begin{bmatrix} a_1 \\ a_2 \end{bmatrix}$$

we write $\mathbf{a}^T = [a_1, a_2]$

for the *transpose* of $\mathbf{a}$, i.e., the *row-vector* with components identical with those of the column-vector $\mathbf{a}$. Likewise if

$$\mathbf{M} = \begin{bmatrix} a & b \\ c & d \end{bmatrix}, \qquad \mathbf{M}^T \stackrel{\text{def}}{=} \begin{bmatrix} a & c \\ b & d \end{bmatrix}.$$

To the theorem above we can add:

The condition, (i) $\mathbf{M}$ is singular, is equivalent to (iii) there exists a non-zero $\mathbf{v}$ such that

$$\mathbf{v}^T\mathbf{M} = \mathbf{o}^T.$$

If $\mathbf{A}$ is any non-singular matrix:

$$(\mathbf{A}^T)^{-1} = (\mathbf{A}^{-1})^T.$$

If $\mathbf{A}$ and $\mathbf{B}$ are any two matrices, and $\mathbf{k}$ any vector, then

$$(\mathbf{AB})^T = \mathbf{B}^T\mathbf{A}^T$$
$$(\mathbf{Ak})^T = \mathbf{k}^T\mathbf{A}^T.$$

In particular $\mathbf{S}_\theta^T = \mathbf{S}_\theta^{-1} = \mathbf{S}_\theta$, $\mathbf{T}_\alpha^T = \mathbf{T}_\alpha^{-1} = \mathbf{T}_{-\alpha}$.

(v) *Uniqueness of* $\mathbf{S}$ *and* $\mathbf{T}$

THEOREM. If under an affine transformation, with fixed origin, distance is invariant, then the transformation is either a rotation or a reflection.

We have to have

$$x'^2 + y'^2 \equiv x^2 + y^2,$$

i.e.,

$$\mathbf{r}^T\mathbf{r} = \mathbf{r}'^T\mathbf{r}' = (\mathbf{Mr})^T\,\mathbf{Mr} = \mathbf{r}^T\,(\mathbf{M}^T\mathbf{M})\,\mathbf{r}.$$

$$\mathbf{M}^T\mathbf{M} = \begin{bmatrix} a & c \\ b & d \end{bmatrix} \begin{bmatrix} a & b \\ c & d \end{bmatrix} = \begin{bmatrix} a^2 + c^2 & ab + cd \\ ab + cd & b^2 + d^2 \end{bmatrix}$$

$$\mathbf{r}^T\mathbf{M}^T\mathbf{Mr} = (a^2 + c^2)\,x^2 + 2\,(ab + cd)\,xy + (b^2 + d^2)\,y^2$$
$$\mathbf{r}^T\mathbf{r} = x^2 + y^2.$$

Thus, if distance is invariant,

$$a^2 + c^2 = 1 \\ b^2 + d^2 = 1 \qquad ab + cd = 0.$$

Take $a = \cos\alpha$, $c = \sin\alpha$, $b = \cos\beta$, $d = \sin\beta$, then $\cos(\alpha - \beta) = 0$, i.e., either either $\beta = \alpha - \frac{1}{2}\pi$ or $\beta = \alpha + \frac{1}{2}\pi$ and the two possible matrices are

$$\begin{bmatrix} \cos\alpha & \sin\alpha \\ \sin\alpha & -\cos\alpha \end{bmatrix} = \mathbf{S}_{\frac{1}{2}\alpha}$$

$$\begin{bmatrix} \cos\alpha & -\sin\alpha \\ \sin\alpha & \cos\alpha \end{bmatrix} = \mathbf{T}_{\alpha}.$$

A matrix with the property $\mathbf{M}^T\mathbf{M} = 1$, i.e., $\mathbf{M}^T = \mathbf{M}^{-1}$, is called orthogonal.

V. DISPLACEMENT, MATRIX ADDITION, AND MATRIX ALGEBRA

If H is (h, k) and $\mathscr{D}_{OH}P = P'$ where P is (x, y) and P' is (x', y'), we have

$$\begin{bmatrix} x' \\ y' \end{bmatrix} = \begin{bmatrix} x + h \\ y + k \end{bmatrix},$$

which we may write as

$$\begin{bmatrix} x' \\ y' \end{bmatrix} = \begin{bmatrix} x \\ y \end{bmatrix} + \begin{bmatrix} h \\ k \end{bmatrix}$$

or as

$$\mathbf{r}' = \mathbf{r} + \mathbf{h}.$$

This suggests an addition operation for *vectors*, and then immediately an addition operation for *matrices*. If

$$\mathbf{A} = \begin{bmatrix} a_{11} & a_{12} \\ a_{21} & a_{22} \end{bmatrix} \qquad \mathbf{B} = \begin{bmatrix} b_{11} & b_{12} \\ b_{21} & b_{22} \end{bmatrix}$$

we define $\mathbf{A} + \mathbf{B}$ by

$$\mathbf{A} + \mathbf{B} \overset{\text{def}}{=} \begin{bmatrix} a_{11} + b_{11} & a_{12} + b_{12} \\ a_{21} + b_{21} & a_{22} + b_{22} \end{bmatrix}.$$

Thus

$$\mathbf{A} + \mathbf{A} = \begin{bmatrix} 2a_{11} & 2a_{12} \\ 2a_{21} & 2a_{22} \end{bmatrix} \overset{\text{def}}{=} 2A;$$

we define similarly $n\mathbf{A}$, then if $\mathbf{B} = n\mathbf{A}$, $\mathbf{A} = (1/n)\,\mathbf{B}$, and so through rational to real scalar multiples of $\mathbf{A}$. For any number k,

$$k\mathbf{A} = \begin{bmatrix} ka_{11} & ka_{12} \\ ka_{21} & ka_{22} \end{bmatrix} = \begin{bmatrix} k & 0 \\ 0 & k \end{bmatrix} \begin{bmatrix} a_{11} & a_{12} \\ a_{21} & a_{22} \end{bmatrix}$$

$$\det(k\mathbf{A}) = k^2 \det\mathbf{A}.$$

Combination of Sums and Products

We can now work out the rules for the combination of sums and products of matrices; we find that they are the same as those for a field except that:

    (1) multiplication is not commutative;

    (2) not every element has a multiplicative inverse.

VI. THE CHARACTERISTIC FUNCTION, EIGENVALUES, EIGENVECTORS, THE CAYLEY-HAMILTON THEOREM

(i) The Characteristic Function of $A = \begin{bmatrix} a & b \\ c & d \end{bmatrix}$ is

$$\det(x\mathbf{1} - A) = x^2 - (a + d)x + \det A.$$

The roots of $\det(x\mathbf{1} - A) = 0$ are the *eigenvalues*; for these values, the vectors such that $(x\mathbf{1} - A)\mathbf{u} = \mathbf{o}$ are the *eigenvectors*.

(ii) *Cayley-Hamilton Theorem*

The matrix A satisfies

$$A^2 - (a + d)A + (\det A)\mathbf{1} = \mathbf{0}.$$

Thus in forming matrix polynomials

    A^2 may be replaced by $(a + d)A - (\det A)\mathbf{1}$;
    A^3 by $(a + d)\{(a + d)A - (\det A)\mathbf{1}\} - (\det A)A$; etc.

In particular, for

$$M = \begin{bmatrix} rr' & rs' \\ r's & ss' \end{bmatrix}$$

$$M^2 = (rr' + ss')M.$$

VII. CHANGES OF COORDINATES

(i) *Parallel Shift of Axes*

$$\mathbf{r}' = \mathbf{r} + \mathbf{h}$$

may be interpreted *either* as a transformation of the plane in which $\mathbf{r}$ is transformed into $\mathbf{r}'$, *or* as a change of coordinates with new origin at $-\mathbf{h}$ and axes parallelly displaced.

(ii) *Rotation, Reflection*

$$\mathbf{r}' = T_\alpha \mathbf{r}, \qquad \mathbf{r}' = S_l \mathbf{r}$$

may be interpreted *either* as transformations of the plane *or* as changes of axes by rotation through $-\alpha$, reflection in l.

(iii) *Affine Change*

$$\mathbf{r'} = \mathbf{Mr}, \qquad \mathbf{M} = \begin{bmatrix} a & b \\ c & d \end{bmatrix}, \qquad \mathbf{M} \text{ non-singular}.$$

Regarded as a change of coordinate systems we have (see Figure 3).

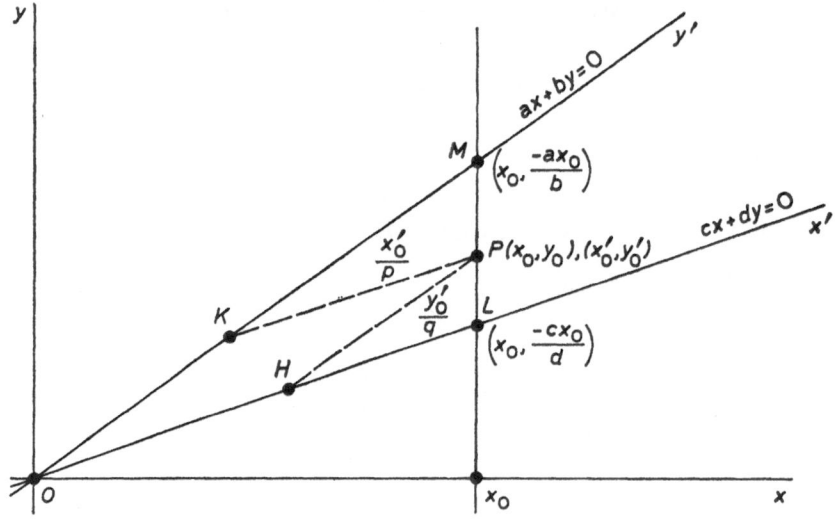

Fig. 3.

$$\mathbf{L} \text{ is } \left(x_0, \frac{-cx_0}{d} \right),$$

$$\mathbf{M} \text{ is } \left(x_0, \frac{-ax_0}{d} \right)$$

$$x_0' = ax_0 + by_0 = p \cdot {}^*KP$$
$$y_0' = cx_0 + dy_0 = q \cdot {}^*HP$$

where p and q are the constants:

$$p = \frac{ad - bc}{\sqrt{c^2 + d^2}}, \qquad q = \frac{ad - bc}{\sqrt{a^2 + b^2}}.$$

The new axes are arbitrary lines, $cx+dy=0$, and $ax+by=0$, through the origin, and the new coordinates are proportional to constant multiples of the distances of the point from the new axes.

VIII. COMBINATION OF CHANGE OF AXES AND TRANSFORMATION: "SIMILARITY" OF MATRICES

Let us take an affine transformation of the plane in which the coordinates of P' and P are connected by

$$\mathbf{r}' = \mathbf{Mr},$$

and a change of coordinates from (x, y), $\mathbf{r}$ to (X, Y), $\mathbf{R}$ where

$$\mathbf{R} = \mathbf{Cr}, \quad \text{and} \quad \mathbf{r} = \mathbf{C}^{-1}\mathbf{R},$$

$\mathbf{C}$ being non-singular. In this new coordinate system, the coordinates (X', Y') of the point with original coordinates (x', y') are also given by $\mathbf{R}' = \mathbf{Cr}'$ so that $\mathbf{R}' = \mathbf{Cr}' = \mathbf{CMr} = \mathbf{CMC}^{-1}\mathbf{R}$. Thus, if in the new coordinates, the transformation from P to P' is represented by

$$\mathbf{R}' = \mathbf{NR},$$

then

$$\mathbf{N} = \mathbf{CMC}^{-1},$$

$\mathbf{N}$ is the transform of $\mathbf{M}$ by $\mathbf{C}$, and the matrices $\mathbf{M}$ and $\mathbf{N}$ are described as similar.

THEOREM. If $\mathbf{L}$ is similar to $\mathbf{M}$ and $\mathbf{M}$ is similar to $\mathbf{N}$, then $\mathbf{L}$ is similar to $\mathbf{N}$. "Similarity" is reflexive, symmetric, and transitive.

The principal algebraic theorem for similar matrices states a condition which is sufficient but not necessary, although it is only in the "exceptional" cases that the converse theorem is not valid. In the usual applications made of the theorem only the sufficiency condition is required.

THEOREM. If two matrices are similar, then their characteristic functions are identical, i.e.,

$$\det(x\mathbf{1} - \mathbf{M}) \equiv \det(x\mathbf{1} - \mathbf{CMC}^{-1}).$$

There is no theorem: if two matrices have the same characteristic function, then they are similar. For example, the set of matrices similar to $\alpha\mathbf{1}$, namely, $\mathbf{C}(\alpha\mathbf{1})\mathbf{C}^{-1}$, consists of the single matrix $\alpha\mathbf{1}$ itself. The characteristic function of $\alpha\mathbf{1}$ is $(x-\alpha)^2$, which is the characteristic function of any matrix of the set

$$\left\{ \begin{bmatrix} \alpha & u \\ 0 & \alpha \end{bmatrix}, \alpha \text{ fixed, } u \text{ arbitrary} \right\}.$$

Such a matrix is similar to $\alpha\mathbf{1}$ only if $u=0$. We shall prove in fact (for 2×2 matrices, there is no corresponding simple result for larger matrices) that the only set of exceptional characteristic functions is $\{(x-\alpha)^2\}$. We have however the following weaker theorem valid for all 2×2 matrices, and in fact for all squares matrices.

THEOREM. If $M = \begin{bmatrix} a & b \\ c & d \end{bmatrix}$ and $N = \begin{bmatrix} a' & b' \\ c' & d' \end{bmatrix}$ have the same characteristic function, i.e., if M and N are such that

$$a + d = a' + d'$$

and

$$ad - bc = a'd' - b'c',$$

then there exists a non-empty set of non-zero matrices $\{C\}$ such that

$$CM = NC.$$

The theorem fails to be the complete converse because if M and N are to be similar, we require also that C should be non-singular. If for a given M and N all members of $\{C\}$ are singular, then M and N are not similar – there is no C for which $CMC^{-1} = N$.

PROOF. Assume that $C = \begin{bmatrix} x & y \\ z & t \end{bmatrix}$ and that $CM = NC$. Write

$$CM - NC = P = \begin{bmatrix} p & q \\ r & s \end{bmatrix}$$

then

$$CM = NC \Leftrightarrow P = 0.$$

That is,

$$p \equiv hx + cy - b'z = 0$$
$$q \equiv bx - ky - b't = 0$$
$$r \equiv -c'x + kz + ct = 0$$
$$s \equiv -c'y + bz - ht = 0,$$

where

$$h = a - a' = -(d - d')$$
$$k = a - d' = a' - d,$$

and

$$bc - b'c' = ad - a'd' = -hk.$$

We have the identities

$$bp + b's \equiv hq, \qquad c'p + cs \equiv -hr,$$

so that, if $h \neq 0$,

$$(p = 0 \text{ and } s = 0 \text{ and } h \neq 0) \Rightarrow (q = 0 \text{ and } r = 0).$$

Similarly

$$(q = 0 \text{ and } r = 0 \text{ and } k \neq 0) \Rightarrow (p = 0 \text{ and } s = 0).$$

Thus for $h \neq 0$,

$$C_1 = \begin{bmatrix} \dfrac{b'z - cy}{h} & y \\ z & \dfrac{bz - c'y}{h} \end{bmatrix}$$

and for $k \neq 0$,

$$C_2 = \begin{bmatrix} x & \dfrac{bx - b't}{k} \\ \dfrac{c'x - ct}{k} & t \end{bmatrix}$$

both satisfy the condition for all pairs of values (y, z) and (x, t), respectively.

The matrix C_1 is singular if

$$bb'z^2 - (bc + b'c' + h^2)\, yz + cc'y^2 = 0$$

and all matrices of the set $\{C_1\}$ are singular if

$$bb' = 0, \qquad cc' = 0, \quad \text{and} \quad bc + b'c' + h^2 = 0.$$

We already have the relation

$$bc - b'c' + hk = 0,$$

so that

$$bc = -\tfrac{1}{2}h(h + k), \qquad b'c' = -\tfrac{1}{2}h(h - k).$$

Since $h \neq 0$, the only two sets of solutions of the equations are:

$$b = c = h + k = 0, \quad \text{which imply} \quad a = d,$$

and

$$b' = c' = h - k = 0, \quad \text{which imply} \quad a' = d',$$

for which respectively

$$\mathbf{M} = a\mathbf{1} \quad \text{and} \quad \mathbf{N} = a'\mathbf{1}.$$

The same result is obtained from the conditions that all matrices of the set $\{C_2\}$ are singular. Thus, provided either $h \neq 0$ or $k \neq 0$, the only sets of pairs of matrices $\mathbf{M}$ and $\mathbf{N}$ having the same characteristic function which are such that *every* matrix $\mathbf{C}$ which satisfies the condition $\mathbf{CM} = \mathbf{NC}$ is singular are

$$\mathbf{M} = a\mathbf{1}, \qquad \mathbf{N} \neq a\mathbf{1}$$

and

$$\mathbf{M} \neq a'\mathbf{1}, \qquad \mathbf{N} = a'\mathbf{1}.$$

We have now to examine the case $h = 0$, $k = 0$, that is:

$$a = d = a' = d' \quad \text{and} \quad bc = b'c'.$$

We have

$$p_0 \equiv cy - b'z, \qquad s_0 \equiv -c'y + bz$$
$$q_0 \equiv bx - b't, \qquad r_0 \equiv -c'x + ct,$$

so that

$$bp_0 + b's_0 \equiv 0, \quad \text{and} \quad c'q_0 + br_0 \equiv 0.$$

Again therefore only two of the four equations are independent and the possible forms of the matrix $\mathbf{C}$ are, for $b \neq 0$,

$$\mathbf{C}_3 = \begin{bmatrix} \dfrac{b't}{b} & y \\ \dfrac{c'y}{b} & t \end{bmatrix}$$

and for $c \neq 0$,

$$\mathbf{C}_4 = \begin{bmatrix} x & \dfrac{b'z}{c} \\ z & \dfrac{c'x}{c} \end{bmatrix}$$

(which are singular for all (y, t) and (x, z) if and only if $b' = c' = 0$) and corresponding matrices for $b' \neq 0$, $c' \neq 0$, which are always singular if and only if $b = c = 0$. Thus again, the only cases in which all matrices of the set $\{\mathbf{C}\}$ are singular are

$$\mathbf{M} = a\mathbf{1}, \qquad \mathbf{N} \neq a\mathbf{1}$$
$$\mathbf{M} \neq a'\mathbf{1}, \qquad \mathbf{N} = a'\mathbf{1}.$$

We have now exhausted all the possibilities.

THEOREM. (i) Matrices for which the characteristic function is not the square of a linear form.

Σ_1 is the set of matrices

$$\left\{ \begin{bmatrix} a & b \\ c & d \end{bmatrix} : \begin{matrix} a, b, c, d \text{ arbitrary subject} \\ \text{only to } (a-d)^2 + 4bc \neq 0 \end{matrix} \right\}.$$

Any two members of Σ_1 are similar if and only if their characteristic functions are identical.

(ii) Matrices for which the characteristic function is the square of a linear form (other than x^2).

(α) Σ_2 is the set of matrices

$$\left\{ \begin{bmatrix} a & b \\ 0 & a \end{bmatrix} : \begin{matrix} a, b \text{ arbitrary sub-} \\ \text{ject only to } b \neq 0 \end{matrix} \right\} \cup \left\{ \begin{bmatrix} d & 0 \\ c & d \end{bmatrix} : \begin{matrix} c, d \text{ arbitrary sub-} \\ \text{ject only to } c \neq 0 \end{matrix} \right\}.$$

Any two members of Σ_2 are similar if and only if their characteristic functions are identical, that is, their leading diagonals are the same.

(β) Σ_2' is the set of matrices

$\{a\mathbf{1} : a \text{ arbitrary}\}$.

Each member of Σ_2' is similar only to itself.

It is necessary to emphasize that the theorem in this form is peculiar to 2×2 matrices, the full statement for larger matrices is much more complicated. It would never of course be expected that students would "learn" the proofs of the converse theorem, but the sufficiency theorem and its method of proof are of considerable importance. The value in this course of the converse theorem lies in the way it illustrates the vagaries of the solutions of linear equations in relation to conditions which are sufficient but not necessary. Another facet of the same algebra is the determination of the matrices which commute with a given matrix, i.e., of the set C of matrices for which, for a given matrix M, $MC = CM$.

IX. QUADRATIC FORMS, CONICS

(The treatment using only 2×2 matrices is clumsier than that using 3×3 matrices, but gives significance to the eigenvalues and eigenvectors, and provides useful practice in matrix manipulation.)

The quadratic form

$$\Gamma(x, y) \equiv ax^2 + 2hxy + by^2 + 2gx + 2fy + c$$

$$\equiv [x, y] \begin{bmatrix} a & h \\ h & b \end{bmatrix} \begin{bmatrix} x \\ y \end{bmatrix} + 2[g, f] \begin{bmatrix} x \\ y \end{bmatrix} + c$$

$$\equiv \mathbf{r}^T C \mathbf{r} + 2\mathbf{u}^T \mathbf{r} + c, \quad \text{where} \quad C^T = C.$$

The objective is to show that by shifting the origin and rotating the (rectangular) axes the form can be reduced either to $\alpha x^2 + \beta y^2 + \gamma$ or to $\lambda y^2 + \mu x$.

(i) $C = \det C \neq 0$. The reduced form is $\alpha x^2 + \beta y^2 + \gamma$, where α, β are the eigenvalues of C (and are real when C is real).

First step. Shift the origin so that new coordinate vectors $\mathbf{r}'$ are given by $\mathbf{r} = \mathbf{r}' + \mathbf{r}_0$, where $\mathbf{r}_0 = -C^{-1}\mathbf{u}$.

$$\mathbf{r}^T C \mathbf{r} + 2\mathbf{u}^T \mathbf{r} + c = (\mathbf{r}' + \mathbf{r}_0)^T C(\mathbf{r}' + \mathbf{r}_0) + 2(\mathbf{r}' + \mathbf{r}_0)^T \mathbf{u} + c$$

$$= \mathbf{r}'^T C \mathbf{r}' + 2\mathbf{r}'^T (C\mathbf{r} + \mathbf{u}) + \mathbf{r}_0^T (C\mathbf{r}_0 + 2\mathbf{u}) + c$$

$$= \mathbf{r}'^T C \mathbf{r}' + 0 - (C^{-1}\mathbf{u})^T \mathbf{u} + c$$

$$= \mathbf{r}'^T C \mathbf{r}' - \mathbf{u}^T C^{-1}\mathbf{u} + c$$

$$= [x', y'] \begin{bmatrix} a & h \\ h & b \end{bmatrix}$$

$$\qquad - [g, f] \begin{bmatrix} b & -h \\ -h & a \end{bmatrix} \begin{bmatrix} g \\ f \end{bmatrix} C^{-1} + c$$

$$= ax'^2 + 2hx'y' + by'^2 + \gamma,$$

where

$$\gamma = \frac{\Delta}{C}$$

and

$$\Delta = abc + 2fgh - af^2 - bg^2 - ch^2.$$

Second step. Rotate the axes, that is, take new coordinate-vectors $\mathbf{r}''$ given by $\mathbf{r}' = \mathbf{T}_\theta \mathbf{r}''$ choosing θ so that $\mathbf{r}'^T \mathbf{C} \mathbf{r}'$ becomes $\mathbf{r}''^T \mathbf{D} \mathbf{r}''$ where

$$\mathbf{D} = \begin{bmatrix} \alpha & 0 \\ 0 & \beta \end{bmatrix}.$$

$$\mathbf{r}'^T \mathbf{C} \mathbf{r}' = \mathbf{r}''^T \mathbf{T}_\theta^T \mathbf{C} \mathbf{T}_\theta \mathbf{r}'',$$

that is,

$$\mathbf{D} = \mathbf{T}_\theta^T \mathbf{C} \mathbf{T}_\theta = \mathbf{T}_\theta^{-1} \mathbf{C} \mathbf{T}_\theta.$$

Thus $\mathbf{C}$ and $\mathbf{D}$ are similar and from their characteristic functions we have

$$a + b = \alpha + \beta, \quad \text{or} \quad ab - h^2 = \alpha\beta.$$

That is, α, β are the roots of $\xi^2 - (a+b)\xi + ab - h^2$. Since $(a+b)^2 - 4(ab - h^2) = (a-b)^2 + 4h^2 > 0$ for all relevant a, b, h (when $\mathbf{C}$ is real), the roots α, β are always real. $\{\alpha, \beta\}$ is the pair of eigenvalues of $\mathbf{C}$ and we may associate them in either order with x'', y''. If we associate α with x'', then under the transformation $\mathbf{r}' = \mathbf{T}_\theta \mathbf{r}''$ we have

$$ax'^2 + 2hx'y' + by'_2 + \gamma \quad \text{becomes} \quad \alpha x''^2 + \beta y''^2 + \gamma.$$

The directions of the (x'', y'')-axes referred to the (x', y')-axes, and therefore also to the (x, y)-axes, are given by the eigenvectors corresponding to the eigenvalues α, β. An eigenvector $\mathbf{e}$ corresponding to α satisfies $\begin{bmatrix} a-\alpha & h \\ h & b-\alpha \end{bmatrix} \mathbf{e} = \mathbf{0}$, and $\mathbf{e}$, that is $y'' = 0$, is therefore parallel to $y = hx/(x-\alpha)$ and the angle of rotation is given by $\tan\theta = h/(a-\alpha)$.

(ii) $C = \det \mathbf{C} = 0$. Without using 3×3 matrices any expression in matrix form of the reduction to $\lambda y^2 + \mu x$ is laboured and artificial, and it is better to follow the standard textbook process.

(iii) No detailed study of conics is required. It will be sufficient to identify the curves $\alpha x^2 + \beta y^2 + \gamma = 0$ and $\lambda y^2 + \mu x = 0$ with those given by the focus-direction definition and to investigate the shapes of these curves.

X. COMPLEX NUMBERS AS MATRICES
(to supplement the usual treatment of the Argand diagram)

$$z = x + iy = r(\cos\theta + i\sin\theta)$$

$$z = \begin{bmatrix} x \\ y \end{bmatrix} = r \begin{bmatrix} \cos\theta \\ \sin\theta \end{bmatrix}$$

$$Z = \begin{bmatrix} x & -y \\ y & x \end{bmatrix} = r \begin{bmatrix} \cos\theta & -\sin\theta \\ \sin\theta & \cos\theta \end{bmatrix}$$

$$z'' = z + z' \leftrightarrow z'' = z + z' \leftrightarrow Z'' = Z + Z'$$

$$z'' = zz' \leftrightarrow z'' = Zz' \leftrightarrow Z'' = ZZ'$$

$\dfrac{1}{z}$ corresponds to Z^{-1}. $\qquad Z\bar{Z} = (x^2 + y^2)\,\mathbf{1}$.

If α is a point with coordinate vector $\alpha = \begin{bmatrix} a \\ b \end{bmatrix}$ on the Argand diagram and

$\lambda = \varrho(\cos\theta + i\sin\theta)$, then $\lambda\alpha$ corresponds to $\varrho \begin{bmatrix} \cos\theta & -\sin\theta \\ \sin\theta & \cos\theta \end{bmatrix} \begin{bmatrix} a \\ b \end{bmatrix}$, i.e., the

point obtained by rotation and multiplication by ϱ in the usual way.

APPENDIX A. NOTE ON THE ORGANIZATION OF HIGH SCHOOL COURSES IN AUSTRALIA (1968)

Each of the six States (populations ranging from four and a half million in New South Wales to 0.4 million in Tasmania) runs its own highly centralized system of education. The organization of school courses in the various States follows much the same pattern, but there is considerable difference in detail. In each State there is a public examination at the end of the high school course which provides the main criteria for entry to the universities and other tertiary institutions. The age up to which attendance is compulsory is 15. The whole system is currently under close scrutiny and the Federal Government seems anxious to become involved.

In New South Wales the high school course is divided into a four-year general course (ages $11+$ to $15+$) followed by a two-year course (ages $15+$ to $17+$). There is, at present, from the first year onwards, separation by subjects into three streams.

In the fifth and sixth years of high school a student takes five or more subjects each weighted at 6 units in the lower two streams and 8 units in the upper stream, a unit being equivalent to one nominal 40-minute period in a nominal 40-period week in a nominal 40-week year. (For calculating how much ground can be reasonably expected to be covered, each nominal 40 is treated as an actual 35.) In the middle stream alternative fuller courses in mathematics and science are offered each counting as $1\frac{1}{2}$ subjects (9 units). In the upper stream only "full" courses in mathematics and science are offered each counting $1\frac{3}{4}$ subjects (11 units). In 1968 the numbers of candidates sitting for the final examinations in mathematics were approximately:

Lower stream	Middle stream		Upper stream	Total	Total for whole exam.
	1-sub.	1½-sub.			
7000	8000	6000	1000	22000	25000

The syllabuses for the mathematics courses are devised so that the middle and upper streams can be taught together for five or six periods a week. The segment of the upper course described in this paper is much the largest connected part of the course which is studied by students in the upper stream and not by those in the middle stream, but the expectation is that as teachers become more familiar with this area of mathematics some of the material may be introduced into the middle stream course.

APPENDIX B. SOME QUESTIONS SET ON GEOMETRY AND MATRIX ALGEBRA IN THE 1967 AND 1968 FINAL EXAMINATIONS

(1) In a plane, in relation to a given coordinate system with origin O, P_i is the point represented by the column-vector $\mathbf{r}_i$. Let $\mathbf{r}' = \mathbf{Mr}$, where $\mathbf{M}$ is a non-singular 2×2 matrix, be the equation of a transformation of the plane onto itself in which P_i is transformed into P_i'. Prove that:

(i) if P_1, P_2, P_3 are collinear, then

(a) P_1', P_2', P_3' are collinear, and

(b) $$\frac{{}^*P_1'P_2'}{{}^*P_1'P_3'} = \frac{{}^*P_1P_2}{{}^*P_1P_3};$$

(ii) area $\Delta OP_1'P_2' = (\text{area } \Delta OP_1P_2) \cdot \det \mathbf{M}$.

(2) $\mathscr{D}$ is the operation of displacement through one (positive) unit parallel to the x-axis in a rectangular coordinate system, origin O; $\mathscr{S}$ is the operation of reflection in the line $y = x$. Draw a sketch to show the points $\mathscr{D}O$, $\mathscr{D}^2O$, $\mathscr{S}\mathscr{D}O$, $\mathscr{D}\mathscr{S}\mathscr{D}O$, $\mathscr{S}\mathscr{D}\mathscr{S}\mathscr{D}O$.

Prove that, for any sequence of non-negative integers r, s, t, ...,

$$\dots \mathscr{D}^t\mathscr{S}\mathscr{D}^s\mathscr{S}\mathscr{D}^rO = \mathscr{D}^m\mathscr{S}\mathscr{D}^nO,$$

where m, n are non-negative integers. Describe the set of points which corresponds to the set of ordered pairs $\{(m, n)\}$.

(3) $\mathscr{S}_a$, $\mathscr{S}_b$, $\mathscr{S}_c$ are the operations of reflection in the lines a, b, c, respectively, where a, b, c contain the sides of an equilateral triangle. Draw a diagram (not on graph paper) which shows the lines e and f, and a segment determining the displacement $\mathscr{D}$, which are such that

(a) $\mathscr{S}_a\mathscr{S}_b\mathscr{S}_a = \mathscr{S}_e$

(b) $\mathscr{S}_b\mathscr{S}_a\mathscr{S}_b = \mathscr{S}_f$

(c) $\mathscr{S}_c\mathscr{S}_b\mathscr{S}_a = \mathscr{D}\mathscr{S}_b.$

(4) The matrices $\mathbf{J}$, $\mathbf{K}$ are

$$\mathbf{J} = \begin{bmatrix} 0 & 1 \\ -1 & 0 \end{bmatrix}, \qquad \mathbf{K} = \begin{bmatrix} 0 & i \\ i & 0 \end{bmatrix}, \quad \text{where } i^2 = -1.$$

Prove that the set of eight matrices $\{\pm 1, \pm \mathbf{J}, \pm \mathbf{K}, \pm \mathbf{JK}\}$ forms a group under the operation of matrix multiplication.

The Open University
Bletchley, Bucks
England

SEYMOUR SCHUSTER

ON THE TEACHING OF GEOMETRY
A POTPOURRI

I. INTRODUCTION

My understanding is that this International Conference on the Teaching of Geometry will address itself to the improvement of geometry courses. Naturally, vitalizing and modernizing the content of the courses will have the strongest impact on the projected improvement. However, content depends on some prior conditions, namely, on the philosophy and orientation of the courses; in simple terms, on the purposes of the course.

I intend to make some critical remarks on the purpose(s) and content of existing geometry courses. In doing so, I shall put forth opinions and judgments – many without documentation – based on personal experience and interaction with teachers and students from various parts of the U.S.A. over a period of approximately 15 years. I hope that these judgments are intelligent and don't raise the ire of too many Conference participants. However, it may be that a bit of abrasiveness on the part of some of us will precipitate more heated discussions of the group, in which case we will all gain. I shall therefore not hesitate to offer my "seat-of-the-pants" appraisals and conclusions.

In addition to criticism, I shall offer some alternative purposes and goals for geometry courses, and some consequences (virtues) of this alternative point of view. Finally, I will put forth specific mathematical suggestions to implement the alternative philosophy and to attain the new goals.

II. THE LAST HALF-CENTURY

The secondary schools of the U.S.A. now have – and have had for most of the twentieth century – Euclidean Geometry as the content of their geometry curriculum. This subject matter often consisted of emasculated versions of Euclid's *Elements*, with dependent (redundant) axioms to make the arguments shorter and circular reasoning of the sort that the Greeks would never have tolerated.

With twentieth century perspective, especially since Hilbert's *Grundlagen der Geometrie*, all mathematicians have been sensitive to the logical weaknesses in Euclid's *Elements*. Those few who were concerned with school geometry were also sensitive to the even greater weaknesses of the textbooks.

But it wasn't until after World War II that mathematicians were sufficiently influential to bring to the high schools versions of Euclidean Geometry that were more in accord with modern standards of logical rigor. The reformers wrote more careful developments, paid attention to the notions of order, treated congruence in some proper manner (without superposition), and enriched the content of the course.

Thus, the current situation is that Euclidean Geometry in some axiomatic form, remains the principal content of school geometry courses.

III. THE PHILOSOPHY

When we ask questions like "Why Euclidean Geometry?" and "What purpose does it serve?", the answer is twofold: (1) For historical reasons and otherwise Euclidean Geometry is regarded as a (the?) means of showing students how to build an axiomatic system, of teaching them the deductive character of mathematics, and training them to write formal proofs; (2) Euclidean Geometry is useful – indeed necessary – for science and engineering.

The first reason is stated first because it has been the most influential in the teaching of geometry. The primary focus of the teaching has been on constructing the formal systems: the postulates, definitions and formal proofs. The important thing to note is that this focus has not been changed by the reform of recent years. The reform assumed the existing philosophy, namely that tenth-year geometry was to be devoted to an axiomatic development of Euclidean Geometry, and exerted its energies to preserve Euclidean Geometry as a model of logical thinking. To be sure, some of the new books have enriched the content of the course in other ways, and many of the mathematicians associated with the reform had broader goals, but the mold was cast and it was hard to break. Thus, the reformers did not change the basic aim of the high school course nor were they successful in altering the spirit in which the geometry was to be studied.

IV. CRITICISM

I wish to argue that focusing on the formal structure of geometry is a serious mistake which has had some unfortunate results.

(1) The program has failed; that is, the goal of teaching formal structure via geometry has not been achieved.

It is my experience and the experience reported by others that few students come away from high school geometry with anything beyond a superficial understanding of, and appreciation for, an axiomatic structure. Paul Kelly is said to have made an informal survey with the conclusion that the thing most people recall from their high school geometry courses is that geometry is the subject in which proofs are written in two columns separated by a line on the page.

Some mathematicians argue that the program was doomed from the start because of the inherent nature of the subject. For example, Albert Wilansky believes that axiomatic Euclidean Geometry belongs in graduate school, Abraham Seidenberg confesses that he once "killed" a Berkeley course by doing parts of Hilbert's *Grundlagen*, and Murray Klamkin asserts that Euclidean Geometry is too rich and complicated to serve for conveying the meaning of formal structure and formal proof.

Although I am not in possession of the data, I have been told by mathematics educators that there is substantial documentation of the fact that large numbers of high school students memorize proofs they don't really understand, come to believe that the only acceptable proofs are those given in "step-reason" form (echoing Paul Kelly's finding), and in spite of the fact that they have heard of non-Euclidean Geometry, still believe that the axioms they memorized are "self-evident truths" and are *the* axioms for Euclidean Geometry rather than *a* set of axioms for the subject.

Klamkin argues, as have others before him, that elementary group theory or Boolean algebra, which require simpler axiom systems, are more appropriate as subjects for conveying the meaning of formal structure and formal proof.

(2) The overemphasis on axiomatics has resulted in an underemphasis on applications.

While lip service is paid to the fact that geometry is used extensively by physical scientists and also by biological and social scientists – indeed, by anyone who mathematizes at all – very little is done about this in the actual teaching of the subject. Klamkin, in his paper 'On the Ideal Role of an Industrial Mathematician and Its Educational Applications' (reprinted in *Educ. Studies in Mathematics*, Vol. 3 (1971), 244–269), states that "the present applications are very few and artificial". Any perusal of current textbooks bears out Klamkin's assertion.

Not only are applications of the results of geometry slighted, but so are the methods of geometric conceptualization. Geometrizing constitutes a very powerful tool for the mathematician and applied scientist alike. It is an instrument for gaining insight and intuitive understanding in problems that may come from mechanics, electrical network theory, thermodynamics, ecology, function theory – almost anywhere. Yet training in geometrization of physical phenomena is either totally absent or barely present in the current teaching of geometry.

(3) Not enough concern is given to three- (and higher-)dimensional concepts.

Admittedly, the reasons for slighting geometry of three-space are perhaps several; but I place much of the blame on the overemphasis on the axiomatic structure. My argument is that since the study of formal structure and proof can be (and has been) satisfied by studying plane geometry, there is little need to go into higher dimensions.

The fact that the study of three-space has been so slighted is further evidence that there was little concern for teaching geometry for its usefulness in science and engineering. There is the obvious fact that three-dimensional concepts are essential for applications; but aside from this, it should be recognized that students are being robbed at a crucial time in their lives of the practice of reasoning about space and developing the intuition necessary for the analysis of higher-dimensional problems of science and mathematics.

(4) Other important topics, appropriate to secondary school work, are slighted. Among these are: combinatorial aspects of geometry, transformations, symmetry, and topological aspects of geometry. Though it is not a *bona fide* topic of geometry, problem-solving is one of the serious casualties in the current teaching of geometry. Klamkin is most adamant and eloquent on this point. (See his 'On the Teaching of Mathematics so as to be Useful', *Educational Studies in Mathematics* 1 (1968), 126–160.)

V. ALTERNATIVE PHILOSOPHY

The tradition of reserving the tenth year for geometry and confining nearly all study of geometry to that period has now been broken to some extent. So when we think about the curriculum in geometry, we can think about it quite broadly, beginning even in kindergarten. In addition to this departure, I suggest that we go a step further and drop using geometry as the main vehicle for teaching axiomatic development of mathematics; and then, decide on a new set of priorities for the curriculum.

Among anyone's top priorities is, quite naturally, the requirement that a body of facts is to be learned; the body of facts currently being taught must be enlarged, and this can easily be done. But the real question is: How should these facts be learned, and in what context? The answer depends on the level of the student's experience. "Experience" is a key word. I believe that we should build a curriculum in which the student gets a sequence of *geometric experiences*. Examples of such experiences, which would begin in the lower grades, would include: handling (and studying) polyhedra, dissection of polygons, measurement, play with mirrors, experimenting with linkages, and counting in geometric problems. These examples indicate that

I believe that high priority consideration should go to relating geometry to science, particularly to the physical world.

In line with relating geometry to the physical world, we should place on our list of priorities *training in geometric conceptualizing* or what I earlier referred to as *geometrizing*. Geometry should be learned as an instrument for interpretation of concepts and problems that arise in other branches of knowledge (including other branches of mathematics). Examples of this sort of work would come from studying reflection and refraction, mechanics of forces and velocities, work, laws of chance (area), elementary combinatorial problems, errors in measurement, ratios, and the basic operations of arithmetic and algebra.

I shall now extrapolate from what has been said thus far and state what I propose as a basic point of view or attitude toward a geometry curriculum.

Geometry should be regarded as a body of knowledge that had its roots in the study of physical space and physical objects, and now concerns itself with abstract concepts such as points, lines, curves, surfaces and volumes. The study of geometry over two-plus millennia had resulted in the development of different techniques for the analysis of geometric problems. Among these are the various methods: synthetic, coordinate, analytic, vector, algebraic, combinatorial, and so on. Some problems admit to solution more readily under attack by one method in preference to the others. Each method has its virtues and each its shortcomings. *The curriculum should be developed to teach the variety of techniques for solving geometric problems.*

Finally, among the techniques should be that of transforming to, or interpreting a geometric problem in some other context. For example, mechanical considerations can be utilized to solve geometric problems (see, e.g., V. A. Uspenskii, *Some Applications of Mechanics to Mathematics*, Blaisdell Publishing Co.), electrical networks can often accomplish such results, optics can be used to solve geometric inequalities, and probability (combinatorial analysis) can solve some geometric problems. In short, if concept X has a geometric interpretation, then it is very likely that concept X can be used to solve some geometric problems as well as geometry can be used to solve problems involving concept X.

Now to return to the question of teaching axiomatic structure. As stated earlier, the burden placed on geometry to do this job is too great and the consequences of carrying the load are many and serious. I offer a compromise solution.

The load should be distributed. Algebra could easily be used at the junior high and high school level to share the burden. Probability can do some of the work, and a good look at other parts of the curriculum will certainly turn up other candidates.

But more important than identifying topics that can be developed axiomatically with ease is that we reflect on what should be taught, how to teach it, etc., because a more pedagogically sound approach is necessary.

Axiomatics is learned slowly, so axiomatic treatment should be spread out over the entire curriculum. Great concentration on formal development, as it currently exists in secondary geometry, is overwhelming to students. The compromise I suggest here is that we make geometry *locally axiomatic* in contrast to the *globally axiomatic* treatment that is now given. That is, I believe that students as they grow to mathematical maturity would benefit more from smaller axiomatic systems within the subject of Euclidean Geometry. This point of view might be regarded as a request to teach mathematics the way it existed among the Greeks prior to Euclid. The theory of congruence no doubt existed as a set of interrelated propositions; the theory of ratio, proportion and similarity was developed as a unit by Eudoxus. An examination of the curriculum topics will surely turn up portions that admit to simple axiomatic development. Perhaps a unit on *transversals*, or one on *circles and angle measure*, or another on *parallelograms*. And some of these could be offered at much lower levels than they are now taught precisely because of their simplicity.

Part of the motivation behind my "locally axiomatic" proposal is that I think it is most important to begin early in inculcating in students the notion that the propositions of mathematics are logically interrelated and dependent upon one another. A great deal of this can be done before axioms are ever introduced, and before we say that a logical system must rest on certain unproved propositions and in terms of some undefined terms.

Another thing that motivates my proposal is that the present manner of teaching formal structure doesn't teach students where the primitive frame comes from or (rephrased) how and why the mathematician creates his deductive system the way he does. If teaching at the early stages gives students experience – physical and logical – with the concepts, then there is a much better basis for teaching the important matters in foundations. For example, it is important to see *where* the definition comes from; the study of interrelationships of propositions will turn up necessary and sufficient conditions which, the student should learn, can often be utilized to define a concept in a convenient manner. Another thing of importantce is the idea of generalizing: How is this done? With a view toward what? Most important are the questions of how to abstract, how to choose axioms, how to mathematize a subject that we already have some knowledge about. This sequence of considerations would be closer to real mathematical development than exists in the current curriculum, because students would then learn that axiomatic treatment of a subject is usually the last stage in the creation of a mathematical subject. Moreover, there would be more training in mathematical creativity than can possibly come from studying the finished and highly polished product, and only filling in the gaps that textbook authors select as exercises.

At long last, I return to questions concerning a "globally axiomatic" treatment of Euclidean Geometry. As to the question of "When ", the answer is that nobody knows without far more experimentation with the curriculum at the lower level. From Piaget we have learned that what may be sacrosanct is the sequence in which certain topics are learned; but as for the pace of learning and the age at which given topics can and should be studied, we are still somewhat ignorant. As to the question of how the "globally axiomatic" treatment should be developed, the answer is, "it's up for grabs"; it certainly depends on the geometry program at the lower level. Most recently, the linear algebraic approach has been championed by a number of prominent mathematicians, among whom are Choquet, Dieudonné, Rosenbloom, Snapper and Troyer. While this method provides a major economy, defining Euclidean space very quickly as a real inner-product vector space, the intricacies and subtleties are hidden behind the powerful algebra. Therefore, if this be the method employed for an axiomatic development of Euclidean Geometry, it must be preceded by perhaps several years of groundwork that would lay the basis for this abstract definition. The geometric basis for each of the axioms of a linear vector space must be understood first, and also the importance of the concept of perpendicularity to Euclidean Geometry must be appreciated before the notion of inner product can be very meaningful.

VI. ENRICHING THE CURRICULUM

To begin with, I offer the proposition that *invariance* is one of the most important ideas in all of mathematics, and that geometry is unquestionably the most natural subject for the demonstration and use of this idea. I would therefore like to see geometry serve the curriculum by maintaining the invariance notion as one of its themes, utilizing it as an instrument for problem-solving as well as for the usual purpose of classifying geometries.

In a sense, this notion should be brought into a child's early mathematical experience. Some of Piaget's astounding experiments show that small children don't realize that the size (cardinality) of a set is invariant under physical rearrangement of the elements. Thus, arithmetic concepts cannot possibly have meaning until this hurdle is passed; that is, until an invariance is recognized. True geometric experiences with invariance are easy to generate. For example, reflections in mirrors are familiar to every child, and these can be studied with some seriousness to lead to important geometric understanding at an early age. (Steps in this direction have been taken by the Educational Development Corporation in their experimental programs.)

Before proceeding further along these lines, I must backtrack to get the horse before the cart. One can talk about invariance only after *transformations*. That is, a set or structure is invariant relative to, or under, a transformation. Hence, the subject of *geometric transformations* is essential in developing new curricula. This would have many advantages: there is the natural tieup with algebra and the function concept; the Euclidean transformations can be taught concretely with mirrors and/or physical translations and rotations; translations serve as a basis for an early study of vectors; and the notion of isometry gives the simplest formulation of the idea of congruence. Incidentally, isometry is an excellent example of the mathematizing of a physical concept, for an isometric transformation is a mathematical abstraction (and generalization) of Euclid's idea of superposition.

Once transformations are available, the *symmetry* concept is apparent. For a structure possesses symmetry if it is invariant under some transformation other than the identity. There are many virtues to reaching this subject early in the curriculum: important properties of figures can be gleaned from their symmetries; symmetry concepts are fundamental in science, form the most elementary work on optics to the most sophisticated in elementary particle theory; symmetry principles constitute an important tool for problem solving; finally, with the modern view of symmetries (see E. Artin, *Geometric Algebra*, Interscience Publishers) as transformations that preserve some structure, all geometries are simply studies of symmetries.

In the early grades, Euclidean symmetry can be introduced in many simple ways. For example, if two congruent squares (or other plane figures) are available, one can be taped to the desk top and the other given to the student to fit on top of the taped one. In how many ways can this be done? Mirrors can be used to verify the axes of symmetry. The group concept is concretely available any time the instructor wants to introduce it. Another possibility would be the shifting of strip patterns or plane patterns. Of course, three-dimensional symmetry should be studied concretely through the regular polyhedra; and the students should be taught to abstract by drawing and otherwise describing the planes of reflectional symmetry (since only the rotational symmetries of a polyhedron can be realized by physical handling).

Incidentally, I think that learning mathematics this way would be sheer fun for youngsters.

As students move on, feeling at home with transformations, it would be appropriate to go beyond the Euclidean isometries. The most natural next step would be to introduce first magnification and contraction and then point reflections (all *central dilatations*). This is the beginning of the study of *affine transformations*, which also relate directly to physical concepts, and should be related to and used in physical problems.

Now is the time to study *parallel projection* and, in particular, *orthogonal projection* by means of shadows and diagrams. "What are the invariants?" should be asked. The answer helps to define affine symmetry, which not only explains why spirals are symmetric in a technical sense, but also turns out to be prevalent in biological studies: e.g., phyllotaxis and conchyliometry (see H. S. M. Coxeter, *Introduction to Geometry*, Wiley).

Since I have asserted that invariance should be an ever-present theme in geometry curriculum, I think it advisable that time now be given to an example of how invariance becomes a tool for problem-solving and mathematical creativity apart from its philosophical (or foundational) use in classifying geometries.

Consider the problem of determining whether it is possible to inscribe an ellipse in a given triangle so that the ellipse touches the triangle at the midpoints of its three sides. I have given this problem to college calculus students, most of whom failed to obtain any worthwhile insight. However, only a small amount of information about affine transformations provides the key:

(1) An affine map, call it f, can be used to transform the given triangle T into an equilateral traingle E.

$$f : T \to E.$$

(2) A circle C inscribed in E touches the sides of E at their midpoints.

(3) Since the affine map f^{-1} *preserves midpoints*, the image of C under f^{-1} is an ellipse that satisfies the desired conditions.

All the facts about affine transformations used in the solution of this problem are things that I can imagine teaching junior high children by means of parallel projections.

(1') Three given non-collinear points of a plane can be mapped onto any other three given non-collinear points of the plane.

(2') Affine images of circles are ellipses. (Perhaps one would define an ellipse this way at the junoir high level.)

(3') Ratios of distances, hence midpoints, are invariants of affine transformations.

The next step in physical motivation of geometry is *central projection*, which can be demonstrated by casting shadows using a light bulb. Again, this leads to more mathematizing of physics when the light bulb becomes a mathematical point, the projection screen becomes a line or plane extending without bound, and the notion of shadow is extended and abstracted in terms of geometric incidences. Of course, this is *projective geometry*, which D. N. Lehner, A. N. Whitehead, and H. S. M. Coxeter (in historical order) have indicated as possible to treat in secondary schools. Coxeter's *Projective Geometry* (Blaisdell) was an attempt to make the subject accessible to advanced high school students. It would seem advisable to introduce some of the notions of projective geometry even earlier because they are so intuitive and they offer so much.

The projective equivalence (invariance) of conic sections is seen immediately when the center of the perspectivity is taken as the vertex of a cone. This will not only serve to delight students, but will be another problem-solving tool as it no doubt was for Pascal when he proved his famous theorem on hexagons inscribed in conics.

Even the invariance of cross-ratio can be derived from physical considerations. I recently discovered that a high-school teacher from this state, a Mr. Robert Wegner of Highland Park High School, developed a unit for projective geometry for sophomore students. In this unit, which is called "Shades and Shadows", the students study light projection, its relation to Renaissance painting , and then work through exercises that lead them to the discovery of the invariants. including cross-ratio.

Throughout the various geometric experiences suggested thus far, collinearity has been an invariant. More general experiences would lead to topological and combinatorial invariants. Although I don't have any clear idea of what might be fruitful topological lessons, I do believe that it would be worthwhile experimenting at the secondary level with a geometric (rather than epsilontic) definition of continuity, such as proposed by N. Steenrod at the CUPM Geometry Conference (see Part II of the Proceedings.) But there is no doubt that combinatorial matters should be brought into geometry training very early, certainly as soon as students have algebra under their belts. There are many problems that are natural, have physical application, and are fun. Of course, Euler's formula relating the vertices, edges and faces of a polyhedron should be seen as a combinatorial result. In fact, this is graph theory, which can be used to provide a geometric formulation of problems from a variety of fields.

I am afraid that I have rambled over many topics, giving the impression that everything belongs in the school curriculum. (I refrained from mentioning vectors, analytics and linear algebra only because these are already regarded as high priority topics.) Although I don't adhere to my position with complete confidence, I am fairly certain that:

(1) Much more geometry could and should be in the curriculum.

(2) Suggested topics (e.g., affine transformations, projective transformations) need not be treated extensively; even Euclidean Geometry should not

get an extensive treatment until students have had several exposures to some of its subject matter.

(3) Much more of what we teach should be motivated by the world of physical, biological and social science.

(4) There should be more emphasis on the creative and problem-solving aspects than on studying the highly polished formal structure; that is, the school curriculum should emphasize geometric techniques, formulations and approaches more than it does the complete development of a geometry.

Carleton College
Northfield, Minn., U.S.A.

HANS-GEORG STEINER

A FOUNDATION OF EUCLIDEAN GEOMETRY
BY MEANS OF CONGRUENCE MAPPINGS

Traditional teaching of geometry has been strongly influenced by the Euclid-Hilbert foundation of geometry. In his refinement of Euclid's axiomatic system Hilbert[1] used as undefined terms: to be a point, to be a line, to be a plane, the incidence relation, the betweenness relation, and the congruence relation (for explicitly defined objects such as line segments and angles). In this approach *congruence* is a basic (undefined) concept and the well known axioms and theorems on congruence play an important role in the whole development. The techniques of congruence proofs consist in a piece by piece comparison of the two figures which are to be proved congruent. In contrast to this piece by piece comparison Euclid also seemed to have in mind the idea of a *motion* of one figure onto another. This idea has been made precise during the last hundred years. One can reduce the cinematic concept of physical motions to that of mappings of the set of all points onto itself: the so called *congruence mappings*. These transformations of the space played an important role in F. Klein's "Erlanger Programm" (1872). In this program the congruence mappings or *euclidean transformations* were specialized among a more general set of transformations, the set of all projective transformations. However Klein's attitude was not directed towards a synthetic foundation of geometry. His underlying structure was that of analytic geometry: the number space.

On the basis of Hilbert's axiomatics for synthetic geometry one can define the *congruence mappings* to be those permutations of all points in space which preserve incidence and betweenness and map every triangle onto a congruent one. These mappings form a group. A good knowledge of significant properties of the congruence mappings and of their groups is a very useful tool in proofs, in addition to the congruence theorems. This is one aspect (a technical one) of the so called *motion geometry*.

A strict motion geometric approach, however, does not introduce the congruence mappings as tools, based on other congruence concepts, such as the congruence relation. It inverts the order: "congruence mappings" becomes a basic concept and the congruence relation is defined by means of congruence mappings. In the following we shall describe one of the various axiomatic systems for a strict motion geometric approach. For the sake of simplicity

[1] D. Hilbert, *Grundlagen der Geometrie*, 1899. First English translation 1902.

we shall restrict on plane geometry. The axioms can be modified in a natural way to form an analogue motion geometric foundation for space geometry.[2]

By expressing the incidence relation between points and lines in set theoretical language we can deal with one sort of primitive objects only: the *set P of all points*. The *plane incidence axioms* (including the *strong parallel axiom*) can then be stated as follows: There is distinguished a nonvoid set Λ of subsets of P, the set of all so-called lines, such that:

(I_1)   For all $L \in \Lambda : L$ is an infinite proper subset of P.

(I_2)   For every $p, q \in P, p \neq q$, there exists exactly one $L \in \Lambda$ such that $\{p, q\} \subseteq L$.

DEFINITION: $L, M \in \Lambda$ are called parallel ($L \parallel M$) iff $L \cap M = \emptyset$ or $L = M$.

DEFINITION: For every line $L \in \Lambda$ the set dir L (read: "direction of L") is the set of all lines $M \in \Lambda$ such that $M \parallel L$.

(I_3)   For every $L \in \Lambda$: dir L is a partition of P.

(*Remark:* (I_3) implies the strong parallel axiom and the fact that parallelity is an equivalence relation on Λ)

Ordering in the plane can be based on the following: On each line $L \in \Lambda$ there are distinguished two linear ordering relations $<_L$ and $>_L$, one being the inverse relation of the other.

DEFINITION: A point r is said to be *between* two points p and q, iff r is on the line L, determined by p and q (according to (I_2)) and

$$p <_L r <_L q \quad \text{or} \quad p >_L r >_L q.$$

DEFINITION: A subset $S \subseteq P$ is called *convex*, iff for all $p, q \in S$ also $r \in S$ for all r between p and q.

The ordering axiom then says:

(O)   Every line $L \in \Lambda$ determines a partition of P into three convex sets L, P'_L, P''_L, such that for all $p \in P'_L, q \in P''_L$ there is a point r on L which is between p and q.

(The sets P'_L, P''_L are called the *halfplanes* determined by L. Since every point $p \notin L$ determines one of the two halfplanes, belonging to L, the halfplane which contains p may be denoted by "$P_L(p)$".)

Some more definitions:

DEFINITION: The set $\{r \mid r = p$ or $r = q$ or r between p and $q\}$ is called the *line segment* of p and q and is denoted by "pq".

DEFINITION: For all lines $L \in \Lambda$ and all points $p \in L$ the sets $\{r \mid r <_L p\}$

---

[2] H. G. Steiner, *Grundlagen und Aufbau der Geometrie in didaktischer Sicht*, Münster 1966.

and $\{r \mid r >_L p\}$ are called the *halflines* on L, determined by p. The halflines on L, determined by $p \in L$, may also be denoted by L'_p, L''_p. If $q \in L$ and $q \neq p$, then the halfline on L, determined by p, such that it contains q, is denoted by "$L_p(q)$".

DEFINITION: A *flag* is a triple (p, L'_p, P'_L) such that $p \in P$, $L \in \Lambda$, $p \in L$, and L'_p is one of the two halflines of L, determined by p, P'_L is one of the two half-planes determined by L.

Preparation for the *axioms on congruence mappings*: Let $A(P)$ be the set of all permutations of P. Then $(A(P), \circ)$ is a group. Let $B(P)$ be the set of all permutations α of P such that:

(1) $L \in \Lambda \Rightarrow \alpha(L) \in \Lambda$

(2) $r, p, q \in P$ and r between p and q, then $\alpha(r)$ between $\alpha(p)$ and $\alpha(q)$.

The mappings in $B(P)$ are called the *automorphisms of the ordered affine structure* on P. It is immediately clear, that $(B(P), \circ)$ is a subgroup of $(A(P), \circ)$. In $B(P)$ we shall identify the congruence mappings by determining a subgroup $(C(P), \circ)$ of $(B(P), \circ)$ which itself is generated by a subset $R(P)$, the set of all reflexions in lines $L \in \Lambda$. The axioms on *congruence mappings* can be stated as follows: The group $(B(P), \circ)$ has a subgroup $(C(P), \circ)$ the group of all so called congruence mappings which is generated by a subset $R(P) \subseteq B(P)$, the set of all so called reflexions in lines, such that:

(CM$_1$):   For every $L \in \Lambda$ there exists a $\varrho \in R(P)$ and for every $\varrho \in R(P)$ there exists a $L \in \Lambda$ such that L is the set of all fixed points of ϱ.

(CM$_2$):   For all $\varrho \in R(P)$: ϱ is involutorial, i.e. $\varrho \neq$ id (the identity mapping of P) and $\varrho \circ \varrho =$ id.

(CM$_3$):   If L'_p and K'_p are halflines, beginning at the same point p, then there exists $\varrho \in R(P)$ such that $\varrho(L'_p) = K'_p$.

(CM)$_4$:   For any $p, q \in P$ there is a $\varrho \in R(P)$ such that $\varrho(p) = q$.

(CM$_5$):   For every flag F and every $\alpha \in C(P)$: if $\alpha(F) = F$, then $\alpha =$ id.

In order to get the full euclidean geometry, one has to guarantee that each line as an ordered set is of the *order type of the real numbers*. This so called *completeness* can be postulated in terms of *Dedekind's cuts*:

(C)   On each line L every partition of L into two sets L_1, L_2 such that for all $p, q \in P$

(1) if $p \in L_1$ and $q <_L p$, then $q \in L_1$

(2) if $p \in L_2$ and $q >_L p$, then $q \in L_2$

has the following property: either L_1 has a greatest element and L_2 has no smallest element or L_2 has a smallest element and L_1 has no greatest element with respect to $<_L$.

So far we have given a complete axiomatic system for plane euclidean geometry, based on the concept of congruence mappings. We shall here not develop any significant part of the theory from the axiomatic system. In the following we shall only prove the *theorem of free mobility* and explain how the concept of congruence relation can be based on $C(P)$.

THEOREM 1 (Theorem of free mobility): For any flags $F=(p, L_p', P_L')$ and $G=(q, M_q', P_M')$ there exists exactly one $\alpha \in C(P)$ such that $\alpha(F)=G$. Every mapping $\alpha \in C(P)$ is the product of at most three reflexions.

Proof: Assume there were congruence mappings α, β which map F onto G. Then $\beta^{-1}(G)=F$ and $(\beta^{-1}\circ\alpha)(F)=\beta^{-1}(\alpha(F))=\beta^{-1}(G)=F$. According to (CM_5): $\beta^{-1}\circ\alpha=\mathrm{id}$, i.e. $\beta=\alpha$.

This shows the uniqueness of α. In order to prove the existence we argue as follows: According to (CM_4) there is a reflexion $\varrho_1 \in R(P)$ such that $\varrho_1(p)=q$. If $\varrho_1(L_p')\neq M_q'$, then we can apply (CM_3): There exists a reflexion $\varrho_2 \in R(P)$ such that $\varrho_2(\varrho_1(L_p'))=M_q'$ and $\varrho_2(\varrho_1(p))=q$. Let be $\varrho=\varrho_2\circ\varrho_1$. If not $\varrho(P_L')=P_M'$, then we consider the reflexion $\varrho_3=\varrho_M$ in M. Since ϱ_M interchanges the halfplanes P_M' and P_M'' and has M as fixed point set (which is an easy consequence of (CM_2) and the fact that all $\alpha \in B(P)$ map lines onto lines and are betweenness preserving), the mapping $\alpha=\varrho_M\circ\varrho=\varrho_M\circ\varrho_2\circ\varrho_1$ has the property that $\alpha(F)=G$. This also shows that every congruence mapping is the product of at most three reflexions, since every congruence mapping maps a given flag onto a flag.

DEFINITION: For any figures Φ_1, Φ_2 (i.e. subsets of P) the *congruence relation* $\equiv$ is defined as follows:

$$\Phi_1 \equiv \Phi_2 \text{ iff there exists } \alpha \in C(P) \text{ such that } \alpha(\Phi_1)=\Phi_2.$$

THEOREM 2: $\equiv$ is an equivalence relation on the set of all figures in P.

Proof: $\equiv$ is reflexive, according to the fact that $\mathrm{id} \in C(P)$ and $\mathrm{id}(\Phi)=\Phi$ for all figures Φ.

$\equiv$ is symmetric, according to the facts that $C(P)$ is closed under inversion, i.e. $\alpha \in C(P)$ implies $\alpha^{-1} \in C(P)$, and that $\alpha(\Phi_1)=\Phi_2$ implies $\alpha^{-1}(\Phi_2)=\Phi_1$.

$\equiv$ is transitive, since $C(P)$ is closed under composition. $\alpha(\Phi_1)=\Phi_2$, $\beta(\Phi_2)=\Phi_3$ implies $(\beta\circ\alpha)(\Phi_1)=\Phi_3$.

The congruence theorems for triangles can now be proved by means of congruence mappings.

University of Erlangen-Nürnberg, PH Bayreuth, F.R.G.

MARSHALL STONE

LEARNING AND TEACHING AXIOMATIC GEOMETRY

INTRODUCTION

Some years ago I prepared for an ICMI conference at Bologna a discussion on "Le Choix d'Axiomes pour la Géométrie", and later published it in L'Enseignement Mathématique [1]. In that paper I examined some of the reasons for continuing to teach axiomatic geometry in the secondary school and went on to explain my own preference for choosing axioms closely related to Artin's treatment of affine geometry [2]. It was pointed out that in this approach one deals separately with the axioms of incidence, the axioms of order, and the axioms of orthogonality, thus dividing and isolating the mathematical difficulties to be overcome. This treatment closely parallels the one proposed by Choquet [3], except that the coordinatization of an affine plane or space has to be carried out abstractly in Artin's manner without assuming the real numbers as already available. The effect of invoking the axioms of order or orthogonality is to impose specific limitations upon the coordinate field. In particular, one sees very clearly how the axioms of incidence and order alone lead to real coordinates, a fact of geometry that, historically speaking, lies behind the invention of the real number system itself.

On the present occasion it will be my purpose to discuss some of the pedagogical problems raised by this approach to axiomatic geometry. I hope that what I have to say will not be so closely tied to a particular choice of axioms as to lack all value in a broader context. Nevertheless there is a definite advantage in being able to focus any general discussion on a particular way of developing Euclidean geometry from a specific set of axioms.

DIFFICULTIES

There seems to be quite general agreement that in all of school mathematics there is no subject more difficult to learn or to teach than axiomatic geometry. Even the most superficial analysis shows that the sources of difficulty are of several kinds.

The most evident troubles are those caused by the relative complexity of the mathematical structure of the subject. In the classical approach as set down by Euclid and in its modern variants the importance attached to the study of special configurations by so-called "synthetic" methods imposes heavy demands on the learner's "geometrical intuition" and mathematical ingenuity. While there is a clear need for developing both intuition and ingenuity on the student's part, the pursuit of these aims by exaggerating the attention paid to synthetic Euclidean geometry was rightly denounced by Dieudonné [4] at Royaumont when he uttered his famous slogan, "Euclid must go!" By choosing suitable axioms and introducing coordinates at an early stage, one is able to pass quickly from synthetic to analytic or algebraic geometry. As has been well known since the time of Descartes, one thus eliminates many of the mathematical difficulties by converting geometric problems into algebraic ones for which algorithmic methods of solution can readily be developed. Indeed, there are many classical problems, such as those concerning the duplication of the cube, the trisection of the angle, and the quadrature of the circle that were not solved until they were thus formulated algebraically. However, the way in which coordinatization is derived from the axioms presents its own mathematical difficulties. In Artin's approach, for example, extensive use is made of geometrical transformations. Here the difficulties are not only technical but also conceptual: the fundamental concept of a transformation has to be taught by the teacher and learned by the student before it can be applied effectively in a technical way.

We are led by this last example to cite as the second general source of difficulties in the axiomatic approach the presence of many concepts that are by no means easy to analyze intuitively or to treat appropriately in the school room. In axiomatic geometry we are engaged in seeking a categorical mathematical description of the physical space in which we live. Each of us brings to this task a rich spatial experience that commenced even before his birth. In the course of this experience each of us has as a child developed unverbalized insights and procedures for dealing with the problems of moving about in physical space. It is from this already partially structured experience that the fundamental concepts and relations of geometry have to be abstracted and characterized. While the concepts of "point" and "straight line" may seem natural enough, they are extremely sophisticated when considered as abstractions from physical reality. We are certainly justified in avoiding at the school level any verbal analysis of the process that leads us to these abstractions. We communicate only by repeatedly exhibiting instances of what we mean by a "point", a "straight line", or the "incidence" of a point with a line. We are literally forced to treat the terms "point", "line", and "incidence" for what they actually are in many axiomatic treatments – undefined or primitive terms. The assumptions we are willing to make about points, lines, and incidence – and eventually to use as axioms – must reflect certain features of our spatial experience with the physical objects to which these terms are intended to refer.

A difficulty deserving most serious consideration is that the experience of one individual may differ widely from that of another and may have led him to quite different geometric insights or inferences. For example, it has been observed that many African children have no intuitive notion of a straight line, possibly because they have experience of so few examples in their daily lives. Probably most American children would not have a similar difficulty. Another interesting example is to be found in the ability to interpret two-dimensional figures as perspective renderings of three-dimensional ones. When a computer-based instructional program was being set up for a selected group, it was necessary to test the participants in advance for this ability. They were found to possess it to an unanticipated degree, but a different group might obviously have shown much less skill. Clearly it is important for us to know more exactly what verbalized or unverbalized geometrical baggage and what unconscious geometrical mastery children bring to their first classes in geometry. Only then can teachers intelligently open up new geometrical experiences for them, filling in the gaps, correcting misapprehensions, and profiting by hidden skills. The geometrical experiences to which children are introduced in school need also to be designed so as to lead gradually to the formation and acceptance of just those concepts that will eventually be emphasized, first in intuitive geometrical thinking and then in the axiomatic approach itself.

A third difficulty in axiomatic geometry is that in order to comprehend adequately its purport the student has to have some knowledge and understanding of logical principles. At the present time the schools have quite generally eliminated logic as a formal subject of instruction. Only in recent years has it begun to reappear in isolated paragraphs or chapters in some modern mathematics texts in which concern is shown for sharpening the students' notions of what constitutes a mathematical proof. Clearly some such preparation needs to be offered before the student is asked to look at geometry from the axiomatic point of view. The challenging question is, "How much?"

LOGIC AND AXIOMATIC GEOMETRY

Let us first examine briefly the matter of the logical difficulties we have just mentioned in our short but by no means unimpressive list of obstacles to the effective teaching of axiomatic geometry. I suggest that we do this forthwith and that we make the examination brief, so that we can pass on quickly to the other difficulties we have noted. The discussion of the latter is likely to be much more fruitful on this occasion, because a thorough discussion of the logical preparation for axiomatic geometry could lead us almost at once into the bigger question of the place of logic in a modern school curriculum. We might then find ourselves irreversibly diverted from the major concerns of this conference. This question is certainly one that deserves far more attention than it has yet received in curricular studies, but is should evidently be set as the sole theme of a major conference devoted exclusively to its discussion.

Under present circumstances we must think of preparing the student for the axiomatic treatment of geometry mainly through repeated exercises in constructing and analyzing mathematical proofs. The current preference for starting with proofs in elementary algebra and arithmetic is undoubtedly sound; but with a view to the student's eventual interest in demonstrative geometry, he should begin quite early to have some experience with simple geometrical proofs as well.

First of all, the student should be led on the basis of physical experience to formulate geometrical propositions that seem to be valid in the physical world. He should then be shown that some such propositions can be proved from others and he should be led to construct other proofs on his own. In the course of his experience with mathematical proofs, whether in algebra, arithmetic, or geometry, he should begin to see that several different proof schemes are used over and over again and even that a limited number of these appears sufficient to build up all the proofs with which he has become familiar. From these observations it should be possible for the student to reach an understanding of two principles: first, that a proof consists in the repeated application of certain specified elementary proof schemes; second, that the logical structure of a set of propositions is described by determining (if possible) what propositions are provable from its various subsets. The problem of axiomatizing a set of propositions can then be defined as that of finding a "small" subset from which every member of the set or its negation, *but not both*, can be proved. The problem of axiomatizing geometry is that of finding a set of basic geometrical propositions, called the axioms, from which "all" others can be proved or disproved. The relevance of this logical structure to the "real" geometry of the physical world should also become plain: namely, that acceptance of the physical validity of the axioms entails a belief in the physical validity of the theorems proved from them.

Evidently what I am proposing is that the student be led to formulate some rather vague principles of logic capable of being made precise, but only at some much later stage in his mathematical education. Perhaps certain moves toward greater precision should already be undertaken in the schools, at least to the extent of making explicit the role of quantifiers and describing some of the proof schemes in which they occur. Even without this additional precision, the student's first notions of what axiomatization means need not be so vague as to throw no light at all on what is taking place in his classes on axiomatic geometry. Even though these initial insights into logic, as well as those into geometry itself, are perforce somewhat vague and imprecise, they may be so directed as to aid in eventually gaining a much more profound understanding. It seems to me futile to hope for a great deal more in the present state of our educational establishments.

THE MATHEMATICAL DIFFICULTIES

Turning now to the mathematical difficulties with which we are confronted in the axiomatic treatment of geometry, we should first observe that many of them could be eliminated if we were willing to break all initial ties between physical geometry and abstract geometry and to postpone the interpretation of abstract geometry as a description or a model of the real world. For instance, we could start with the real numbers, introduce real vectors as algebraic objects, and then attach to each ordered pair of "points" in an abstract "space" a real vector. As Artin pointed out in the report of the Dubrovnik Conference, a few simple axioms and definitions concerning this structure serve to characterize an m-dimensional affine space [5]. This procedure is analogous to that used in setting up a metric space, though to characterize Euclidean spaces among all metric spaces happens to involve some very heavy algebra. With real affine space at our disposal, we have many simple ways of imposing on it a Euclidean structure. For instance, we can suppose that each vector x has a length or real norm $|x|$ satisfying the identity

$$|x + y|^2 + |x - y|^2 = 2|x|^2 + 2|y|^2,$$

which expresses the condition that "the sum of the squares on the two diagonals of a parallelogram be equal to the sum of the squares on the four sides", a variant on the Pythagorean theorem. This is more or less the sort of procedure Dieudonné would have us adopt "after Euclid has gone".

In my opinion this procedure is one that has to be rejected for both philosophical and pedagogical reasons. It prematurely divorces mathematics from reality, and it deprives the child of the personal experience of gradually conceptualizing and formalizing in meaningful mathematical terms his deep intuitions of space. If one accepts the view that physical space is known to us primarily in terms of light, an electromagnetic phenomenon, then it may appear "right" and "natural" to introduce the "point" and the "straight line" as the primitive abstract concepts respectively associated to the light corpuscle (or photon) and the light ray in the real world. Historically, we should remark, the corpuscular theories of light were the first to come upon the scene (as proposed by Lucretius and Newton) and are now once more at the center of the stage, albeit after fundamental quantum-theoretical revisions. In other words, in trying to "analyze" or "explain" our inseparable experiences of light and space, the first and perhaps simplest approach is to put everything in terms of photons or "points" on the one hand, and of light rays or "straight lines" on the other, with incidence as the significant relation connecting them. If we are to start with points, lines, and incidence as the primitive terms in formal geometry, subject of course to appropriate axioms, then everything else has to be defined in terms of them. If we want to introduce vectors, then vectors have to be suitably defined and their properties derived from the axioms. If we want to introduce a field and use it to coordinatize space, then we have to define the field elements in terms of points, lines, and incidence and to derive the field properties of these elements from the definitions and the axioms. For a fairly long time the mathematical difficulties involved in these derivations were too severe to be admitted to our school rooms. Eventually, as is so often the case in mathematics, these difficulties were eliminated – as a result of Emil Artin's discovery of a new line of attack [2]. The price that has to be paid for using Artin's strategy is the effort involved in introducing and applying the concept of a transformation or mapping and various related concepts from group theory. Since Felix Klein enunciated his Erlanger Programm, it has been recognized that these concepts are basic for geometry and therefore need to be taught for their own sake. That they are equally important for other branches of mathematics and physics as well has been established by the researches of the past fifty years and more. Thus the price for using Artin's insights into the structure of affine geometry is one we should be very glad to pay. In fact, we would be delinquent if we failed to introduce these dominant concepts of modern mathematics into the school curriculum. It hardly needs to be said in this gathering that most modern school mathematics programs give place to many, if not all, of them.

In outline, Artin's development of affine geometry starts from an elementary analysis of parallelism based on simple axioms of incidence, similar to those of Euclid. However, it quickly appeals to the transformation concept and focusses attention on the affine transformations as those that preserve parallelism. A link with group theory and algebra is thus established at a very early stage. It permits the continual interplay of geometric and algebraic reasoning in all further study of affine planes or spaces. Among the affine transformations the dilatations are then characterized geometrically and classified (by an examination of their fixed points) into homotheties and translations. The translations are shown to form an abelian or commutative group. Now from abstract algebra it is known that the endomorphisms of any abelian group constitute an associative ring. Artin's theory next singles out certain special endomorphisms – the "trace-preserving endomorphisms" – of the translation group, correlates them with certain dilatations, and thereby establishes that they determine a division subring K of the full endomorphism ring. In order to eliminate uninteresting degeneracies, it is now necessary to establish the existence of "enough" translations and "enough" homotheties. Thus one is confronted with the problem of constructing, if possible, a translation of a homothety moving one given point into another. A geometrical analysis easily leads to the conclusion that the desired constructions hinge upon the validity of certain incidence relations for various simple instances of the so-called Desargues configuration. It turns out that in space these relations can be proved from the axioms but that this is not the case in the plane. Hence in plane geometry these crucial incidence properties have to be postulated as *additional* axioms. Once this point has been reached, coordinatization in terms of the division ring K is immediate, and affine geometry becomes linear algebra over K. In this whole development the geometrical demonstrations are relatively simple, while the algebraic arguments are all standard proofs of standard elementary results in group-theory and linear algebra. Indeed, it seems safe to say that there is hardly any mathematical feature of the whole discussion that has not already been handled, though perhaps in a different context, in one or another of the more advanced experimental secondary school mathematics programs.

Similar remarks apply to the study of the consequences of adjoining the axioms of order and orthogonality. We have already noted that these axioms entail limitations upon the division ring K. It is here that the characterization of the real field as a certain kind of ordered division ring emerges in its natural connection with geometry. Of course, one can no longer avoid the question as to the *existence* of such a field. This question now presses for some sort of answer – and one [6] should certainly be given with as little further delay as possible, even though tradition has it that this is not altogether easy!

THE CONCEPTUAL DIFFICULTIES

In a sense Artin's approach has thus shifted the burden of difficulty onto those basic concepts that provide the key to the whole strategy. They are, as we have seen, the concepts of point, straight line, incidence, transformation (mapping, function, or operation), group, morphism, order, and orthogonality. Associated with these, of course, are many derived concepts, such as those of collinearity, parallelism, endomorphism, automorphism, ring, division ring, field, and so on. The problem is to arrange the child's school mathematics program so that he is confronted with a variety of concrete situations out of which in due course he may spontaneously abstract the particular concepts we expect him to learn. Clearly this is primarily a psychological rather than a mathematical problem, although the selection of appropriate concrete situations evidently requires the participation of mathematicians.

In my opinion the concepts of point, straight line, and incidence can best be developed through having the child experiment actively with pencil and straight edge constructions carried out on paper, or with other materials not too difficult to manipulate. By making compasses available as instruments of construction the child's geometrical experience can be greatly broadened and the groundwork prepared for him to learn the concepts of order, parallelism, and orthogonality. Simultaneously he should learn to verbalize his experience by gradually acquiring a standard descriptive vocabulary and by attempting to use it in stating general features or properties of the configurations with which he becomes familiar. From the work of Suppes and Hawley [7] with large numbers of first-grade children in Palo Alto, California, some years ago it may be inferred that this approach to geometry through constructions can be very successful at a surprisingly early stage. Of course, as the geometry of physical space is Euclidean, at least to a first approximation, it would be both premature and futile at this early stage to try to separate out from one another the affine, ordinal, and metric properties. Indeed, for a long time the child should be expected to deal actively, intuitively, and in the end verbally with physical geometry which he continually explores with the help of the material aids put at his disposal. How far this approach can be carried with quite ordinary materials is indicated by the geometrical parts of the SMSG elementary mathematics texts for grades K through 6.

At some stage, probably a good deal earlier than many might suppose, experiences with the transformation concept should be initiated. Some years ago, I prepared a paper on 'Learning and Teaching the Function Concept' [8] in which many passages are presumably relevant to the present somewhat narrower discussion. However, in geometry one does not often need or at first appreciate the full generality of the function concept, with which I was there concerned. Consequently, in the context of elementary school geometry the function or transformation concept should be developed slowly from a relatively small number of special cases. There is an extensive literature [9] dealing with many different ways of introducing children to geometrical transformation theory. I believe that enough is known from the numerous experiments described there to warrant confidence in the possibility of achieving substantial progress with children still in the later elementary grades, 4–6 and a fortiori with children in the earlier secondary grades, 7–10. It seems to me that this passage from a vague intuitive notion of a transformation to a more precise concept cannot be made until the child has been confronted with the problem of describing just how a particular transformation acts on each point of the plane or space. For instance, a proper translation can be described in Artin's terminology as a transformation without fixed point, moving every line into a parallel line. The problem becomes that of finding out how such a transformation acts in general when its action on a single point is known or prescribed. As we noted in the preceding section, this problem leads inevitably to an existence question hinging on the validity of certain incidence relations in simple instances of the Desargues configuration. The discussion of these matters does not involve anything beyond simple facts about parallelism and can therefore be initiated at a fairly early stage and carried on at first without appeal to formal proofs. In fact some preliminary notions about proofs could well emerge from the early attempts at discussing the problem.

When the notion of a transformation has been grasped, the composition of two transformations appears quite naturally on the intuitive level and leads quickly into group theory. It is not easy to say whether the group concept should be introduced independently in an algebraic or arithmetic setting before it is brought in from geometry. Probably both geometric and algebraic backgrounds should be built up in parallel and then used together as a basis for abstracting the group concept. The important algebraic notions needed in group theory, such as that of morphism, are probably somewhat easier to introduce from arithmetic and algebraic considerations than from geometry. Ways for doing this have been rather thoroughly tried out in some of the modern experimental school mathematics programs, and the study of geometric symmetries has also been tested for the same purpose.

To summarize, we see that the conceptual difficulties raised by Artin's approach to axiomatic geometry can be identified and can be handled pedagogically in a variety of ways that have already been explored in some depth. We have not gone into detail about the concepts of order and orthogonality, thinking that the basic concepts involved are not so difficult pedagogically as to make this necessary. The question that has to be asked is, "How can we use, refine, and extend the psychological methods already known to us so as to make the basic concepts more readily comprehensible?" It seems to me that part of the answer lies in the vigorous pursuit of psychomathematical researches, but another part is to be found in a much more careful organization of the school geometry program about some clearly defined approach such as Artin's to an axiomatic treatment of the subject.

THE ORGANIZATION OF THE SCHOOL GEOMETRY PROGRAM

From all that has been said so far it should be evident that I do have in mind, at least in broad outline, a school geometry program designed to culminate in an axiomatic treatment of affine and Euclidean geometry in the 10th or 11th grade, when the average student would be around 16 or 17 years of age. For students showing a special interest in geometry the program could be topped off by the study of various additional geometrical subjects such as projective spaces, non-Desarguesian planes, non-Euclidean geometries, finite geometries, and so on.

First of all, geometry should be presented as a subject of study throughout the school program up to the point where the student has acquired a good intuitive working knowledge of the basic concepts of geometry and their application to elementary problems of physical geometry. Beyond this point, geometry should be presented in all programs intended for precollege students and should be developed so far as to include some discussion of axiomatics. It is only for mathematically able pre-college students, regardless of their eventual specialization, that a program culminating in a full axiomatic treatment of affine and Euclidean geometry can reasonably be recommended in the present state of the physical, biological, and behavioral sciences. It is not altogether clear where the suggested terminal points could reasonably be fixed for these different classes of student, but I would guess that they might occur at the end of the 9th, 10th, and 11th grades respectively. The corresponding programs would presumably differ in pace, in depth, and to some degree in content, though they would all be guided by a common approach to their subject matter.

At the beginning of the program, whether in kindergarten or grade 1, and also at suitable later times, an inventory should be taken of each student's geometrical concepts, vocabulary, and skills, as we have already suggested in our general discussion of the difficulties that have to be overcome. Until we have more experience with such inventories, they may prove to be a somewhat crude instrument but one capable of improvement through future psycho-mathematical investigations. Even now these inventories should give some guidance as to useful ways of individualizing instruction as the student commences and then continues his study of geometry.

During the elementary school years, in grades 1–6, the geometry program should be concerned with ruler and compass constructions, transformations, and mensuration, as described more fully in our discussion of the conceptual difficulties involved. In grades 4–6 the student should learn about the various kinds of transformation of interest in affine and Euclidean geometry and should begin to formulate and test general geometrical propositions suggested by his acquaintance with concrete situations. The concepts of length, area, and volume have an undeniable claim to be treated at this stage as part of metric – that is, Euclidean – geometry. In other words, the time for separating affine from metric phenomena is yet to come.

In grades 7–9 the study of geometry on the intuitive and descriptive levels in terms of the fundamental concepts should be deepened, and a start made toward developing an increasing awareness of the logical connections among the geometrical propositions that have emerged from this study. Perhaps some of the very simplest proofs even belong in the earlier grades, but there seems to be no urgent need for forcing the child's efforts toward mathematical understanding to this point quite so early. Naturally, if it should happen that a particular child or group of children should show signs at any stage whatsoever of being interested in these logical questions, the teacher should be ready to turn this interest to good advantage. In any case, by the end of the 10th grade the student should have sufficiently clear notions of logic and the nature of proof that he can understand both the possibility and the significance of an axiomatic treatment of geometry. Not only that! He should also be psychologically prepared by this piece-meal approach to the study of the logic of geometry to separate out the affine properties of space for independent consideration, leaving their synthesis with the ordinal and orthogonality properties to follow. Mathematically talented students could undoubtedly complete their study of axiomatic geometry by the end of grade 10, when most of them would be around 15 or 16 years of age. However, all students who have followed the proposed program for the first nine grades should be expected in grade 10 to become acquainted with the axiomatic approach, to grasp the general strategy adopted for carrying it out, and to undertake some typical mathematical exercises involved in implementing that strategy. On the whole it would seem best to leave for grades 11 and 12 the completion of the proposed axiomatic treatment and possible extensions to other kinds of geometry. Not only would this offer students who are not motivated by rather strong needs or inclinations or interests an excellent chance of knowing and understanding the axiomatic method without too much involvement in technicalities as one of the very great achievements of the human intellect. It would also permit greater flexibility in combining the more advanced parts of the geometry program with the high points in algebra and analysis that should be included in a modern secondary school mathematics program.

It hardly seems necessary to elaborate upon the remark that this geometry program needs to be properly articulated with the other components of the school mathematics curriculum. It is evident that there are conceptual, technical, and pedagogical connections between what we would like to see done in geometry and what is already being done in set-theory, algebra, and arithmetic at the various stages. By identifying these connections and properly organizing their treatment in the class-room, we can help the student to make more rapid progress in his study of mathematics and to recognize the basic unity of the subject.

SUMMARY

To sum up, we have examined the various difficulties-mathematical, conceptual, and logical – involved in learning and teaching axiomatic geometry. Taking as a typical axiomatic approach one that couples Emil Artin's treatment of affine geometry with Choquet's handling of the basic order and orthogonality phenomena of Euclidean geometry, we have tried to identify these difficulties in some detail. We have noted that this particular approach isolates the graver difficulties, separating them from one another in such a way that they can be mastered by comparatively simple techniques. For both student and teacher the burden is shifted to the fuller understanding of certain fundamental concepts, such as that of a geometrical transformation, which permit making an early contact with algebra – specifically, with group theory – in the spirit of Descartes and Felix Klein. We have proposed an organization of the school program in geometry from grade K or grade 1 through grade 12 aimed at gradually acquainting the student with these basic concepts and their technical uses as they emerge from the persistent contemplation of the physical, mathematical, and logical facts. We have stated our conviction that by the 10th grade students should be prepared to understand the strategy and the significance of the axiomatic approach, and by the 11th and 12th grades to carry through that approach in adequate detail. We have taken explicitly into account the need to arrange the school geometry program in such a way that students can reach terminal points appropriate to their capacities, needs, and interests. We must emphasize that what has been presented here is no more than a rough outline, to be filled out on the basis of further psychological, mathematical and curricular investigations. I am confident that these suggestions can mature into a school geometry program not only far more significant intellectually and technically than those now commonly offered but also one easier to learn and to teach. This is not just a matter of personal belief, because many elements of the proposed program have actually been tried out in one form or another with success. What is now needed is a grand effort to assemble the tested and untested parts into an experimental whole.

University of Massachusetts
Amherst, Mass., U.S.A.

BIBLIOGRAPHY

[1] Stone, M. H., *L'Enseignement Mathématique* (2) **9** (1963), 45–55.

[2] Artin, E., *Geometric Algebra*, Interscience Publishers, New York (1957), 51–103.

[3] Choquet, G., *L'Enseignement de la Géométrie*, Hermann et Cie, Paris (1964).

[4] Dieudonné, Jean, 'New Thinking in School Mathematics', *OEEC* (1961), 35–39.

[5] Artin, E., 'Synopses for Modern Secondary School Mathematics', *OEEC* (No date) 202–219.

[6] There are other possibilities than the traditional one due to Dedekind – references and discussion may be found in Stone, M.H., *L'Enseignement Mathématique* (2) **15** (1969), 261–267.

[7] Hawley, Newton and Suppes, Patrick, *Geometry for Primary Grades*, Book I, and accompanying *Teachers Manual*, Book I (private edition).

[8] Stone, M. H., *Monographs of the Society for Research in Child Development*, Serial No. 99 (1965), Vol. 30, No. 1 (1965), 5–11 and 143–150, the first part being reprinted in a book, *Cognitive Development in Children*, University of Chicago Press, Chicago (1970), 455–461.

[9] See for example *Report on Methods of Initiation into Geometry* (H. Freudenthal, Editor), J. B. Wolters, Groningen (1958), 120 pages.

HERBERT E. VAUGHAN

THE DEVELOPMENT OF EUCLIDEAN GEOMETRY
IN TERMS OF TRANSLATIONS

The following is a description of a two-year course (grades 10–11 or 11–12) encompassing plane and solid geometry, trigonometry and considerable additional material.

The course is based on the fact that Euclidean geometry may be thought of as the theory of a linear space with an inner product – the space of translations – operating on a set of points. We deal, then, with three sets – the set $\mathscr{E}$ of points, the set $\mathscr{I}$ of translations, and the set $\mathscr{R}$ of real numbers. We shall use 'A', 'B', 'C',... as variables whose domain is $\mathscr{E}$, '$\vec{a}$', '$\vec{b}$', '$\vec{c}$',... as variables whose domain is $\mathscr{I}$, and 'a', 'b', 'c',... as variables whose domain is $\mathscr{R}$.

After a short review of functions there is a considerable amount of exploratory work in which students discover various properties of translations – for example, that a translation is determined by a single point and its image and that the set of translations is closed under function compostion. The next problem is to devise a way of expressing these facts in a convenient notation. After some discussion we adopt '$B-A$' as a notation for the unique translation which maps A on B; '$A+\vec{a}$', (rather than '$\vec{a}(A)$') as a notation for the image of A under the mapping $\vec{a}$; and '$\vec{a}+\vec{b}$', (rather than '$\vec{b}\circ\vec{a}$') for the resultant of $\vec{a}$ followed by $\vec{b}$. This leads to the rather elegant postulates:

(1a)	$B - A \in \mathscr{I}$	(b)	$A + \vec{a} \in \mathscr{E}$
(2a)	$A + (B - A) = B$	(b)	$\vec{a} = (A + \vec{a}) - A$
(3)	$(B - A) + (C - B) = C - A$		
(4)	$\mathscr{I}$ is a three-dimensional inner product space over $\mathscr{R}$.		
(5)	$\mathscr{R}$ is a complete ordered field.		

Postulates (4) and (5) are introduced slowly. That $\mathscr{R}$ is an ordered field is essentially review material for our students and is reviewed in Chapter 4. The completeness of $\mathscr{R}$ is not treated until Chapter 18, late in the second year.

Postulate (4) is dealt with much more circumspectly. In Chapter 3 it turns out that $\mathscr{I}$ is a commutative group; in Chapter 5 it turns out that $\mathscr{I}$ is a vector space over $\mathscr{R}$; $\mathscr{I}$ becomes three-dimensional in Chapter 10, at the end of the first year; and an inner product for $\mathscr{I}$ is introduced in Chapter 11. Other chapters in the first year deal with linear dependence, lines, triangles and quadrilaterals (affine properties) and planes. Other chapters in the second year deal with perpendicularity, orthonormal coordinate systems, distance, angles, triangles and quadrilaterals (metric properties), circles, oriented planes and sensed angles, and the circular functions. For lack of time, volumes of solids are discussed from an elementary point of view in an appendix.

Throughout the course, but especially in Chapters 2, 4, and 6, considerable attention is devoted giving rigorous paragraph proofs based on a systematization of logic derived from Gentzen's system of natural inference.

The notion of a group is introduced in Chapter 3, that of a field in Chapter 4, and that of a vector space in Chapter 5.

The convenience of our notation is brought out in Chapter 3 (where $\mathscr{I}$ is known to be an abelian group). At this point any sentence is a theorem if and only if it makes sense in our notation and has the property that, if all variables are taken to have $\mathscr{R}$ as a domain, it can be deduced from the group properties of addition of real numbers. For example:

$$A + (B - C) = B + (A - C) \quad \text{and:} \quad \vec{a} + (B - A) = (B + \vec{a}) - A$$

$$A + (B - C) = B + (A - C) \quad \text{and:} \quad \vec{a} + (B - A) = (B + \vec{a}) - A$$

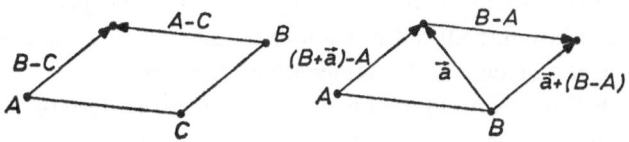

Fig. 1.

are both theorems. Their content is illustrated in Figure 1. Students are encouraged to draw such figures to illustrate the theorems they prove. Another thing which is provable at this stage is that, given a point O, there is a canonical correspondence between points and translations given by the mappings defined by:

$$P_O(\vec{a}) = O + \vec{a}, \qquad T_O(A) = A - O.$$

Another testament to the interest of the notation is that Postulates (1)–(3) and the postulate that $\mathscr{I}$ is an abelian group are equivalent to the following six postulates:

$$\vec{a} - (\vec{b} - \vec{c}) = \vec{c} - (\vec{b} - \vec{a}) \qquad \vec{a} - (\vec{a} - \vec{b}) = \vec{b}$$
$$A - (B - \vec{c}) = \vec{c} - (B - A) \qquad A - (A - \vec{b}) = \vec{b}$$
$$A - (B - C) = C - (B - A) \qquad A - (A - B) = B$$

together with three definitions and three closure postulates:

$$-\vec{a} = (\vec{a} - \vec{a}) - \vec{a}, \qquad \vec{a} + \vec{b} = \vec{a} - {}^{-}\vec{b}, \qquad A + \vec{a} = A - {}^{-}\vec{a}$$
$$B - A \in \mathscr{I}, \qquad \vec{b} - \vec{a} \in \mathscr{I}, \qquad A - \vec{a} \in \mathscr{E}.$$

As in the case of all UICSM texts, students are led to discover much of the material through doing appropriate exercises. This accounts in part for the length of the text. I shall take this as understood and restrict myself to outlining the mathematical content of the course.

After linear dependence is discussed with some care in Chapter 6 we proceed to more conventional geometric concepts and theorems. To begin with, in Chapter 7 we adopt the definitions:

A, B, and C are collinear $\Leftrightarrow (B-A, C-A)$ is linearly dependent.

A line is a set l such that

(i) l is nondegenerate, and

(ii) $(\{A, B\} \subseteq l$ and $A \neq B) \Rightarrow [C \in l \Leftrightarrow A, B,$ and C are collinear]

$$\overleftrightarrow{AB} = \{X : \exists_x X = A + (B - A) x\}$$

$$\overleftrightarrow{A[\vec{a}]} = \{X : \exists_x X = A + \overrightarrow{ax}\}.$$

It is then possible to prove that $A \neq B$ and $\vec{a} \neq 0$ then $\overleftrightarrow{AB}$ and $\overleftrightarrow{A[\vec{a}]}$ are lines – the line through A and B, and the line through A in the direction of $\vec{a}$, respectively. (The direction of a translation is the subspace of $\mathscr{J}$ which it spans.) Planes are introduced in a similar manner and parallelism of lines and of planes can be defined by sameness of direction.

However, before taking up planes it is convenient te develop some of the affine theory of triangles and quadrilaterals. This is done in Chapter 8. Here parallelism and ratio play the main roles. The theorems of Menelaus and Ceva are among those proved in this chapter.

The developing of the concept of a plane in Chapter 9 suggests the introduction of dimension in Chapter 10. Here we settle that $\mathscr{J}$ (and, with it, $\mathscr{E}$) are three dimensional. This leads to the discussion of bases for $\mathscr{J}$ and coordinate systems for $\mathscr{E}$. The chapter closes with a discussion of the problem of representing lines and planes by equations. This ends the first year of the course.

The second part of the course begins with a rather long chapter in which the postulates for an inner product are suggested by intuitive work with the notions of orthogonal projection, perpendicularity, and distance. Once the inner product postulates are adopted we reverse this process and give formal definitions of our intuitive notions. On this basis we discuss, in Chapter 12, perpendicularity of planes and lines, of lines, and of planes.

In Chapters 13 and 14 we return to the introduction of coordinate systems in $\mathscr{E}$ – based either on general bases for $\mathscr{J}$ or on orthonormal bases for $\mathscr{J}$. This discussion involves developing the properties of second and third order determinants.

In Chapter 15 we define distance in terms of norms of translations:

$$d(A, B) = \|B - A\|.$$

In this chapter we prove the Cauchy-Schwartz inequality and the triangle inequality. We prove standard theorems about triangles, introduced the notion of an isometry, prove that reflections in planes are isometries, and, using reflection, prove that there are the expected isometries. Also, each isometry of $\mathscr{E}$ is the resultant of at most four reflections.

In Chapter 16 we introduce angles and their cosines and sines. In Chapter 17 we apply these ideas to proving the usual congruence theorems for triangles and to the study of squares, rectangles, rhombuses, isosceles, trapezoids, etc. The notion of similarity and a study of the areas of polygonal regions come in here.

In Chapter 18 we deal with spheres and circles and, on an intuitive notion of arc measure (bolstered up somewhat by a discussion of the completeness of the real numbers) introduce angle measures. We also discuss areas of circular sectors and segments.

In Chapter 18 we introduce sensed angles – ordered pairs of covertical rays – and discuss orientation of lines and of planes. We see how to assign a positive or negative measure to a sensed angle in an oriented plane. This is the basis for discussing the winding function which maps a line onto a circle.

The winding function is used in Chapter 20 to introduce the circular functions, and the chapter is devoted to exposing the analytic theory of these functions and their inverses. The inverse circular functions are, of course, used in solving "trigonometric equations".

It may be appropriate at this point to give two examples illustrating the organizing power of this approach to geometry. To do so in a reasonable space some knowledge of vector spaces must be assumed. Our first example deals with the possible intersection of two diagonal lines of a plane quadrilateral.

Consider the plane quadrilateral $ABCD$ (Figure 2) and let $B - A = \vec{a}$ and

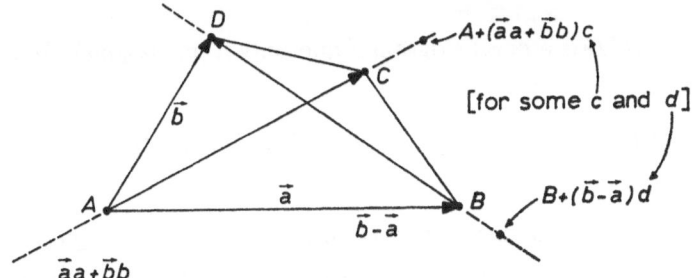

Fig. 2.

$D - A = \vec{b}$. Since $ABCD$ is a quadrilateral, A, B, and C are noncollinear and, so, $\vec{a}$ and $\vec{b}$ are linearly independent. Since A, B, C, and D are coplanar, there are numbers a and b such that $C - A = \vec{a}a + \vec{b}b$. Note that $C - D = \vec{a}a + \vec{b}(b-1)$ and, so, since $\vec{a}$ and $\vec{b}$ are linearly independent, $\overleftrightarrow{CD} \parallel \overleftrightarrow{AB}$ if and only if $b = 1$. We wish to find the point of intersection, if any, of the diagonal lines $\overleftrightarrow{AC}$ and $\overleftrightarrow{BD}$. To this end we consider the equation:

$$A + (\vec{a}a + \vec{b}b)\, c = B + (\vec{b} - \vec{a})\, d\,.$$

On substituting 'A + $\vec{a}$' for 'B' this simplifies, successively, to:

$$A + (\vec{a}a + \vec{b}b)\, c = (A + \vec{a}) + (\vec{b} - \vec{a})\, d$$
$$A + (\vec{a}a + \vec{b}b)\, c = A + (\vec{a} + (\vec{b} - \vec{a})\, d)$$
$$(\vec{a}a + \vec{b}b)\, c = \vec{a}(1 - d) + \vec{b}d$$
$$\vec{a}(ac + d - 1) + \vec{b}(bc - d) = 0$$

and, since $\vec{a}$ and $\vec{b}$ are linearly independent, to:

$$ac + d - 1 = 0 \quad \text{and} \quad bc - d = 0$$

$$c = \frac{1}{a+b} \quad \text{and} \quad d = \frac{b}{a+b}\,.$$

Note, in particular that $c = d$ if and only if $b = 1$ and recall that $b = 1$ if and only if $\overleftrightarrow{CD} \parallel \overleftrightarrow{AB}$. It follows that $\overleftrightarrow{AC}$ and $\overleftrightarrow{BD}$ intersect at a point which divides the segment from A to C and the segment from B to D in the same ratio if and only if $\overleftrightarrow{AB}$ and $\overleftrightarrow{CD}$ are parallel.

From this quite special consequence of our result follow several standard theorems. In case $a > 0$ we have:

> $ABCD$ is a trapezoid if and only if the diagonals intersect at a point which divides them in the same ratio.

In particular, with $a = 1$:

> $ABCD$ is a parallelogram if and only if its diagonals bisect each other.

In case $a < 0$ (you may wish to draw a different figure from the one given above) we have:

> The line through C and D on two sides of a triangle is parallel to the third side if and only if C and D divide the sides on which they lie in the same ratio.

In particular, with $a=\frac{1}{2}$ we have:

> A line through the midpoint of one side of a triangle contains the midpoint of a second side if and only if it is parallel to the third side.

Our second example deals with the dot product of two translations. Recall that such a dot product is 0 if and only if the directions of the two translations are orthogonal. Recall, also, that the distance between points A and B is $\sqrt{(B-A)\cdot(B-A)}$. It follows easily from the postulates for the dot product that

$$(\vec{a} - \vec{b})\cdot(\vec{a} + \vec{b}) = \vec{a}\cdot\vec{a} - \vec{b}\cdot\vec{b}.$$

If we apply this result as suggested in the first of these two figures we obtain the well known result according to which a parallelogram is a rhombus if and only if its diagonals are perpendicular. Applying the same identity as suggested in the second figure we obtain three theorems (Figure 3):
$\triangle ABC$ has a right angle at C if and only if the length of the median from C is half the length of side $\overline{AB}$.

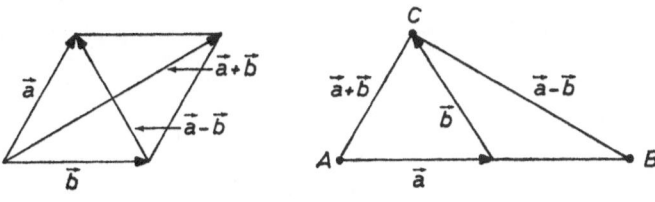

Fig. 3.

If A and B are endpoints of a diamater of a circle then $\angle ACB$ is a right angle if and only if it is inscribed in the circle.
A parallelogram is a rectangle if and only if its diagonals are congruent.

On first hearing of this course one is likely to feel that it is "too algebraic" and that the spirit of geometry has been lost. This, however, need not be the case. For, after enough basic theorems have been proved algebraically, students may be led to derive others by standard "synthetic methods". It is also worth noting that it is very easy to introduce coordinates at any point from Chapter 10 on and then use "analytic methods". Of course, it is usually easier to deal directly with translations rather than with their components with respect to some coordinate system.

Summarizing, the course teaches the standard theorems of plane and solid geometry, and of trigonometry; it introduces the student to groups, fields, and vector spaces; and it makes use of a function space operating on a space of points. It also deals with varous bits of algebra: determinants, complex numbers, and mathematical induction, to name a few.

UTCSM, Urbana, Ill., U.S.A.

MARIO VILLA

A LOGICAL APPROACH TO THE TEACHING
OF GEOMETRY AT THE SECONDARY LEVEL

[1] In this conference I shall present a few personal ideas on what I feel might be a sensible approach to the teaching of geometry at the secondary level (Section III), with special reference to a strictly logical treatment of it which might be suitable in the senior year. I shall go into details about the latter only (Section IV).

The orientation I suggest happens to break with a very deeply rooted tradition in Italy, a country where geometry has always occupied a rather prominent place among the mathematical sciences at all levels.

It will help my presentation if I dwell briefly, first, upon the teaching of mathematics (and of geometry in particular) in the Italian secondary schools under present-day established curricula (see [2] below); secondly, upon the axiomatic approach to Euclidean geometry in the strictly classical sense [3], thirdly, upon the results of present-day teaching of Euclidean geometry in Italy [4].

I shall then mention briefly [5–8] the vastly significant steps that were taken in Italy after the 1961 Bologna International Conference on the updating of the teaching of mathematics at the secondary level; significant as these steps were, they have not yet brought about a genuine reform of the standard programs, but they did help to formulate all the premises for just such a radical reform.

I

[2] *The teaching of mathematics in the Italian secondary schools according to the present standard curriculum*

The teaching of mathematics at the secondary level in Italy follows two distinct cycles. The lower cycle comprises the first three years of what is known as the "Scuola Media" (ages 10–14). The second cycle takes in the next 5 years (from "ginnasio" to "liceo", whether classical or scientific; also technical institutes and teachers colleges).

Lower cycle teaching is essentially of an intuitive and experimental character; one does not, however, exclude some deductive work that may throw light on facts and rules that are less intuitively evident. Much time is allotted to numerical calculations, arithmetical problems, geometric applications of arithmetic and geometric drawing. Emphasis is placed on the formal operational properties as they constitute the foundations of algebra. Since the 1961 Bologna Conference this lower cycle teaching has been acquiring a more modern orientation, as I shall have occasion to mention later [8].

At the second cycle ("ginnasio-liceo") one no longer dwells on the elements of arithmetic; algebra and geometry, however, are developed from their foundations according to a method which is intentionally rigorous. Thus, in geometry, one lists explicitly the undefined terms and the axioms; one allows the widest freedom of choice, but every assertion must be logically justified. However, to avoid boring and lengthy discussions, one usually handles very lightly, if at all, especially at the very beginning, demonstrations of facts of lesser importance or intuitively evident.

The teaching is always accompanied by exercises and problems, both algebraic and purely geometric (constructions and demonstrations); also, exercises dealing with trigonometry and with applications of algebra to geometry (one does not, however, at this level, give much importance to numerical calculations). It is a fact that, because it is the long established tradition of the Italian school and, I might say, its very cornerstone, we have always tended to give the greatest importance to the intellectual formation value of the mathematical sciences: a value which is derived from correct reasoning, from the strict logical dependence of propositions, and from precision of language. One hoped, of course, that this intellectual formation would also benefit those students who, after their "liceo", might abandon completely all scientific pursuits. I shall have occasion shortly [4] to express some personal reservations and criticism of this educational program.

To appreciate the rationale behind the methods of exposition, especially in geometry, one should remember what Gaetano Scorza had already pointed out back in 1908 that in Italy there exists a deeply ingrained taste for precise formulations and systems, an indelible consequence of centuries of Euclidean teaching; one should perhaps point out that this taste for purism and precision may be due to needs and inclinations that are more aesthetic in nature rather than strictly philosophical. These cultural needs may explain the decisions taken by Cremona*, Betti and Brioschi. It was in 1867 that F. Brioschi prescribed for the "liceo" the original text of Euclid, and this with the idea of exorcising from the Italian school the arithmetical methods of Legendre and his followers. It was not long before this rather rigid position became considerably more flexible; following the constructions of Peano, Veronese, and Hilbert, methods of exposition were adopted that were rather removed from the Euclidean model and acquired a definite scientific value; but, in spite of this, they still persisted in the notion of introducing measure theory much later and, for the most part, kept adhering

---

* Luigi Cremona (1830–1903) is considered the founder of the Italian School of Geometry.

to a purely synthetic approach to geometry. The Euclidean theory of proportion was practically abandoned; one chose rather to rely upon the arithmetical concept of ratio and upon the theory of real numbers which was often developed much too prematurely.

With regard to the logical development of arithmetic, its teaching at the secondary level was all but abandoned after the *Gentile* Reform* of 1923, except in teachers colleges ("istituti magistrali"**) where one can still detect some of its basic principles.

The characteristic features of the classical school are still evident and unchanged in the "liceo scientifico", with much greater importance given to drills and problems. The elements of analysis, a required topic in the "liceo scientifico", are taught only in a rather weak and unpretentious program where intuition is frequently invoked at the most critical and delicate stages of development.

Less importance is given to the formative value of mathematics in the technical and teachers institutes where the method of teaching is similar, for the most part, to that of the "liceo". Here one appeals with greater frequency to intuitive arguments to avoid minute demonstrations of easily predictable results; clearly, students here realize the inherent weakness of such "proofs".

In industrial schools and naval academies one finds that the infinitesimal calculus is dealt with much more extensively than in the scientific lyceums. Due to recent decrees, students may transfer from technical and teachers institutes to the university; but this requires, besides an understandable updating of programs and curricula, a radical revision of the academic offerings of those institutes. Such revision is long overdue.

[3] *The logical systematization of Euclidean geometry in the classical style*

The last decades of the last century and the first decades of the present one have witnessed the birth of a great number of critical studies of Euclidean geometry. The problem was to develop it rigorously through what is technically known as a hypothetico-deductive system.

Clearly, however, the problem of organizing Euclidean geometry into such a system does not have a unique solution: many are the ways of selecting the primitive concepts and the axioms that link them together. Hence it is that the problem is, in a certain sense, still open, in spite of the seriousness of the research and of all the studies that have been directed to it.

The serious studies I am referring to have been carried out in Italy and in

* G. Gentile, philosopher and politician, instituted a radical reform in secondary education in 1923.
** These institutes were set up to train elementary school teachers especially.

Germany especially. In Italy, side by side with the strictly scientific research work of Veronese (1854–1917), of Peano (1858–1932), and of his students (Vailati, Padoa, Pieri, and Cassina, to single out a few), a lively movement made itself felt, more perhaps than in any other country, in the didactic field, with regard to the systematic logical method of teaching elementary geometry at the secondary level. From this movement there came a flowering of textbooks, all of them truly worthwhile under several aspects; suffice it to mention the texts by Faifofer, Sannia, d'Ovidio, De Franchis, Rosati, Benedetti, Enriques, Amaldi, Severi, and many others. In actual practice, at the secondary level, the strictly axiomatic approach is appropriately toned down depending on the ability of the students. Often, for instance, one would permit the use of redundant axioms; such axioms, that is, that could strictly be derived from other accepted axioms, but which the teacher preferred to add to the list of primitive axioms for purely pedagogical reasons, even if the independence of the axiom system had to be sacrificed; again, one often had recourse to the handy expedient of using, even in a strictly logical presentation, words already possessed of an intuitively clear physical content.

In Germany, after the deep research work of Von Staudt, Dedekind, and Pasch, the studies of Hilbert, gathered in the classical volume *Grundlagen der Geometrie* (with seven editions between 1899 and 1930!) became for a long time the standard textbook. Hilbert's work has deeply influenced Italian scholars and writers.

In the Anglo-Saxon countries one should point out the research work of Moore and Veblen; in the field of didactics – as far as I am able to ascertain – the teaching of elementary geometry is kept at an intuitive and empirical level much longer than is done in Italy; for this reason one has not heard of much significant work done with regard to the organization of geometry from a strictly logical point of view.

In France, even though geometry in the secondary school has always been held in high regard, the prevailing orientation in teaching assumes very early the nature of a mixed approach, with simultaneous development of elementary geometry, algebra and analytic geometry, so that, also in France the problems in question have been of far less relevance than elsewhere.

[4] *Results of present day teaching of geometry in Italy*

The orientation given to the teaching of geometry in Italy, which I have briefly outlined above, has yielded, unfortunately, results that are rather negative and certainly different from what its proponents and advocates had expected.

It is with sadness that I make this assertion, both because of the damage that this has caused to the mathematical culture of my country, and also because so many truly outstanding geometers were involved in the establishment of such a program and had advocated its adoption.

That was a mistake which one must now have the courage to admit to the end that the teaching of mathematics in the Italian secondary schools may take a radically different course. What I have stated is not a matter of opinion but an incontrovertible fact confirmed by many years of experience and a question which, in this age of mass education, rings with even more ominous overtones.

The Hilbert postulates, for instance, represent an abstract structure so delicately balanced and complex that the great mass of students 15 years of age, or thereabouts, are absolutely incapable of mastering it, even with the help of an exceptionally good teacher. Wouldn't, for instance, the abstract group structure be far simpler to comprehend?

A young man who has just completed cycle one (11 to 14 years of age), where he was taught geometry from a purely intuitive and experimental viewpoint, and who goes on to the upper cycle and a strictly reasoned approach to geometry, is incapable – in almost every case – of adjusting to such a profound and sudden change. Efforts at tempering the rigor of a strict axiomatic approach (such as the expedient I have just mentioned [9] of making use of words that are rich in intuitive and physical connotations in the context of a logical demonstration) are very tempting but counterproductive in that they completely confuse the mind of a student. He has no way of telling, for instance, when the teacher uses the word "point", whether it is an empirical point he is talking about, or whether it is the "point" defined by Hilbert's postulates; he cannot see the purpose of the axioms themselves nor of certain demonstrations (which, from an experimental point of view, he correctly judges to be self evident and hence trivial). Since he is as yet incapable of grasping such a high level of abstraction, he does not understand what is being discussed and remains confused and bewildered.

It is of course not surprising that students can't develop a taste (much less a passion!) for such a subject; this type of teaching will not only fail in its primary stated objective (that of intellectual formation), but is fraught with utterly negative aspects. One is left with bitter memories, witness the successful lawyer who proudly claims he has never seen much point in mathematics.

It is also a fact that such a method tends to be time-consuming; the teaching of geometry is then forced into extremely tight boundaries when one ought instead to give it the free scope which analytic geometry easily permits and which often becomes such an engrossing challenge for young minds.

Let us not forget to add one further modern problem, namely, the critical shortage of good teachers. I should have said, more realistically, the critical shortage of teachers, period.

II

[5] *The 1961 Bologna conference and the national pilot study*

Italy is today one of the countries of the world most committed to the improvement of the teaching of mathematics at the secondary level. The most important step was undertaken by the Department of Public Education in collaboration with the Organization for Economic Cooperation and Development.

An international conference on the teaching of modern mathematics at the secondary level (organized by the International Commission on Mathematical Instruction) was held at the University of Bologna in October 1961. After this conference, and as an immediate consequence of it, the Department of Public Education set in motion in 1962 a vast experimental study for the updating of mathematics education in the Italian secondary schools.

A great number of teachers institutes were organized to prepare teachers for the pilot study planned for the following year. Each institute program consisted of two parts: the first was mainly cultural in scope; the second had a strictly didactic character. In the second part of the program the teachers would carefully go over, and discuss, the pedagogical implications of the topics pertaining to the experimental content which had been covered during the first part of the institute.

In every region of Italy many pilot classes were set up by the Department of Education to test the new program. Several teachers' handbooks and students' texts were prepared expressly for the pilot classes:

Per un insegnamento moderno della matematica nelle Scuole secondarie, 1962;
Per un insegnamento moderno della matematica nei Licei classici, nei Licei scientifici e negli Istituti magistrali, 1963;
Per un insegnamento moderno della matematica negli Istituti Tecnici, 1963;
Per un insegnamento moderno della matematica negli Istituti Tecnici, 1964;
Per un insegnamento moderno della matematica nella Scuola media, 1964.

Two other books were published later:

Matematica moderna nella Scuola media, 1965;
Matematica moderna nelle Scuole secondarie superiori, 1966.

Many seminars were held with the pilot test teachers, and one can now testify to the positive worth of the experiment. The pilot classes very soon became centers of great interest also outside their immediate location, so that a veritable "movement" began forming around the various pilot centers.

[6] *The pilot study program*

One should keep in mind that present secondary mathematics curricula in

Italy list topics which, by and large, belong to the body of mathematics that was familiar to the Greeks of the third century B.C. Clearly one could not possibly tolerate this situation much longer. One just had to take a bold stand, even at the risk of stepping on some toes. This we did. One had just to sell the idea that, at the secondary level, a veritable storehouse of topics was available, far richer in intellectual content and of greater relevance to modern applications than those offered by the traditional program.

The characteristic features and topics covered in the pilot study were mainly these: elements of set theory; elements of abstract algebra; elementary geometric transformations (with particular stress on their group structure); a rather extensive treatment of analytic geometry beginning from the earliest grades.*).

We all agreed that a good measure of traditional mathematics had naturally to be kept, but its orientation became more modern; we all agreed on a balanced fare of set theory and abstract algebra. In outlining the experimental program one had always to strive for realistic moderation, common sense and a sound balance between the old and the new.

In the pilot classes there was no increase in the length or the number of class meetings: one just presented certain parts of the traditional program in a more concise manner, thus leaving ample time for the new topics.

[7] *The clearly positive nature of the experiment with the pilot classes*

The national experiment, as I have already observed, yielded genuinely positive results.

Many teachers who at first were rather skeptical about introducing elements of set theory and of abstract algebra at the secondary level ended up becoming so enthusiastic about the program that one now is almost forced to check and restrain their zeal.

The concepts of set theory and of abstract algebra are easily accessible to the young and – I want to add – to the very young. Of this we have ample proof. They appreciate this type of mathematics for its high intellectual content and they find it to a certain extent entertaining and enjoyable. This was confirmed over and over both by the teachers who conducted the pilot classes and at many conventions dealing with "the new math". At the secondary level held in Italy during the past several years.

I must add that in Italy several experiments have been carried out on introducing some elements of set theory at the elementary level (ages 6 to 10). The results were likewise highly positive.

---

* This describes the basic preparatory math. courses. Later, and depending on the type of schools, one covers Fourier series and transformations, linear programming, elements of computer science, elements of probability theory and of actuarial mathematics.

[8] *Toward a reform of the teaching of mathematics in Italian secondary schools*

One can, first of all, declare that all Italian mathematicians are in agreement as to the wisdom and practicality of the decisions taken in this field, starting with the pilot classes. There are differences of opinion, but they do not concern essential or substantive questions.

The new programs which the Department of Education will make mandatory for all Italian secondary schools will be deeply influenced by the pilot classes experiment.

I should add that, at the intermediate level (Scuola Media), a partial reform has already taken place with the decision that permitted the introduction of many elements of the "new math".

III

[9] *A proposal for the teaching of geometry prior to the senior year at the secondary level*

I hold that a strictly rigorous presentation of Euclidean geometry should not be introduced before the last year of secondary school; the 17 or 18 year old student will then be sufficiently mature.* I shall give a detailed proposal for such a rigorous approach to geometry in a later section. I would like to dwell briefly here on what might be a feasible teaching plan for the preceding few years (ages 11–17).

After introducing the real numbers through their decimal representation, I would as soon as possible begin to develop analytic geometry, and this as extensively as possible, together with an adequate dose of traditional algebra and of the key concepts of calculus.

Having introduced from the very start the first notions of set theory, I would dwell particularly on elementary mappings (or pointwise transformations).

Experience with elementary mappings on a line (or from one line to another), and on the plane (or from one plane to another) will make it possible to dwell through concrete instances on the fundamental mathematical notion of correspondence.

Elementary transformations naturally lead to groups and rings, and serve as concrete illustrations of these abstract algebraic structures. By means of groups of transformations one is naturally led to a classification of the various geometries, and is thus given a glimpse of the boundless field of geometrical research.

---

* The two present cycles in Italian secondary education would be automatically eliminated.

The study of elementary transformations (affinities, similitudes, congruences, inversions) would be carried out mainly with the aid of analytic geometry.

As a plan for the study of geometry during this period (ages 11–17) I would definitely suggest to develop geometry and algebra simultaneously.

A PROPOSED APPROACH TO THE TEACHING OF
SECONDARY (Senior Year) GEOMETRY

IV. EUCLIDEAN LINE AND EUCLIDEAN PLANE

[10] *Basic set theory notions*

For the sake of completeness let us recall some notions of set theory which we assume are already familiar to students in the senior year.

Let I be a set. We write '$e \in I$' to denote that the element e is one of the elements of I; that is, e *belongs* to I. Given any two sets I and F, if every element of F is an element of I, we say that F is *contained* in I, or I *contains* F, and we write $F \subset I$ (or $I \supset F$); we say that F is a *subset* of I.

The set formed by the elements belonging to a set I *or* to a set F is called the *union* of the two sets and is denoted by $I \cup F$. The set of elements belonging to a set I *and* to a set F is called the *intersection* of the two sets and is denoted by $I \cap F$. "Nothingness" as the object of thought is called the *empty set* and is denoted by $\emptyset$.

[11] *Correspondence between sets*

Let's recall also some notion of correspondence between sets. A one-to-one correspondence between two sets I_1 and I_2 may be thought of as being formed by two *operations*, one of which "changes", so to say, the elements of I_1 into the elements of I_2, the other, on the contrary, the elements of I_2 into the elements of I_1. Each operation is called the *inverse* of the other, and if one of them is denoted by ω, the other is denoted by ω^{-1}.

Let I_1, I_2, I_3 be sets; let ω_1 be a correspondence between I_1 and I_2; let ω_2 be a correspondence between I_2 and I_3; let us call corresponding the two elements of I_1 and I_3 that are mapped to the same element of I_2 under ω_1, ω_2, respectively; a transformation Ω between I_1, I_3 arises which is called the *composite* (alternatively, their *composition*) of the two transformations ω_1, ω_2 and is denoted by $\omega_1 \circ \omega_2$.

It is clear that this notion of composition may be carried over to more than two correspondences at once. In fact, if we consider the composition of Ω and another correspondence ω_3, we have a new correspondence Ω_1 which is called the composition of ω_1, ω_2, ω_3 and is written $\Omega_1 = \Omega \circ \omega_3 = \omega_1 \circ \omega_2 \circ \omega_3$, and so on.

If the three sets I_1, I_2, I_3 are the same, we remark that the composition of the transformations ω_1, ω_2 *does not*, in general, have the commutative property. That is to say, the two transformations $\omega_1 \circ \omega_2$ and $\omega_2 \circ \omega_1$ are, in general, different. The correspondence from a set to itself in which every element corresponds to itself is called the *identity transformation* on the set.

If ω is a one-to-one transformation between two sets I_1, I_2 and $\mathscr{I}$ is the identity on I_2 we have, obviously, $\omega \circ \mathscr{I} = \omega$.

It is clear, too, that the composition of a transformation ω and its inverse transformation ω^{-1} is an identity, that is, $\omega \circ \omega^{-1} = \mathscr{I}$.

[12] *Topological spaces*

Let I be a non-empty set, and let Σ be a set of subsets of I. We say that Σ defines a *topological structure* or a *topology* on I, if the following conditions are satisfied:

1)   The arbitrary union of any sets of Σ belongs to Σ; the empty set belongs to Σ, and $I \in \Sigma$.

2)   The finite intersection of any sets of Σ belongs to Σ.

The sets of Σ are called the *open sets* of the topology defined on I, which is now called a *topological space*; the elements of I are called the *points* of the topological space.

Any non-empty set I may be always endowed with a topological structure: for instance, we have a topology on I by considering, as subsets of I, I itself and the empty set.

Some examples of topological spaces will be given later.

If P is a point of I, any subset including an open set A which contains P is called a *neighborhood* of P.

Two topological spaces S and S' are called *homeomorphic* if there is a one-to-one transformation from S onto S' which maps the open sets of S into the open sets of S' and conversely; such a transformation is called a *homeomorphism*.

[13] *Continuous transformations*

Given two topological spaces S and S' (which may be the same), let us consider the one-to-one transformation $\mathscr{C}$ under which to every point P of S corresponds a point P' of S'. The transformation $\mathscr{C}$ is said to be *continuous* if (and only if) for every neighborhood γ' of P' the set corresponding to γ' under $\mathscr{C}^{-1}$ is a neighborhood of P. The transformation $\mathscr{C}$ is called *continuous* in S if it is continous at every point of S.

[14] *Homeomorphism*

A one-to-one transformation $\mathscr{C}$ from a topological space S onto a topological space S' is called *bicontinuous* if both $\mathscr{C}$ and its inverse $\mathscr{C}^{-1}$ are continuous. It may be proved that:

> A one-to-one transformation from a topological space S onto a topological space S' is a homeomorphism [12] if and only if it is bicontinuous.*

We have then:

> Homeomorphism is an equivalence relation.**

It is obviously reflexive and symmetric. But it is transitive, too; that is, if the topological space S is homeomorphic to the topological space S', and S' is homeomorphic to the topological space S'', then S is homeomorphic to S''. It is sufficient to observe that the composition of two one-to-one transformations is still a one-to-one transformation, and it is also bicontinuous if the two transformations are bicontinuous.

Homeomorphism is the main equivalence relation in topology. In topology two homeomorphic spaces are regarded as different only in the name given to their elements, and therefore the object of topology is the study of the properties of topological spaces up to a homeomorphism. Topology – in short – treats definitions, properties, operations, numbers, etc., which are invariant under homeomorphisms.

[15] *Metric spaces*

A set I is called a *metric space* when to every pair P, Q of elements of I (called points), different or the same, corresponds a real number $|PQ|$, called their *distance*, satisfying the two following conditions:

1) $|PQ| = 0$ if and only if P and Q coincide, and conversely.
2)   (1) $|PQ| \leqslant |PR| + |QR|$,

where P, Q, R are any three points of I. This relation is called the "triangle inequality".

We have then:

> The distance between two points is non-negative.

In fact, if $P=Q$, it follows from condition (1) that $2|PR| \leqslant 0$, that is, the distance of two points is positive or zero.

From this last result and condition (1) it follows that the distance between two different points is positive. We have also: $|PQ|=|QP|$; that is, the distance between two points does not depend on their order. In fact if

---

* This theorem will not be proved.
** This notion will be recalled, of course.

$R=P$, it follows from (1) that $|PQ| \leqslant |QP|$. It follows also from (1) that $|QP| \leqslant |QR| + |PR|$, and if $R=Q$, $|QP| \leqslant |PQ|$. Then, from $|PQ| \leqslant |QP|$ and $|PQ| \geqslant |QP|$, we have $|PQ| = |QP|$. If P is a point of the metric set I, the set of points Q of I such that

$$|PQ| < \varrho$$

is called an open sphere with center P and radius ϱ (ϱ being a real number >0).

This concept allows us to define a topological structure on the set I [12]. If we consider the subsets A of I, such that for any point $P \in A$, there is an open sphere with center P with all its points belonging to A, it is possible to verify that these subsets A satisfy conditions 1) and 2) of [12] so that the class of the subsets A defines a topology on I in which the sets A are the open sets.*

[16] *The real line*

Consider a variable x and the set of real values which may be given to this variable. Each of those values is called a *point* and also the *coordinate* of the point. The point with coordinate 0, that is, the point O, is called the *origin*. Every set of points P with coordinate x, such that

$$a < x < b \qquad\qquad\qquad (2)$$

is called an *interval*; (a, b two real numbers with $a<b$). In the set of all real numbers a topology is determined by the requirement that the open sets be intervals, unions of intervals, or the empty set. In fact, it is immediately seen that conditions 1) and 2) of [12] are satisfied. Then the sets of all such points is a topological space called the *real line*. Any of the topological spaces [12] which is homeomorphic [14] to the real line is also called the real line. A *segment* of the real line is the set of the points P of coordinate x, such that

$$a \leqslant x \leqslant b$$

(a, b two real numbers with $a<b$). If on the real line the number

$$|p - q| \qquad\qquad\qquad (3)$$

is associated with the pair of points P, Q with coordinates p, q respectively, a metric space, [15], is obtained because conditions 1) and 2) of [15] are satisfied. Such a metric space – in which the distance between points P, Q is determined by (3) – will be called the *real line with Euclidean metric*. The topology defined by the distance (3) [15] on the real line is, as we can easily verify, the same topology defined above.

---

* This proof, even though it doesn't present difficulties, will not be made.

[17] *The real plane*

Let us consider two ordered variables x_1, x_2 and the set of all ordered pairs of real values that may be given to these two variables. Each of these pairs of values is called a *point*; the numbers forming one pair are called the *coordinates* of that point. The point $(0, 0)$ is called the *origin*. By an *interval* we mean a set of points P of coordinates x_1, x_2 such that

$$a_1 < x_1 < b_1,$$
$$a_2 < x_2 < b_2,$$

where a_1, b_1 and a_2, b_2 are two pairs of real numbers such that $a_1 < b_1$ and $a_2 < b_2$. In the set of all the points considered, a topology is determined by taking as open sets the intervals, the unions of intervals, and the empty set. Indeed, it is immediately seen that conditions 1) and 2) of [12] are verified. The set of all the given points is then a topological space [12] which is called the *real plane*. Any topological space [12] which is homeomorphic [14] to the real plane is called the real plane also.

If in the real plane the number

$$|PQ| = \sqrt{(p_1 - q_1)^2 + (p_2 - q_2)^2} \tag{4}$$

is associated with the pair of points $P(p_1, p_2)$, $Q(q_1, q_2)$, we obtain a metric space [15] because the first and second conditions of [15] are satisfied. Such a metric space – in which the distance between two points P, Q is given by (4) – will be called the *real plane with Euclidean metric*.

It is possible to prove that the topology of the real plane defined by the distance (4) [15] is the same topology we have considered above.

[18] *The group of translations on the real line with Euclidean metric*

Let us consider the relation

$$y = x + k, \tag{5}$$

k being a (real) given constant. Calling *corresponding* any two points x and y of the real line when their coordinates satisfy (5), a transformation between the points of the real line arises. This transformation is called a *translation**. Equation (5) is called the *translation equation*. It can be verified that the translations on the real line form a group****. It is also possible to verify

---

* During the preceding school years the elementary correspondences have already been studied (among them the translations on the line) by the methods of analytic geometry. Essentially, therefore, we return now to already known concepts and properties.
** The concept of the group of transformations has already been acquired during the earlier years of school.

immediately that, given any two points $P(p)$, $Q(q)$ of the real line with Euclidean metric,

$$|p - q|$$

that is the distance between the two points, is invariant under translation. Translations are homeomorphisms of the real line onto itself [14].

[19] *The Euclidean line*

The real line with Euclidean metric in which we consider only the translations as transformations or, in other words, the real line with the structure of the group of translations [18], is called the *Euclidean line* and is denoted by E_1. The points of the real line with a Euclidean metric are called *points of E_1*; the coordinate of any point P of the real line is called the *coordinate* of the point P of E_1; the translations on the real line are called the *translations* on E_1; the distance between two points P, Q of the real line with a Euclidean metric is called the *distance* between the points P, Q of E_1; etc.

[20] *Euclidean geometry (on the Euclidean line)*

The translations of a Euclidean line E_1 onto itself form a group [18, 19]. More precisely, the geometry of this group, according to the usual meaning*, is called the *Euclidean geometry* of E_1.

[21] *The group of isometries in the real plane with Euclidean metric*

Let us consider the relations

$$y_1 = ax_1 - bx_2 + c_1 \tag{6}$$
$$y_2 = bx_1 + ax_2 + c_2,$$

a, b, c_1, c_2 being (real) constants, such that

$$a^2 + b^2 = 1. \tag{7}$$

Any two points of the real plane are called corresponding when their coordinates x_1, x_2 and y_1, y_2 satisfy (6). A one-to-one transformation then arises, called a *direct isometry* between the points of the real plane**.

Equations (6) are called *equations of isometry*. It happens that the direct isometries of the real plane onto itself form a group. It happens also that, given any pair of points $P(p_1, p_2)$, $Q(q_1, q_2)$ of the real plane with a Euclidean metric,

$$\sqrt{(p_1 - p_2)^2 + (q_1 - q_2)^2}$$

---

* Klein's Erlangen Program will have naturally come up while treating groups of transformations in the lower classes.
** We assume that isometries have also been studied in the lower classes.

(that is, the distance between the two points [17]) is invariant under isometries. The isometries are homeomorphisms of the real plane onto itself [17].

[22] *The Euclidean plane*

The real plane with a Euclidean metric in which we confine our attention solely to direct isometries as transformations or, in other words, the real plane endowed with the structure of the group of isometries [21], is called the *Euclidean plane*, and is denoted by E_2. The points of the real plane with a Euclidean metric are called *points* of E_2; the coordinates of any point P of the real plane are called the *coordinates* of the point P of E_2; the direct isometries in the real plane are called the *direct isometries* of E_2; the distance between two points P, Q of the real plane with a Euclidean metric is called the *distance* between the two points P, Q of E_2; etc.

[23] *Euclidean geometry (on the Euclidean plane)*

The isometries in the Euclidean plane form a group [21, 22]. More precisely, the geometry of this group of transformations, according to the usual meaning, is called the *Euclidean geometry of E_2*.

V. (EUCLIDEAN) GEOMETRY ON THE EUCLIDEAN LINE

[24] *Direction and distance on the Euclidean line*

Let us consider the Euclidean line [19]. Now, the coordinate of any point is called, more properly, the *abscissa*. Two directions, *opposite* one to the other, can be distinguished on the real line. We say that a point *moves* along the line in one direction when its abscissa is always increasing (from $-\infty$ to $+\infty$), a point moves along the line in the other direction when its abscissa is always decreasing (from $+\infty$ to $-\infty$). The first of the two will be called the *positive direction*, the second the *negative direction*.

A line is divided into two half-lines, with origin at A, by any point A on the line: the positive half-line is the one traced out by a point moving from A in the positive direction; the negative one is the half-line traced by a point moving from A in the opposite direction.

A segment AB [16, 19] on the Euclidean line is considered *positive* or *negative* depending on whether its tracing point going from A to B follows the positive or the negative direction. The distance between two points A, B with respective abscissas a and b is given by [16, 19]

$$d = |a - b|.$$

This distance will be given the sign $+$ or $-$ depending on whether the segment AB is positive or negative. If the distance between the two points A and B is still denoted by AB we have

$$AB = b - a. \tag{8}$$

Thus, *the directed distance between two points is given by the difference between the abscissa of the second and the abscissa of the first.* The following relation (of *Chasles*) holds:

$$AB + BC + CD + \cdots + MN + NA = 0, \tag{9}$$

where A, B, C,..., M, N are a finite number of points arbitrarily chosen on the line. In fact from (8) we have:

$$b - a + c - b + d - c + \cdots + n - m + a - n = 0,$$

where $c, d, ..., m, n$ are the abscissas of the points $C, D, ..., M, N$.

[25] *Meaning of abscissas*

Let us consider the origin O [16, 19], that is, the point of abscissa 0, and a point A of abscissa a. We have:

The abscissa of a point A is the distance OA.

In fact, from (8),

$$OA = a.$$

If we consider the point U of abscissa 1 (unit point) on the line we can say also that the abscissa of a point A is equal to the ratio OA/OU.

[26] *Properties of translations*

The equation of a translation [18, 19] is

$$y = x + k, \tag{5}$$

where k is a given constant. We have:

A translation is an order preserving transformation.

In fact, if a, b are the abscissas of any two points A, B, and a', b' the abscissas of their corresponding points A', B', under (5) we have

$$a' = a + k$$
$$b' = b + k.$$

Hence,

$$a' - b' = a - b. \tag{10}$$

It follows that $a' < b'$ if $a < b$, and $a' > b'$ if $a > b$, and this establishes the theorem. Hence also:

The distance between two points is invariant under translations.

The distance between two points A, B [24] is given by $b-a$; likewise the distance between points A', B' is $b'-a'$. Now, from (10) these two numbers are equal. Furthermore:

A translation is determined by one pair of corresponding points.

In fact, if a, a' are the abscissas of two corresponding points A, A', from (5) we have

$$k = a' - a,$$

so that the translation is completely determined.

A translation may be defined also as a one-to-one transformation between the points of the line under which to the point P corresponds a point P' such that the distance PP' is equal to the distance AA', called the *amplitude* of the translation, where, from (5) the amplitude is given by k.

[27] *Transformation of abscissas*

Formula (5) may be also interpreted as the formula for a transformation of the abscissas, corresponding to a change of their origin. For all new y abscissas, the new origin is the point which had abscissa $-k$ in the original x abscissas.

VI. (EUCLIDEAN) GEOMETRY OF THE EUCLIDEAN PLANE

[28] *Lines of the Euclidean plane*

In the Euclidean plane [22] the coordinates of a point, x and y, are called *Cartesian coordinates* of the point; more precisely, the first (x) is called the *abscissa*, the second (y) the *ordinate* of the point.

The set of all those points of the Euclidean plane whose coordinates satisfy the equation

$$ax + by + c = 0, \qquad (11)$$

where a, b, c are arbitrary constants (a, b not both equal to zero) is called a *line* of the Euclidean plane; (11) is called an *equation* of the line.*

The lines having the equations $x=0$, $y=0$ are called Cartesian *axes*, $y=0$ being called the *abscissa axis* (or x-axis), and $x=0$ the *ordinate axis* (or y-axis). The point of intersection of the axes is the origin of the Cartesian coordinates [17, 22].

---

* At this point we shall not go into details about the concept of parallelism of two lines, about the condition for parallelism of two lines (the numbers b and $-a$ defined up to a factor of proportionality, are called *direction coefficients* of line (11), and their ratio $-a/b$ is called the *direction ratio* of line (11)), about the point of intersection of two lines, about bundles of lines and their equations.

[29] *Transformation of Cartesian coordinates*

Equations (6) and (7), which we write

$$x' = ax - by + c_1$$
$$y' = bx + ay + c_2, \tag{12}$$

where

$$a^2 + b^2 = 1, \tag{13}$$

may be interpreted also as formulas of a transformation of the Cartesian coordinates, in conformity with a change of Cartesian axes.

In the new Cartesian system the axes x', y' are the lines which had equations

$$bx + ay + c_2 = 0, \qquad ax - by + c_2 = 0$$

in the first Cartesian system.

[30] *Directed (oriented) lines*

In the Euclidean plane – as we have already learned – any pair of points $P_1(x_1, y_1)$, $P_2(x_2, y_2)$ has an invariant under isometries, given by

$$d = \sqrt{(x_1 - x_2)^2 + (y_1 - y_2)^2} \tag{14}$$

and called the distance between the two points [17, 22].

A line in the Euclidean plane is a Euclidean line [19] because the number $d = |x_1 - x_2|$ is determined for every pair of points $P_1(x_1)$, $P_2(x_2)$. In fact, when a transformation of coordinates [29] is effected under which the new x-axis is the given line, from (14) it follows that

$$d = |x_1 - x_2|$$

for every pair of points $P_1(x_1, 0)$, $P_2(x_2, 0)$ of this axis.

There are two directions on the line of the Euclidean plane. We say that a point *moves* along the line in one direction when its abscissa, for instance, is always increasing (from $-\infty$ to $+\infty$), while we say that a point moves along the line in the other direction when its abscissa is always decreasing (from $+\infty$ to $-\infty$). However an isometry in the plane (that maps the line onto itself) can change one direction into the other, contrary to what happens on the real line when we consider translations; so that the choice of the *positive direction* is arbitrary. A line is called *directed* when the positive direction (and hence, also the negative one) has been chosen. A directed line is divided by one of its points A into two half-lines having A as the origin. We call the *positive half-line* that which is traced out in the positive direction by a point moving from A; the negative half-line is the other. A segment AB [16, 19] of the directed line is called positive or negative depending on whether a point which traces it out follows the positive or the negative direction. Depending on whether the segment AB is positive or negative we shall give the distance between the two points the sign $+$ or $-$. Two parallel lines are called *similarly directed* when the positive direction on both of them is the direction of their increasing or decreasing abscissas.*

[31] *Cosine of the angle between two directed lines*

Let r, s be two directed, non-parallel lines in the Euclidean plane; let P_1 be their point of intersection and let P_2, P_3 be any two points of r and s respectively; by the *cosine of the angle of the two directed lines r, s* we mean the number

$$\cos rs = \frac{P_1P_2^2 + P_1P_3^2 - P_2P_3^2}{2 \cdot P_1P_2 \cdot P_1P_3}, \tag{15}$$

where the directed distance between the two points P_1 and P_2, for instance, is denoted by P_1P_2 [30].**

If the two lines are parallel and similarly directed [30], we get $\cos rs = +1$; if the two lines are the same or parallel but not oppositely directed we get $\cos rs = -1$.

Given lines r, s having equations [28],

$$\frac{x - x_1}{x_1 - x_2} = \frac{y - y_1}{y_1 - y_2}, \qquad \frac{x - x_1}{x_1 - x_3} = \frac{y - y_1}{y_1 - y_3}, \tag{16}$$

the cosine of the angle between the two given lines (14), (15) is

$$\cos rs = \cos sr = \frac{(x_1 - x_2)(x_1 - x_3) + (y_1 - y_2)(y_1 - y_3)}{\pm\sqrt{(x_1 - x_2)^2 + (y_1 - y_2)^2}\sqrt{(x_1 - x_3)^2 + (y_1 - y_3)^2}}.$$

The ambiguity in the choice of the sign is removed by directing the lines, while equations (16) represent the non-directed lines. If l, m and l', m' are the direction coefficients of lines r, s [28], we have:

$$\cos rs = \frac{ll' + mm'}{\pm\sqrt{l^2 + m^2}\sqrt{l'^2 + m'^2}}. \tag{17}$$

(17) shows that cos rs depends on the two directed lines only, not on the points P_1, P_2, P_3 chosen on them.

---

* If the two lines are parallel to the y-axis they are called similarly directed when the positive direction chosen for both of them is the direction of the increasing or decreasing ordinates.
** If the two directed lines coincide [24, 30], $P_2P_3 = P_1P_3 - P_1P_2$, and then from (5) we obtain cos $rs = +1$.

If the two lines r, s have, respectively, equations

$$ax + by + c = 0,$$
$$a'x + b'y + c' = 0,$$

the direction coefficients of line r are [28] b, $-a$, and the direction coefficients of line s are b', $-a'$. From (17) we have:

$$\cos rs = \frac{aa' + bb'}{\pm \sqrt{a^2 + b^2} \sqrt{a'^2 + b'^2}}. \tag{18}$$

If r' is a line which is parallel to r', s' a line parallel to s, and if the parallel lines are similarly directed [30], since r, r' have the same direction coefficients (as also s, s'), from (17) we deduce $\cos rs = \cos r's'$.

[32] *Condition for perpendicularity of two lines*

Two lines are called perpendicular, or orthogonal, or normal, if the cosine of their angle is zero.* From (18) it follows that

a necessary and sufficient condition that the lines

$$ax + by + c = 0,$$
$$a'x + b'y + c' = 0$$

are perpendicular is that

$$aa' + bb' = 0.** \tag{19}$$

[33] *Meaning of Cartesian coordinates*

From (19) it follows that the two Cartesian axes x, y are orthogonal [28]. On the x-axis [28] the positive direction is given by the increasing of x, and on the y-axis, the positive direction is given by the increasing of y.

Given a point P in the plane, let P_1 be the point of intersection of the x-axis with the line passing through P and parallel to the y-axis, and P_2 the point of intersection of the y-axis and the line passing through P and parallel to the x-axis. Then

the Cartesian coordinates $\bar{x}$, $\bar{y}$ of a point P are the signed distances OP_1, OP_2, respectively (O being the origin of the coordinates).

In fact, the line passing through P and parallel to the y-axis has equation $x - \bar{x} = 0$, and this line intersects the x-axis at the point P_1 with coordinate $\bar{x}$.

---

* When the lines P_1P_2, P_1P_3 are perpendicular, Pythagoras' Theorem follows from (15).
** It must be remarked that the two lines $bx + ay + c_2 = 0$, $ax - by + c_1 = 0$ [29] are orthogonal, too.

Similarly, the line passing through P and parallel to the x-axis, has the equation $y - \bar{y} = 0$, and this line intersects the y-axis at the point P_2 with coordinate $\bar{y}$.

On the x-axis, which is a Euclidean line [30], the points O, P_1 have coordinates 0, $\bar{x}$, respectively, and hence [25] $OP_1 = \bar{x}$. Similarly, on the y-axis the points O, P_2 have coordinates 0, $\bar{y}$, respectively, and hence $OP_2 = \bar{y}$.

[34] *Polygonal paths and their orthogonal projections on a line*

The *orthogonal* (or *normal*) projection of a point P on an oriented line r is P', the intersection between r and the perpendicular from P to r.

Given a segment AB on an oriented line s in the plane, A', B' being the orthogonal projections of A, B on r, the segment $A'B'$ (with its own sign) is called the *orthogonal projection* of the segment AB on r.

But if r, s are orthogonal, the orthogonal projection of every segment on s becomes a point. One can verify that equal segments, placed on parallel and equally oriented lines, have equal projections on the same line.

Now, the following relation holds between a segment AB on an oriented line s and its orthogonal projection $A'B'$ on an oriented line r:

$$A'B' = AB \cos rs. \tag{20}$$

To prove this relation we can suppose that the point A belongs to the line r, because the projection segment does not change if one replaces segment AB with an equal segment on a line parallel to s and equally oriented. Then

$$A' = A.$$

From the triangle ABB', it follows from (15) that

$$\cos rs = \frac{AB^2 + A'B'^2 - BB'^2}{2 \cdot AB \cdot A'B'}. \tag{21}$$

But, from Pythagoras' Theorem [31]

$$AB^2 = A'B'^2 + BB'^2. \tag{22}$$

From (21) and (22) we have (20).

If A, B, C,..., M, N are any points in the plane, taken in a specified order, a *polygonal path* is the set formed by these points (*vertices*) and the segments AB, BC,..., MN (*sides*) determined by the first and the second points, by the second and third, etc., according to the specified order. If $N = A$, the path is *closed*. If $N \neq A$, the path is called *open*: A and N are the *extremes* and the segment AN is called the *closure segment*. By the orthogonal projection of a path $ABC...MN$ on the line r we mean the sum $A'B' + B'C' + \cdots + M'N' = A'N'$, and $A'N'$ is the orthogonal projection of the closure segment AN of the polygonal path. So we have that

> *The orthogonal projection of a polygonal path on a line is equal to the orthogonal projection of the closure segment.*

[35] *Direction cosines of an oriented line*

The direction cosines of an oriented line r are the cosine of the angle between line r and the x-axis and the cosine of the angle between line r and the y-axis [31].

[36] *A theorem on isometries*

We shall just state the following theorem, which in the proposed plan will be applied repeatedly:

> *There is exactly one direct isometry under which a point A' corresponds to a point A, and a half-line t' with origin at A' corresponds to a half-line t with origin at A (the point and the half-lines being given arbitrarily).*

[37] *Equation of an oriented line*

Let r be an oriented line; draw line n perpendicular to r and passing through the origin O; call N the point of intersection of r and n. Let us consider the direct isometry under which N corresponds to O, and the positive half-line of r (with origin at N) to the positive half-line of the x-axis (with origin at O) [36].

Under this isometry a well-defined half-line (assumed positive) of line n (with origin at N) corresponds to the positive half-line of the y-axis (with origin at O). If $P(x, y)$ is a point of r, let P_1 be the orthogonal projection of P on the x-axis. The sides of the polygonal path OP_1P are given by the coordinates of P. Projecting the path OP_1P perpendicularly on line n, by setting $ON = P$, we have [34]

$$x \cos xn + y \cos yn - p = 0. \tag{23}$$

The latter equation is the equation of the line r, and because of its form it is called the *normal equation* of r.

If a line is represented by Equation (11), from (23) we conclude:

> *The direction cosines of a line n normal to the line $ax + by + c = 0$ are*
>
> $$\cos xn = \frac{a}{\pm \sqrt{a^2 + b^2}}$$
> $$\cos yn = \frac{b}{\pm \sqrt{a^2 + b^2}}, \tag{24}$$
>
> *where the upper or the lower signs must be taken together.*

[38] *Some remarks on the direction cosines of a line*

From the latter result it follows that *the direction cosines of line r with equation $ax + by + c = 0$ are given by*

$$\cos xr = \frac{b}{\pm \sqrt{a^2 + b^2}}, \qquad \cos yr = \frac{a}{\mp \sqrt{a^2 + b^2}} \tag{25}$$

where the upper or the lower signs must be taken together.

In fact, draw lines r' and n' through O and parallel to r and n, respectively (parallel lines being equally oriented) [30]. Consider the direct isometry under which the positive half-line of the y-axis (with origin at O) corresponds to the positive half-line of the x-axis (with origin at O) [36]. Under this isometry the positive half-line of n' (with origin at O) corresponds to the positive half-line of r' (with origin at O). Hence [31],

$$\cos xr = \cos xr' = \cos yn' = \cos yn. \tag{26}$$

Let the direct isometry be considered under which the positive half-line of the x-axis (with origin at O) corresponds to the positive half-line of the y-axis (with origin at O). Under the last isometry the negative half-line of n' (with origin at O) corresponds to the positive half-line of r (with origin at O). Therefore, [31]

$$\cos yr = \cos yr' = -\cos xn' = -\cos xn. \tag{27}$$

From (24), (26), (27) we obtain (25). From the last result it is easily deduced that

> The cosine of the angle between two oriented lines is equal to the sum of the products of the direction cosines of one line and the corresponding direction cosines of the other.

[39] *Sine of the angle between two oriented lines*

Let r, s be two oriented lines in the plane; let A be their point of intersection: if r and s are distinct and not parallel (in the opposite case, let A be any point of r); let n be the line which is perpendicular to r and passing through A. Consider the direct isometry under which A corresponds to O and the positive half-line of r (with origin at O) corresponds to the positive half-line of the x-axis. Under this isometry, to the positive half-line of the x-axis (with origin at O) corresponds a well-determined half-line of the line n (with origin at A) which we assume positive. Having made this definition, by the sine of the angle between two oriented lines r, s we mean the number:

$$\sin rs = \cos sn.$$

I shall forego here the full development of the program since I believe the above presentation suffices as an illustration.

Università di Bologna
Bologna, Italy

JOHN WILLIAMS

PROBLEMS AND POSSIBILITIES IN THE ASSESSMENT AND INVESTIGATION OF MATHEMATICS LEARNING

I. INTRODUCTION

It is very apparent to most of those who are concerned with the teaching of mathematics that some kind of attempt should be made to establish whether the results of a teaching endeavour are more desirable than those of its alternatives. The curriculum evaluator addresses himself to this kind of task, and, although the utility of the outcome of his efforts is, more often than not, highly questionable, we are likely to accept that the aims of this kind of exercise are relevant to the teacher.

Of more doubtful relevance, in the eyes of the mathematics educator, are the investigations carried out by the psychologist, under "artificial conditions" in his laboratory. His aims, it seems, coincide only by accident with those of the teacher, for *he* is primarily after general laws of learning, which may, or may not, be of direct use in the classroom.

It is difficult for me to believe that those who are concerned with the development of curricula can fail to be interested, and to some extent, involved, in the problems of assessing the products and investigating the processes of the learning activities that they design. Presumably, *any* organism that needs to adjust to a reality needs also to receive feedback relating to the success of his attempts to make this kind of adjustment. There is no reason why the curriculum developer should be an exception to this rule.

However, the way in which such feedback is sought and the form in which it is obtained, are most certainly matters of hot contention. Attempts to cope with the problem of obtaining feedback have to date fallen far short of success, and their results should not be taken at face value, for they are likely to promise more than they yield.

In this paper I shall do my best to survey some aspects of both product evaluation and process investigation insofar as they impinge on mathematics education. My objectives are the following:

(a) To give some idea of the role of these activities, and to urge that, in some form or other, they are indispensable to curriculum development;

(b) More particularly, to show some of the pitfalls they offer;

(c) To suggest that the feasible version of each differs greatly from the version that we assume in much of our discussion; occasionally I shall indicate directions that I think attempts to increase their feasibility might take.

II. CURRICULUM EVALUATION

First let us consider curriculum evaluation.

It is not inconcervable that a new course, while giving everyone the impression that it is excellent, should be inferior to an alternative, or inferior to what it might be. For this reason we are prompted to look for objective proof of the course's worth. How do we obtain such proof? The obvious steps to take would be first to state clearly and in objective terms the details of the outcomes we demand of the curriculum and then to construct instruments and procedures that we might apply to the task of establishing whether these outcomes have been achieved.

It is usual to talk of two kinds of evaluation:

(1) *Summative evaluation.* This consists of the assessment of the final effects of a course. It enables us to compare the effectiveness of one course with that of another, or with that required by standards that we have set, and thus to make decisions concerning either whether we should use it or in what form we should use it.

(2) *Formative evaluation.* This consists of the assessment of the effectiveness of a course while it is being developed. This enables us to detect how the course might be improved in such a way that its parts become more compatible both with one another and with the objectives that we have set for it.

A. *Some Difficulties*

Both of these kinds of evaluation are easier to conceive in the abstract than they are to implement. We shall examine just a few of the obstacles to their realization.

(a) *Which criteria should we use?* Let us first consider this at the level of the actual tests that we might use in evaluating a course. Suppose we wished to choose or construct a test for comparing the effectiveness of two mathematics courses, in one of which traditional topics were taught and in the other of which modern mathematical topics were taught. Of what should our test consist? If it covered only traditional topics it would favour one course while if it covered only modern topics it would favour the other.

An apparent way out would be for us to state clearly what we wanted from *any* course and then to devise a test which was appropriate for the measurement of this *whether or not* the particular courses under evaluation catered for it. That a course did *not* cater for it would be a criticism of that course. We shall consider a third way out later.

(b) *At which level should we judge a course?* Should a new course in the primary school be assessed for the effects it produces at the end of the primary stage, or should we assess it in terms of its effects at a *later* stage – say, when the pupil takes a public examination, or when he actually has to use his education to earn a living in the adult world? Obviously we are not interested only in the immediate effects of a course; but obviously, on the other hand, the further we get from the impact that the curriculum has on the pupil, the more likely it is that these effects will have become diluted by intervening influences.

Two of the devices that might be used for surmounting this problem are as follows: The obvious means is that of assessing the products of the course both immediately and then subsequently through a series of follow-up evaluations. A less obvious means is to use, in the immediate asesssment of a course, instruments whose relationship to the long-term effects in which we are interested is known or may reasonably be assumed. Where, in our assessment, we can use "experienced" instruments, whose relationship to various kinds of educational achievement has been established, this solution is possible. But are "experienced" tests likely to be relevant to the content of newly-developed courses?

(c) *The atypicality of the conditions of innovation.* Unless we wish to close the stable door after the horses have bolted we wish also to evaluate a course *before* it has become established. But our need is for information about what the course will be like *once it has become* established. Moreover, in an evaluation, we frequently judge such a course by comparing it with one that is established and which it might replace.

But consider some of the obstacles that might prevent us from obtaining information that is relevant to our purposes and some of the difficulties in making a fair comparison of two courses under these circumstances:

If we use those instances of the employment of the new course that already exist, we are likely to find that these "pioneer cases" have occurred in circumstances that are by no means representative of the circumstances under which the course will eventually need to be implemented if it is generally adopted. For example, the teacher who adopts a new and relatively untried course is likely to be exceptionally talented or exceptionally keen.

If, on the other hand, we introduce the course anew to teachers and classes that we have chosen for their representativeness, we are faced with other problems: the new course may be at a disadvantage because the teacher who uses it is inexperienced in its use, or it may be at an advantage because the teacher and pupils who use it are fired with great enthusiasm over its novelty.

Is there a way out of this dilemma? *Relatively* practicable devices for avoiding some of the difficulties are, on the one hand, to concentrate our evaluation on those aspects of the courses that are unlikely to be affected by novelty and, on the other hand, to assess and allow for the effect of increasing the teacher's familiarity with the new course. However, this is a difficulty of evaluation which is seldom tackled by the evaluator.

(d) *Diversity and Change.* The validity and usefulness of the usual form in which summative evaluation is carried out are threatened where courses are diverse and changeable. In the middle of a mathematics teaching revolution, courses tend to be both of these. Since it cannot be hoped that an evaluation will do justice to more than very few of the existing courses, we cannot be sure that unfairness does not derive from the fact that the best contenders among these courses are not entered in the competition. For the same reason we cannot be sure that the results of our evaluation will have more than a very limited application. Since evaluation often takes an appreciable length of time, and mathematics courses are in a state of change, we cannot be sure that the results of our evaluation will remain relevant to a course; these results may become obsolete because new rivals appear on the scene.

(e) *Formative Evaluation and External Criteria.* In its present form, *formative* evaluation of teaching courses also labours under certain difficulties. Some of these difficulties derive from the lack of criteria that are external to such courses.

Many of the decisions that need to be taken in the development of a teaching-course amount to choices such as these: whether to use one teaching method rather than another; which of two topics to include, when these are rivals for time; how long to allot to a particular topic; in which order to introduce topics. The answer to most of these questions about parts of a course will depend upon the nature of the remainder of the course. Theoretically, one can conceive that, by making one's decisions to include parts in a way that is dependent only upon the *other* parts of a course, one could arrive at a course which, although its parts had been tested and proved compatible, was a very bad course. Therefore, reference to the "performance" of parts in *other* courses is desirable – especially at the stage at which the foundations of a new course are being laid.

Again, while it is possible to devise a situation which would enable the curriculum evaluation to compare the effectiveness of method A with that of method B as means of teaching topic X, difficulties arise in situations in which it is necessary to compare the effectiveness of method A in teaching topic X with that of method B in teaching topic Y. Situations of the latter kind frequently arise in formative evaluation and cannot really be dealt with meaningfully unless reference is made to some kind of external standard by means of which the *usual* "performance" of these methods can be assessed.

Finally, without reference to outside standards, it is often difficult to know whether or not a part of a course is being taught as effectively as it might be taught. Even where learning has been achieved successfully, we cannot be sure that greater success could not have been achieved, or equal success in less time.

B. *A Proposed Solution: Course-component Evaluation*

A means of coping with these difficulties might be some such system as the following:

(i) The parts of mathematics courses are classified in terms of topics taught and teaching objectives.

(ii) A set of test items appropriate to these topics and objectives is devised and standardised.

(iii) Information about students attainment in terms of mastery of these topic/objective components is collected over a wide variety of mathematics courses. This information is classified in terms of teaching method used, time spent in teaching, ability and age of pupils, preceding learning experiences, etc.

(iv) Teachers needing evaluative information for purposes of formative or summative evaluation specify the components of their course in terms of topics and objectives, administer to their pupils the appropriate set of test items, and, for each component, compare their pupils' results with those already amassed in the system; naturally, each use of the system adds to the amount of information it contains.

(v) In addition to the information that it provides to the teacher who wishes to evaluate his particular course, this kind of system can act as a source of information for general curricular recommendations; for example, for a particular topic/objective it might yield an indication of the best teaching method; again, it might be used to establish which is most likely to be the best *order* of topics.

Let us enumerate some of the advantages of such a "course-component evaluative-data bank:"

(i) By assessing *parts* of courses rather than whole courses, we can obtain limited information that covers many different courses and is both *derived from* a wide range of situations and *generalisable to* a range of great width.

(ii) While courses change, some parts of these courses are likely to remain constant. The relevance of information collected about such parts is therefore likely to endure.

(iii) A readily available "bank" of information about the components of courses is needed by those "formative evaluators" who need external standards in terms of which to assess the effectiveness of parts of a course. Likewise, such a bank can be used to guide the initial design of a course.

III. INVESTIGATING THE LEARNING PROCESS

Now let us consider some of the problems involved in a completely different kind of exercise – that of finding out how learning works, rather than what it does.

An assumption underlying this paper is that the laboratory psychologist can be invaluable to the educational practitioner. After a brief attempt at justifying this assumption, I shall go on to examine some of those variables that might usefully be investigated in the psychological laboratory and to highlight some of the problems of experimental design that are likely to be encountered in investigations of this kind. Throughout, I shall illustrate my points by examples taken from a class of investigations into the learning of group structures. Among my reasons for selecting this kind of example is the fact that you will already have encountered a case of it in Malcolm Jeeves' talk. Other reasons will, I hope, (b) become apparent later.

A. *Basic Research and the Educationist*

The point at which the educationist turns to the laboratory psychologist is the point at which he asks the question, "Why?" He knows, for example, that such and such a way of teaching topic A is better than another way, but wonders whether, because of this, it will also be the better way of teaching topic B. Before he can generalise his knowledge, he needs to know *why* this method is better in the case of A – *what it is* that accounts for its superiority, *how* it achieves its effect. It is at this point that he begins to look for higher-level generalisations – generalisations from which he might *derive* those which he uses in the usual course of his teaching, *and* others like them.

It is partly in order to establish generalisations of this order that the psychologist must resort to conditions of investigation that are so frequently dismissed by the educator as "artificial". Consider a simple illustration of this none-too-subtle point. The educationist might make the statement, "Approach X is better than approach Y for achieving objectives of kind A." The referents of this statement – which we should need to consult if we wished to verify the statement – are clearly available in the everyday teaching situation. Now, if we wished to *explain* the truth of this statement, we should find ourselves asking questions like, "Is it because of aspect (i), rather than a aspects (ii), (iii), etc., of X and Y that X is better than Y?" Of course, once we came to verify such statements as *these*, we should need to begin to concern ourselves with restricted aspects of learning situations.

For example, we might vary aspect X (i) and see what effects this had on Y. Once we are doing this kind of thing, we must use artificial situations.

Please note that this kind of unravelling of the mechanisms of learning – entailing, as it does, the manipulation of limited aspects of the learning situation – is on the one hand *essential* for any but the shallowest of understanding of how things happen in learning, and, on the other hand, sufficient to get the laboratory psychologist stigmatised as a dealer in irrelevancies.

However, we do not need to search very far for examples of psychologists who have done more than just this to earn their stigmata. Many of the aspects of learning studied by educational psychologists are relevant only to a special phantasy world that some members of this fraternity appear to share. For example, investigations involving the endless repetition of responses as a condition of learning or the learning of nonsense syllables as a task could quite legitimately be regarded as having only tangential relevance to the teacher's problems.

In the present paper we shall explore the possibilities of investigating the learning of group structures. The relevance of the study of this subject-matter to the mathematical education of the child is not likely to be disputed by a roomful of geometry specialists. But its pertinence is by no means confined to the fact that groups have a pervasive relevance in mathematics. Piaget, and many other students of cognitive processes, consider that the axioms of the group form the cornerstones of relational and logical thinking.

Hence, in studying the learning of groups, we can kill a mathematical bird *and* a psychological bird. If we can stretch our metaphor just a little to include a third bird, we might point out that even the simpler groups constitute structures of unusual complexity for the laboratory psychologist to use as subject matter in his learning experiments, and that, in this respect, also, they tend to introduce reality into the laboratory. Malcolm Jeeves has pointed out some of the merits of group structures as the subject-matter of investigations into learning.

B. *The Experimental Situation*

At this point, I shall insert a parenthetic comment to the effect that the experiments that are envisaged in this paper would usually involve apparatus that is very similar to that which Malcolm Jeeves has already described and Zoltan Dienes has demonstrated.

Four windows can be used to display the twenty-four permutations of a maximum of four coloured lights. The subject's task is to press buttons (up to three are available), each of which effects a different operation leading to a particular kind of change (or no change) in the display of lights. The apparatus can be programmed for different kinds of operation and display. All displays occuring during the experiments are automatically recorded on punch tapes.

It should be noted that the recording system of this apparatus gives us a complete history of the learner's responses – from which we can obtain information about his successes, his failures and the strategies that inform his responses.

C. *Interesting Variables*

In order to provide some idea of the purposes and scope of this kind of investigation we shall consider some of those aspects of the investigation of complex structures that are likely to be of interest to the educationist and that are pertinent to the study of this kind of subject-matter. We shall classify such variables and briefly explain the point in studying them. Variables whose values we manipulate are known as *independent* variables, while those whose values might be affected by changes in the independent variables are known as *dependent* variables. A particular variable might well be used either as an independent or as a dependent variable.

D. *Independent Variables*

First, let us look at some independent variables:

(1) *Opportunity to Learn:* The obvious ways of varying this are by varying the number of times the learner is presented with the learning situation and the duration for which he is presented with it.

(2) *Information Provided:* One of the perennial issues in education concerns the degree to which the learner should be helped in his task by the provision of guidance. Clearly, the provision of guidance can take a variety of forms.

For example, we can vary the *kind* of information provided – in the present case we could provide kinds of information ranging from prompts concerning correct particular moves for the learner to make, to presentation of parts or the whole of a structure.

Again, we can vary the *embodiments* of the information provided. Some obvious kinds of embodiment are: mechanical, physical, graphical; symbolic. Naturally, the significance of any such embodiment would depend upon the relation it bore to the embodiment in terms of which the learning task was being presented.

Yet another kind of variation of information provided would be in terms of *quantity*. We shall postpone our tussle with the problems of quantifying such information, but may it suffice here to say that the value of information will vary with the *position* in the learning activity at which the information is introduced. This last variable will therefore be yet another way of characterising the information we provide the learner.

It must be pointed out that the introduction of information into the learning situation is most certainly a process that we should do well to study closely. The indications from a wide range of kinds of research are that the optimal conditions for effective learning are not likely to be those in which the learner must rely entirely upon his ability to discover what he is to know.

(3) *Memory Aids.* As Malcolm Jeeves has pointed out, the problems of storage of information in learning tasks of this kind are considerable. In cases in which information exposure is strung out over time (and in the case of most of our forms of communication this is so) some way of reducing the memory-load involved in gaining access to earlier-presented information will be necessary. Two general kinds of aid suggest themselves: devices for *recording* information, and devices for *organising* information.

Devices for recording information would seem to be as plentiful and various as one's imagination can be productive. A very important minimal device would be one which fulfilled the need expressed by Professor Jeeves' learners for a record of the state from which they derived the state at present on display. It would seem a shame not to provide the forgetful learner with the wherewithal to appreciate the nature of single transformations. Presumably, the maximally informative aid would be one which made all states and all operator moves (buttons pressed) available to the learner. If we wished to study the learner's storage strategies we could provide him with some choice concerning the information that was recorded. We might allow him to use some device (pencil and paper, for example) when he felt he needed it.

The *medium* or *embodiment* of the recorded information would very probably be crucial to its utility. Some kinds of embodiment might well restrict the learner's access to the information, or the versatility with which he could scan it for possible relationships. For example, recording in the form of a string of symbols might render information both less accessible and less amenable to scanning for relationships than recording in graphical form. Incidentally, from the relative showing of different kinds of recording, we might obtain indications concerning the most effective ways for the learner to process the information he receives.

Degree of recording permitted is clearly an important variable. For example, copious recording might well reduce the subject's need to search for structures – an effect on learning strategy similar to that which Jeeves discovered in the case in which subjects were given simple tasks to cope with.

Just as important as aids to the *recording* of information are aids to its organisation into a form in which it can readily be stored. Perhaps the simplest of such aids is in the form of periods allowed between trials, in which the subject is given *time* to process the information he has received. Experiments using verbal materials suggest that spacing of this kind can serve such a function. Characteristics of the learning situation that encourage the learner to structure his experiences, such as the "deep-end" sequencing mentioned by Professor Jeeves – or even simply in the form of instructions to do so – are also likely to reduce the memory load.

It should be noted that where a recording device is used, it can be devised
so as to influence the subject's organisation of the information he receives.
As I have pointed out, a two-dimensional, graphical, form of recording,
could well provide more opportunity for organising information than would
a one-dimensional sequence.

(4) *Structures Studied:* Since groups constitute important components of
the curriculum, it will certainly be useful to establish, in the case of particular
groups, the optimal conditions for learning. In addition to this we should
seek information that would enable us to decide on the best order in which
to introduce various groups.

But as seekers after general laws of learning, we might also concern
ourselves with the study of certain *aspects* of group structures.

For example, can we make *generalisations* about the *order* in which kinds
of groups are learnt? The experiment reported by Malcolm Jeeves suggests
that we should be able to do so. Again, in studying the learning of sequences
of groups, or in studying the effect that the learning of certain groups has on
the learning of others, certain intergroup structural relations, such as inclusion of one group in another, exclusion, and overlap, would be likely to
prove important. The studies that Dienes and Jeeves have already carried
out suggest that certain characteristics of groups are likely to affect the ease
with which they are learnt: commutative groups may be easier to learn than
non-commutative groups; the relation between the number of generators and
the number of elements may also be a determinant of a group's difficulty.

(5) *Embodiment:* Whenever an organism is learning, it is abstracting, that
is, acquiring information concerning certain aspects rather than others.
Likewise, when it is *applying* what it has learnt, it must sort out to which
aspects of a situation to respond. Because of this, a variable which is of
pervasive relevance in learning situations is that of the *embodiment* of the
information to be abstracted. One way of enabling the learner to abstract a
structure is to provide him with a variety of embodiments of the structure,
so that he can recognise what these have in common. This teaching technique
not only helps the learner to acquire structures, but provides him with the
necessary discriminations for recognising these structures in a new embodiment. And not only this: it can also provide him with generalisable skills of
abstraction which will help him to detach *classes* of structure from *classes*
of embodiment.

One of the problems of embodiment to be solved is that of the order in
which to use various embodiments. For example, in the case of some of the
embodiments used by Professor Dienes, should the learner encounter the
rotations of the triangle before the lattice games, these before graphical
representation, and this before symbolisation? Another such problem is that

of when to reduce the number of embodiments provided for a structure. Once the learner has discovered which characteristics of an embodiment are likely to be irrelevant to the "message" it conveys, he will perhaps be able to abstract the message with very little recourse to alternative embodiments. Again, embodiments will almost certainly have their individual learning characteristics. For example, learning will be easier with some than with others, and this will probably be affected by various embodiment-structure relationships. Some embodiments are clearly more appropriate for the teaching of certain structures rather than others.

Since the machine is only one of several kinds of embodiment, it is perhaps not surprising that our discussion of multiple embodiment has led us beyond its confines. However, studies of a more detailed kind than the foregoing can be carried out using the machine alone. First, the structure that the learner is required to abstract can be "embodied" in a more complex structure, much of which is irrelevant to the criterion tasks that the learner is required to perform. Second, irrelevant stimulus characteristics can be added to the displays that represent states.

Such "noise" could function in a variety of ways. For example, it could positively mislead the learner into a wrong conception of the structure to be grasped. On the other hand, by providing the learner with an alternative source of information about the structure to be learnt, it could promote learning. Characteristics of an embodiment that are superfluous to those that are strictly necessary for the communication of its "message" can easily be manipulated in such a way that they will help the learner to organise this message. Thus, in one of the experiments we have planned, "essential" stimuli, representing states in the depiction of a group structure, are accompanied by "ancillary" stimuli, which give a clue to the nature of this structure by depicting in a simple way one of its sub-groups.

E. *Dependent Variables*

Having examined some of the possible *independent* variables, let us now look at some of the *dependent* variables that we might expect to be affected by these.

(1) *Measure of Learning Difficulty:* Three obvious choices as measures of difficulty are: time taken to reach criterion (to reach required level of performance, that is); number of trials taken to reach criterion; number of errors made before reaching criterion. For many experimental set-ups the last two measures would be equivalent.

(2) *Learning Strategies:* As Professor Jeeves has so convincingly shown, we can answer some very crucial questions once we know how the learner goes about a complex learning task. To obtain this information we can either

ask him or record his responses. The machine described earlier permits the latter approach.

(3) *Transfer:* The obvious way of establishing whether the learner has abstracted a structure from the embodiment(s) in which he has encountered it is to test his ability to recognise this structure in a new embodiment, in the form of a transfer task. Among the dimensions along which we might qualify results on a transfer task are: kind of embodiment used, in relation to that which is used in the learning task(s); degree of difference between transfer and learning task embodiments.

Since our primary interest in the learner's activity is presumably in the degree to which he is subsequently able to use the skills he has learnt, we should perhaps give more care than seems to have been given hitherto to those conditions, at the transfer stage, that determine the success with which transfer takes place. For example, just as there are successful and less successful strategies for learning, so are there likely to be more or less successful strategies for application of what is learnt.

(4) *Other Learning Outcomes:* I have mentioned the foregoing outcomes mainly because, later in the paper, I mean to make methodological comments that concern them. However, there are many kinds of outcomes that might be studied in such experiments as these. For example, we might examine the learner's ability to recognise rather than recall; to relearn a structure after a long, rather than a short, period; to modify a structure before applying it; to do any of these after having had enough experience with learning situations of this kind to have developed learning sets of a reasonably high order of sophistication.

F. *Problems*

Having outlined some of the major variables that are likely to be involved in the investigation of the learning of mathematical structures, I am now in a position to introduce some of the problems that beset an investigation that is composed of these ingredients. As in the case of course evaluation, it will become clear that, although solutions of a kind are available for these problems, to settle for such solution in some cases transforms our exercise to such an extent that we might well wonder whether our original interests are still being pursued.

(1) *The Time Variable:* Suppose we wish to expose two sets of subjects to different learning conditions before testing them in order to establish which of these conditions is the more effective. In order to give both learning conditions an equal chance, our impulse would be to allow the same amount of time for both sets to learn their structures. But what if subjects in one set learnt faster than those in the other set? (This certainly is not an outside

possibility, especially where we have chosen two conditions for their likely difference in effect on learning.) In this case we are going to be faced with such embarrassing circumstances as the following:

(a) The faster learners must do *something* with the time that is left them, and this is quite likely to cause a difference between the two sets of learners which may be confused with effects due to the official learning conditions. Consider what they might be asked to do:

(i) Just wait. How can we be sure that some kind of covert rehearsal is not taking place during this period? Even if we could be sure of controlling such unamenable aspects of the learner's behaviour, some kind of con-solidation of learning or, alternatively, forgetting, could be taking place.

(ii) "Repeat" their learning task. What does this mean? Once they have grasped a structure, to go through the motions of learning it would be a strange kind of activity indeed.

(iii) Learn further structures. This would be one of the more reliable of possible ways of confusing the issue.

(b) An alternative to leaving the faster learners with spare time is to restrict the learning of the slower learners. This would lead to difficulties with most kinds of learning task, but would be disastrous with a task like the learning of a group structure, for to learn only a part of such a structure would be a completely different matter from learning the entire structure. Usually we should be interested only in subjects who had learnt the entire structure.

In order to avoid such difficulties as these, we might allow all learners to master their learning task. But then, of course, we are faced with the problem that the learners in the slower set will have spent more time in learning, and that this difference, rather than the official difference in learning conditions, might be responsible for the outcome of the experiment.

What should we do about this problem? I cannot think of a watertight solution.

Of course, we can always hope that it will not arise, and that either our two sets of learners will take equal periods of time or the set taking less time will prove to have learnt more effectively. Various devices have been sug-gested for coping with the problem – for example, matching pairs of subjects from different sets, and then considering the results only of those pairs whose members took equal lengths of time. However, such devices usually sub-stitute new problems for old.

It should be noted that this problem would be less acute if we were not concerned with a learning task that is of an all-or-none-and-never-again nature. In the case of some learning tasks, it would make sense to qualify results by saying that the subjects in one set had learnt less or more of a skill in a given time than subjects in another set. But in our case we cannot usefully talk of the learning of a particular structure unless it is completely learnt, and once it *is* learnt, to learn more is to learn something *else* about which we cannot talk usefully.

(2) *The Fair Criterion:* Suppose we wish to compare the effect of learning structure A with that of learning structure B and we wish to assess these effects in terms of transfer to the learning of structure C. Since A and B will have different relationships to C, (for example, A may be more like C) it could be argued that any transfer effects discovered were influenced by this difference. In order to surmount this difficulty, we might decide to compare the effect of A on Ca with that of B on Cb, where the relationship of A to Ca is the same as that of B to Cb. However, in this case, differences could be explained by the *absolute* difference between Ca and Cb.

(3) *Order Effects:* The importance of multi-embodiment, and the concomitant use of transfer tasks as a criterion of success in learning, is likely to lead us to carry out experiments in which the subject learns a structure in two or more different embodiments before being required to "transfer" it.

In such a case, the effect of the order in which tasks are learnt may be or may *not* be of interest to us. Suppose, first, that we are *not* interested in this effect. In this case, we may wish to eliminate it. Accordingly, we might try to balance the effect of following task 1 by task 2, by a condition in which we follow task 2 by task 1. It should be realised that there is no *a priori* reason for assuming that the effect of the "task 1–task 2" order will be the same as that of the "task 2–task 1" order.

Now, let us suppose that we *are* interested in the order effect – for example, as in the case of the study reported by Professor Jeeves (which, incidentally, was *not* designed in the way we are about to discuss). Where the "task 1–task 2" order (simple–complex, say) has a different effect on performance on the transfer task 3 from the effect of the "task 2–task 1" order (complex–simple, say), are we justified in assuming that this difference is not real because, in the one case, this task is immediately preceded by task 1, while in the other, it is immediately preceded by task 2? Again, we are reminded of the wisdom of seeking, as Professor Jeeves sought, information that helps us to understand *why* we obtain the results we do.

(4) *Studying Structural Variables:* We have seen that it is important to examine those *aspects* of a structure that may affect the way in which, or the ease with which, it is learnt. Such aspects, for example, as the number of generators, the number of elements, the number of automorphisms it contains, whether it is commutative, or cyclic. Now the most effective way of studying such aspects would be systematically to vary one while others remained constant. However, this is plainly not possible for, to vary one,

entails – because of the very nature of the structures with which we are concerned – varying others. The fact that these variables cannot be independently varied, seriously limits the degree to which we can subject them to systematic study.

(5) *Studying Embodiment:* It would be interesting to study general characteristics of embodiment, especially as embodiment is such a crucial component in the abstraction and generalization processes. However, it would be difficult to arrive at *general* conclusions about embodiments, for different embodiments are appropriate to the learning of different structures.

IV. CONCLUSIONS

In this paper I have tried to unfold some of the possibilities and limitations of two kinds of investigation to which the curriculum developer might subject his courses.

In conclusion I should like to emphasize that both kinds of investigation have an extremely important part to play, and that, in some form or other, such means as these of obtaining objective information about the behavior of teaching courses are indispensable.

However, I should also like to underline a point that must have become apparent during the course of this paper – that we have a long way to go before either of these kinds of investigation can be made to provide us with the information that we so badly need.

HANS ZASSENHAUS

GENETIC DEVELOPMENT OF THE
CONGRUENCE AXIOMS

The development of axiomatic group theory by Huntington, Dickson and Moore in the last decade of the previous century has opened our minds to a deeper understanding of Euclid's Common Notions 4:

"Things which coincide one with another are equal to one another."

As the conceptual understanding of each generation of mathematicians is increasing, new light is cast on the origins of geometric speculation in Euclid's Elements.

What is really involved in shaping and understanding the concepts of space, plane, straight line, point, angle, area, volume and their mutual relationship by way of incidence, congruence and similarity?

In the light of 20/20 hindsight we see more than the dead letter of an inherited script, important as that is, and we shall continue to restore it as closely as possible to its authentic form as insight increases.

In this light, a new understanding of the groping thoughts of past generations of geometers arises. Each living generation of geometers, by revolving to the original problems of geometric speculation, gathers new strength to venture beyond the limits that have shackled the imagination in the past.

The Greek sentence, "Καὶ τά ἐφαρμόζοντα ἐπ'ἀλλήλα ἴσα ἀλλήλουδ ἐστίν," has always constituted a tough problem of translation since it was coined in antiquity, very probably by Euclid himself.

The temptation here is to interpret ἐφαρμόζοντα as figures which are moved about in the plane or in space in some mechanical way. If we submit to this temptation, then we make geometric truth contingent on the outcome of experiments in mechanics. That is inherently repulsive to the geometric mind. The development of modern quantum mechanics and of modern cosmology has eroded the foundation of deducing geometric truth from mechanical experiments. It rather suggests a common development of the foundations of theoretic understanding within specified limits.

If ἐφαρμόζοντα means those figures which fit by means of a one-to-one correspondence preserving relation-like congruence, then the question arises whether this is a local phenomenon (as early Greek commentators of Euclid thought) or a global phenomenon.

Are there one-to-one mappings of the whole space (or of the whole plane) restricted to distance-preserving one-to-one correspondences (congruent mappings) of two given figures which are supposed to "coincide one with another?" If they exist, what are the properties of the system formed by those mappings which may be called the isometries of the whole space (or of the whole plane)?

Going one step further, to which extent are the geometric properties of the whole space (or of the whole plane) determined by the properties of its isometries?

These are the questions which have been studied by application of 20/20 hindsight by Sr. Louis Goldstein in her Notre Dame University thesis "An Historical Investigation of the Axiom of Congruence in Euclid's Elements: From the First Printed Edition of the Elements until the Development of the Group Theoretic Concept of Congruence" (Department of Education, June 1967).

A close reading of more than 150 Euclid Commentaries yielded two observations:

(1) Most commentators copied one from another;

(2) A few tried to explain anew the undefined and the unexplained in Common Notions 4. Among them very few developed new concepts which carried them to (or even beyond) the threshold of group theory. Finally the nineteenth century saw the full affirmation of group-theoretic foundation of elementary geometry which was climaxed by F. Klein's famous "Erlanger Programm."

The converse way (from group to geometry) was first explored by D. Hilbert in one of the appendices of the *Foundations of Geometry*. His approach was freed from its real plane topological background by Reidemeister and developed in full strength by G. Thomson. Thirty years later it entered the textbook literature in Bachmann's basic book on the geometry of reflections.

In her Ph. D. thesis *A Group-Theoretic Characterization of the Ordinary and Isotropic Euclidean Plane* (Notre Dame University, 1964) Sr. Mary Justin Markham has given a group-theoretic characterization of the ordinary and isotropic Euclidean planes over Pythagorean fields which is as close to elementary geometry as she felt was compatible with the group-theoretic program referred to above.

One starts with a group G generated by certain elements of order two (involutoric elements) called lines (strictly speaking, reflections into straight lines). A point is a product of two distinct commuting lines. Two distinct lines are said to be orthogonal if they commute. A point and a line are said to be incident if they commute. Points are said to be collinear if they are incident with the same line. A set of lines is said to be concurrent if there is a point which is incident with every line in the set. Two lines are called parallel if either they are coincident or if they are distinct and incident with no common point.

It is assumed that the generating set of lines is invariant under the inner automorphisms of G. This implies that the inner automorphism by means of a line (transformation with a line) maps points into points, lines into lines and amounts to an automorphism of the system Γ formed by the points and lines and related by means of incidence, parallelism, orthogonality, etc. Similarly, the transformation with a point is an automorphism of Γ. In this way the line and point reflections of Γ are obtained. The whole group G is homomorphic to a subgroup of the automorphism group of Γ.

The axioms required for making G the full automorphism group of the Pythagorean plane Γ are the following:

(I) There exist at least three non-collinear points.

(II) Given two distinct points, P and Q, there is one and only one line incident with both of them.

(III) The product of three concurrent lines is a line.

(IV) Given a point and a line there is at most one line incident with the point and parallel to the given line.

(V) The product of three points is a point.

(VI) Given two points there exists a midpoint, i.e., a point which reflects one of the two given points into the other.

A set of axioms for the space is simpler in many respects than a set of axioms for the Euclidean plane because then the Desarguean property of the planes is a consequence of the most basic assumptions already.

The strength of the group theoretic approach appears if it is not used merely for the purpose of an exercise in logic (to do this is the purpose and the limitation of the Hilbert school of geometers) but for a better understanding of given geometrical configurations of the Euclidean plane. Then it appears that the group theoretic approach amounts to a dynamic co-ordinatization as was first pointed out by Thomson.

Ohio State University, Columbus, Ohio, U.S.A.

BIBLIOGRAPHY AND LIST OF FILMS

The following list of books and articles has been compiled by the conference chairman, the list of films has been composed by S. Schuster and T. J. Fletscher.

1. MATHEMATICAL BACKGROUND

1.1. Books

Alexandroff, P. S., Markuschewitsch, A. I., and Chintchin, J. (Editors): *Enzyklopädie der Elementarmathematik. IV: Geometrie*, Berlin 1969 (transl. from the Russian).

Artin, E.: *Geometric Algebra*, New York 1957.

Bachmann, F.: *Aufbau der Geometrie aus dem Spiegelungsbegriff*, Berlin, Göttingen, Heidelberg, New York (Springer) 1959.

Bachmann, F. and Schmidt, E.: *n-Ecke*, Mannheim, Wien, Zurich, 1970.

Baer, R.: *Linear Algebra and Projective Geometry*, New York 1952.

Baldus, R. and Löbell, F.: *Nichteuklidische Geometrie*, Berlin 1953.

Beck, H.: *Elementargeometrie* 1 and 2, Leipzig 1929 and 1930.

Behnke, H. Bachmann, F., Fladt, K., and Süß, W.: *Grundzüge der Mathematik. II: Geometrie*, Göttingen (Vandenhoeck and Ruprecht), 1960. New edition in 2 parts A, B (1971).

Behnke, H., Bertram, G., and Sauer, R.: *Grundzüge der Mathematik. IV: Praktische Methoden und Anwendungen der Mathematik*, Göttingen 1966.

Berge, C.: *The Theory of Graphs and its Applications*, New York 1962.

Behzad, M. and Chartrand, G.: *Introduction to the Theory of Graphs* (Allyn and Bacon), 1970.

Bieberbach, L.: *Theorie der geometrischen Konstruktionen*, Basel (Birkhäuser), 1952.

Blaschke, W.: *Kreis und Kugel*, Berlin 1956.

Blumenthal, L. M.: *Theory and Applications of Distance Geometry*, Oxford 1953.

Blumenthal, L. M.: *A Modern View of Geometry*, San Francisco 1961.

Blumenthal, L. M. and Menger, K.: *Studies in Geometry*, San Francisco 1970.

Bonnessen, T. and Fenchel, W.: *Theorie der konvexen Körper*, Berlin 1934.

Borsuk, K. and Szmielew, W.: *Foundations of Geometry*, Amsterdam 1960.

Brückner, M.: *Vielecke und Vielflache*, Leipzig 1900.

Bumcrot, R. J.: *Modern Projective Geometry*, New York 1969.

Busemann, H.: *Projective Geometry and Projective Metrics*, New York 1952.

Coxeter, H. S. M.: *Regular Polytopes*, London 1948.

Coxeter, H. S. M.: *Non-Euclidean Geometry*, Toronto 1957.

Coxeter, H. S. M.: *The Real Projective Plane*, Toronto, London 1949.

Coxeter, H. S. M. and Greitzer, S. L.: *Geometry Revisited*, New York 1967.

Dembowski, P.: *Finite Geometries*, Berlin, New York 1968.

Dieudonné, J.: *La Geométrie des Groupes Classiques*, Berlin, Göttingen, Heidelberg 1955.

Efimow, N. W.: *Höhere Geometrie*, Berlin 1960 (transl. from the Russian).

Eggleston, H. G.: *Convexity*, Cambridge 1958.

Einstein, A.: *Geometrie und Erfahrung*, Berlin 1921.

Fladt, K.: *Elementar-Geometrie*, I–VI, Stuttgart 1957–1963.

Forder, H. G.: *Geometry*, Hutchinson Univ. Library, 1950.

Freudenthal, H. (Editor): *Algebraic and Topological Foundations of Geometry*, Oxford, Paris 1962.

Griffiths, H. B. and Hilton, P. J.: *Classical Mathematics*, Princeton 1970.

Hadamard, J.: *Leçons de Géometrie Élémentaire*, I, II, Paris 1937.

Hadwiger, H.: *Altes und Neues über konvexe Körper*, Basel, Stuttgart 1955.

Hadwiger, H.: *Vorlesungen über Inhalt, Oberfläche und Isoperimetrie*, Berlin, Göttingen, Heidelberg 1957.

Hadwiger, H. and Debrunner, H.: *Kombinatorische Geometrie in der Ebene. Monographie de L'Enseignement Mathématique*, Genève 1959; Engl. transl.: *Combinatorial Geometry in the Plane* (transl. and extended by V. Klee), New York 1964.

Heffter, L.: *Grundlagen und Analytischer Aufbau der Projektiven, Euklidischen, Nichteuklidischen Geometrie*, Stuttgart 1958.

Henkin, L. (Editor): *The Axiomatic Method with Special Reference to Geometry and Physics*, Amsterdam 1959.

Hessenberg, G.: *Ebene und Spärische Trigonometrie*, Berlin-Leipzig 1926.

Hessenberg, G. and Diller, J.: *Grundlagen der Geometrie*, Berlin 1967.

Heyting, A.: *Axiomatic Projektive Geometry*, Groningen 1963.

Hilbert, D.: *Grundlagen der Geometrie*, Stuttgart (Teubner) 1962.

Hilbert, D. and Cohn-Vossen, S.: *Anschauliche Geometrie*, Berlin 1932, New York 1944; Engl. transl.. *Geometry and the Imagination*, New York 1956.

Holden, A. and Singer, P.: *Crystals and Crystal Growing*, Garden City N.Y. 1960.

Jaglom, J. M. and Boltjansky, W. G.: *Konvexe Figuren*, Berlin 1956.

Kelly, P. J. *Geometry*, Chicago 1965.

Kerekjarto, B.: *Les Fondements de la Géométrie*. I., Budapest 1955.

Klein, F.: *Elementarmathematik vom Höheren Standpunkt aus II: Geometrie*, Berlin 1925 (Reprint 1968); Engl. Transl.: *Elementary Mathematics from an Advanced Standpoint. II: Geometry*, Dover Publ. 1939.

Klein, F.: *Vorlesungen über Höhere Geometrie*, Berlin 1926 (Reprint 1968).

Klein, F.: *Vorlesungen über Nichteuklidische Geometrie*, Berlin 1928 (Reprint 1968).

Lenz, H.: *Grundlagen der Elementarmathematik*, Berlin 1961.

Lenz, H.: *Vorlesungen über Projektive Geometrie*, Leipzig 1965.

Lenz, H.: *Nichteuklidische Geometrie*, Mannheim 1967.

Levi, F. W.: *Geometrische Konfigurationen*, 1929.

Levi, H.: *Foundations of Geometry and Trigonometry*, Englewood Cliffs 1960.

Levi, H.: *Topics in Geometry*, Boston 1968.

Lyusternik, L. A.: *Convex Figures and Polyhedra*, Boston 1966.

Maeda, F.: *Kontinuierliche Geometrien*, Berlin 1955.

Meschkowski, H.: *Ungelöste und Unlösbare Probleme der Geometrie*, Braunschweig 1960; Engl. transl.: *Unsolved and Unsolvable Problems in Geometry*, New York 1966.

Norden, A. P.: *Elementare Einführung in die Lobatschewskische Geometrie*, Berlin 1958 (transl. from the Russian).

Ore, Oystein: *Graphs and Their Uses*, New York 1963.

Perron, O.: *Nichteuklidische Elementargeometrie der Ebene*, Stuttgart 1962.

Pickert, G.: *Projektive Ebenen*, Berlin, Göttingen, Heidelberg (Springer) 1955.

Pickert, G.: *Analytische Geometrie*, Leipzig 1967.

Prenowitz, W. and Jordan, M.: *Basic Concepts of Geometry*, New York 1965.

Redei, L.: *Foundation of Euclidean and Non-Euclidean Geometries*, Oxford 1963.

Reidemeister, K.: *Grundlagen der Geometrie*, Berlin 1968.

Reidemeister, K.: *Topologie der Polyeder*, Leipzig 1953.

Room, T. G.: *A Background to Geometry*, Cambridge 1967.

Schur, F.: *Grundlagen der Geometrie*, Leipzig 1909.

Schwan, W.: *Elementare Geometrie*, Leipzig 1929.

Schwerdtfeger, H.: *Geometry of the Complex Numbers*, Toronto 1962.

Seidenberg, A.: *Projective Geometry*, Princeton 1962.

Tarski, A.: *A Decision Method for Elementary Algebra and Geometry*, Berkeley 1951.

Toth, L. F.: *Lagerungen in der Ebene, auf der Kugel und im Raum*, Berlin, Göttingen, Heidelberg 1953.

Veblen, O. and Young, W.: *Projective Geometry* 1, 2, Boston 1910, 1916.

Weber, H. and Wellstein, J.: *Enzyklopädie der Elementarmathematik II: Elementare Geometrie*, Berlin/Leipzig 1915.

Weyl, H.: *Raum, Zeit, Materie*, Berlin 1923 (reprint 1961).

Yale, P. B.: *Geometry and Symmetry*, San Francisco 1968.

1.2. Articles

Ahrens, J.: 'Begründung der absoluten Geometrie des Raumes aus dem Spiegelungsbegriff', *Math. Zeitschrift* **71** (1959) 154–185.

Brossard, R.: 'Birkhoff's Axioms for Space Geometry', *Am. Math. Monthly* **70** (1963) 593ff.

Dembowski, P.: 'Endliche Geometrien', *Math.-Phys. Semesterber.* **XIII** (1966) 32ff.

Freudenthal, H.: 'Der orientierte Raum des Mathematikers', *Die Naturwissenschaften* **50** (1963) 199–205.

Freudenthal, H.: 'Die Orientierung des Raumes', *Math.-Phys. Semesterber.* **X** (1969) 161ff.

Freudenthal, H.: 'Die Geometrie in der modernen Mathematik', *Physikal. Blätter* **10** (1964) 352ff.

Hjelmslev, J.: 'Neue Begründung der ebenen Geometrie', *Math. Ann.* **64** (1907) 449–474.

Hodge, W. V. D.: 'Changing Views in Geometry', *The Math.-Gazette* **39** (1955).

Lorenzen, P.: 'Das Begründungsproblem der Geometrie als Wissenschaft der räumlichen Ordnung', *Philosophia nat.* **6** (1961) 415–431.

Mac Lane, S.: 'Metric Postulates for Plane Geometry', *Am. Math. Monthly* **66** (1959) 543ff.

Nastold, H. J. and Th.: *Begründung der euklidischen Geometrie mit Hilfe von Abbildungen*, Schriftenreihe Math. Inst. Univ. Münster, Münster 1962.

Pejas, W.: 'Die Modelle des Hilbertschen Axiomensystems der absoluten Geometrie', *Math. Annalen* **143** (1961) 212ff.

Pickert, G.: *Die Stellung der geometrischen Grundlagenforschung in der heutigen Mathematik*, Mathematikunterricht an deutschen Universitäten und Schulen, Göttingen 1967.

Schwabhäuser, W.: *Metamathematical Methods in Foundations of Geometry*, Proc. 1964 Int. Congr. Logic, Math. and Phil. of Science, Amsterdam 1965.

Sperner, E.: 'Die Ordnungsfunktionen einer Geometrie', *Math. Anm.* **121** (1949) 107–130.

Thomson, G.: 'The Treatment of Elementary Geometry by a Group-Calculus', *Math. Gazette* **17** (1933) 230ff.

2. DIDACTICS, PEDAGOGY, PSYCHOLOGY, SPECIAL BACKGROUND FOR TEACHERS

2.1. Books

Abbott, E. F.: *Flatland. A Romance of Many Dimensions*, Barnes & Noble, 1963.

Behnke, H. *et al.* (Editors): *Lectures on Modern Teaching of Geometry and Related Topics*, Matematisk Institut Aarhus, 1960.

Castelnuovo, E.: *Didattica della Matematica*, Firenze 1964.

Castelnuovo, E.: *La Via della Matematica, La Geometria*, Firenze 1966.

Cundy, H. M., Rollett, A. P.: *Mathematical Models*, Oxford 1968.

Choquet, G.: *L'enseignement de la Géométrie*, Paris 1967; Engl. transl. *Geometry in Modern Setting*, 1969.

Delessert, A.: *Une construction de la géométrie élémentaire fondée sur la notion de réflexion*, Monogr. de l'Enseign. mathématique, Geneve 1964.

Diénès, Z. P.: *Exploration of Space and Practical Measurement*, Ed. suppl. Ass., 1966.

Diénès, Z. P. and Golding, E. W.: *Geometry of Congruence*, New York 1967.

Diénès, Z. P. and Golding, E. W.: *Geometry of Distortion*, New York 1967.

Diénès, Z. P. and Golding, E. W.: *Groups and Coordinates*, New York 1967.

Dieudonné, J.: *Algèbre Linéaire et Géometrie Élémentaire*, Paris 1969; Engl. transl.: *Linear Algebra and Geometry*, Boston 1969.

Ehrenfest-Afanassjewa, T.: *Didactische Opstellen Wiskunde*, Zutphen 1960.

Ehrenfeucht, A.: *The Cube Made Interesting*, New York 1964.

Fletcher, T. J. (Editor): *Some Lessons in Mathematics*, Cambridge 1964.

Freudenthal, H.: *Report on Methods of Initiation into Geometry*, Groningen 1958.

Guggenheimer, H. W.: *Plane Geometry and its Groups*, San Francisco 1967.

van Hiele-Geldof, D.: *De Didaktiek van het Meetkunde in de Eerste Klas van het VHMO*, Leiden 1957.

Jeger, M.: *Konstruktive Abbildungsgeometrie*, Luzern 1964; Eng. transl.: *Transformatian Geometry*, New York 1969.

Meschkowski, H.: *Nichteuklidische Geometrie*, Braunschweig 1965; Engl. transl.: *Non Euclidean Geometry*, New York 1966.

Moise, E. E.: *Elementary Geometry from an Advanced Standpoint*, Reading (Mass.) 1963.

Papy, G.: *Mathématique Moderne* 1, 2, 3, 6, Bruxelles 1964/67.

Papy, G.: *Géometrie plane et nombres réels*, Bruxelles, Paris 1962.

Papy, G.: *Initiation aux espaces vectoriels*, Bruxelles, Paris 1963.

Papy, G.: *Le premier enseignement de l'analyse*, Bruxelles 1968.

Piaget, J. and Inhelder, B.: *La représentation de l'éspace chez l'enfant*, Paris 1949.

Piaget, J., Inhelder, B., and Szeminska, A.: *The Child's Conception of Geometry*, New York 1970.

Pickert, G.: *Ebene Inzidenzgeometrie*, Frankfurt, Hamburg 1958.

Steiner, H. G.: *Vorlesung über Grundlagen und Aufbau der Geometrie in didaktischer Sicht*, Münster 1966.

2.2. Articles

Castelnuovo, E.: 'Les transformations affines dans le 1er cycle de l'école secondaire', *Educ. Stud. Math.* **1** (1969) 279ff.

Birkhoff, G. D.: 'A Set of Postulates for Plane Geometry, Based on Scale and Protractor', *Ann. of Math.* **33** (1933) 329ff.

Delessert, A.: 'Gibt es Darstellungen der euklidischen Geometrie, die sich wesentlich voneinander unterscheiden?', *Math.-phys. Semesterber.* **XIII** (1966) 165ff.

Dieudonné, J.: 'Winkel, Trigonometrie, komplexe Zahlen', *Der Mathematikunterricht* **12** (1966) 5–15.

Dieudonné, J.: 'Moderne Mathematik und Unterricht auf der Höheren Schule', *Math.-phys. Semesterberichte* **VIII** (1962) 166ff.

Dieudonné, J.: *Geometrie in den Gymnasien und in der modernen Forschung*, Braunschweig 1966.

Fletcher, T. J.: 'Finite Geometry by Coordinate Methods', *Math. Gazette* **37** (1953) 34–38.

Freudenthal, H.: *Report on a Comparative Study of Methods of Initiation into Geometry*, Euclides 1958.

Freudenthal, H.: *Un cours de géométrie. New Trends in Mathematics Teaching II*, UNESCO, Paris 1970.

Glayman, M.: 'A Geometry on the Cube', *Educ. Stud. Math.* **3** (1970) 89ff.

Papy, G.: 'Taximétrie', *Nico* **5** (1970) 87ff.

Pickert, G.: 'Axiomatische Begründung der ebenen euklidischen Geometrie in vektorieller Darstellung', *Math.-phys. Semesterberichte* **10** (1963) 65ff.

Pickert, G.: 'Deduktive Geometrie im Gymnasialunterricht', *Math.-phys. Semesterberichte* **10** (1963) 202ff.

Schneider, E.: 'Spiegelungsgeometrie auf der Oberstufe', *Der math.-naturw. Unterricht* **16** (1963–64) 388ff., 442ff.

Servais, W.: *Axiomatisation et géométrie élémentaire. Les Répercussions de la Recherche Mathématique sur l'Enseignement*, Luxembourg 1965.

Steiner, H. G.: 'Explizite Verwendung der reellen Zahlen in der Axiomatisierung der Geometrie', *Der Mathematikunterricht* **9** (1963) 66ff.

Steiner, H. G.: *Geometry in School Programmes*, Regional Seminar Cairo, UNESCO, Paris 1969.

Willers, H.: 'Die Spiegelung als primitiver Begriff im Unterricht', *Zeitschrift f. den math.-nat. Unterricht* **53** (1922) 68ff and 105ff.

3. HISTORY, CULTURAL ASPECTS

Becker, O.: *Zur Geschichte der griechischen Mathematik*, Darmstadt 1964.

Bourbaki, N.: *Éléments d'histoire des mathématiques*, Paris 1969.

Boyer, C. B.: *History of Analytic Geometry*, New York 1956.

Coolidge, J. L.: *A History of Geometrical Methods*, New York 1963.

Dürer, A.: *Underweysung der Messung mit Zirkel und Richtscheyt*, Nürnberg 1525; ed. by A. Peltzer, München 1908.

Euclides: *The Thirteen Books of Euclid's Elements*, Introd. and comm. by T. L. Heath, New York 1956.

Freudenthal, H.: 'Zur Geschichte der Grundlagen der Geometrie', *Nieuw Arch. Wiskunde, III Ser.* **5** (1957) 105–142.

Freudenthal, H.: 'Die Grundlagen der Geometrie um die Wende des 19. Jahrhunderts', *Math.-Phys. Semesterberichte* **7** (1960) 2–25.

Heath, Th.: *A History of Greek Mathematics I, II*, Oxford 1921.

Hofmann, J. E.: *Geschichte der Mathematik* 1, 2, 3, Berlin 1953, 1957.

Kline, M.: *Mathematics in Western Culture*, Oxford 1953.

Kline, M.: *Mathematics – A Cultural Approach*, Reading (Mass.), Palo-Alto, London 1962.

Smith, D. E.: *History of Mathematics I, II*, Dover Publ.

Steiner, H. G.: 'Frege und die Grundlagen der Geometrie', *Math.-Phys. Semesterberichte* **10** (1964) 175–186 and **11** (1964) 35–47.

Struik, D. J.: *A Source Book in Mathematics* 1200–1800, Cambridge, Mass. 1969.

Szabo, A.: *Anfänge des euklidischen Axiomensystems*, Archive for the History of Exact Sciences, Vol. I, Berlin 1960, 37ff.

Van der Waerden, B. L.: *Erwachende Wissenschaft*, Basel, Stuttgart 1956.

Weyl, H.: *Symmetry*, Princeton 1966.

White, G.: *Perspective*, Batsford 1968.

Wilder, R. L.: *Evolution of Mathematical Concepts*, 1968.

Zeuthen, H. G.: *Die Mathematik im Altertum und im Mittelalter*, Berlin 1912.

4. FILMS

Abbreviations:

I.F.B. = International Film Bureau, Inc., 332 So. Michigan Ave., Chicago, Ill. 60604, U.S.A.

M.L.A. = Modern Learning Aids, 1212 Avenue of the Americas, New York,
        N.Y. 10036, U.S.A.
B.F.I. = British Film Institute, 81 Dean Street, London W1, England.

4.1.

Possibly So, Pythagoras! by Bruce & Katherine Cornwell, 16 mm, 14 min.,
color. Distributor: I.F.B.

The Seven Bridges of Königsberg by Bruce & Katherine Cornwell, 16 mm,
4 min., color. Distributor: I.F.B.

Trio for Three Angles by Bruce & Katherine Cornwell, 16 mm, 7 min., color.
Distributor: I.F.B.

Newton's Equal Areas by Bruce & Katherine Cornwell & Alfred Bork,
16 mm, sound and color, 8 min. Distributor: I.F.B.

4.2.

The following series of short films by J. L. Nicolet are silent, 16 mm, black
and white, each approximately 3 min. They are distributed by Educational
Solutions, Inc., 821 Broadways, New York, N.Y. 10003.

Three Points Determine a Circle
A Given Line Seen at a Given Angle
Angles at the Circumference
Internal Bisectors of a Triangle
External Bisectors of a Triangle
The Construction of the Regular Pentagon
The Golden Section and the Regular Pentagon
Triangle Formed from Sides of Regular Polygons
Hypocycloid Motion with Circles in a Ratio of One or Two
Two Given Circles seen Under Equal Angles
The Strophoid and the Golden Section
Poles and Polars in the Circle
Generations of an Ellipse I and II
Generation of a Hyperbola
Generation of a Parabola
Another Generation of a Parabola
Common Generation of Conics
Locus of Vertex of Right Angles Tangent to Ellipse

(Also in the series are the following 3 which are in color.)
Circles Tangent to Two Concentric Circles
Contact Point of Parallel Tangents to Circles
Subtended Arc

4.3.

Let Us Teach Guessing, Mathematician G. Polya, 16 mm, sound and color, 61 min., produced by Mathematical Association of America Committee on Individual Lectures under the direction of A. N. Feldzamen. Distributor: M.L.A.

Notes on a Triangle by René Jodoin at National Film Board of Canada, 16 mm, sound and color, 7 min. Distributor: I.F.B.

Dance Squared by T. J. Fletcher, 16 mm, sound and color, 4 min. Distributor: I.F.B. and Contemporary Films (McGraw-Hill) and National Film Board of Canada.

Four Line Conics by T. J. Fletcher, 16 mm, sound and color, 13 min. (British Film Academy reward 1962). Distributor: Nat. Film Board of Canada, Montreal.

4.4.

Plucked Strings by T. J. Fletcher, 16 mm, 5 min. 1952. (Animation of two diagrams showing the vibrations of plucked strings.) Distributor: B.F.I.

The Simpson Line by T. J. Fletcher, 16 mm, 13 min. 1953. (A sequence of animated geometrical diagrams showing properties of the Simson line and its envelope, the three-cusped hypocycloid. Teaching notes.) Distributor: B.F.I.

The Cardioid, by T. J. Fletcher, 16 mm, 15 min. 1954. (A sequence of animated geometrical diagrams showing properties of epicyclic curves. Teaching notes.) Distributor: B.F.I.

Four Point Conics, by T. J. Fletcher, 16 mm, 13 min. (Animated diagrams showing properties of four-point systems of conics, including the eleven-point conic. Teaching notes.) Distributor: B.F.I.

The Heat Equation, 16 mm, sound, color, 10 min. (Animations of the heat distribution in bars, given various boundary and initial conditions.) Distributor: B.F.I.

Hypocyclic Motion by Fairthorne and Salt. Distributor: B.F.I.

4.5.

The following five films were produced by the Calculus Film Project of the M.A.A. under the direction of H. M. MacNeille.

Volume of a Solid of Revolution by George Leger, 16 mm, sound and color, 8 min. Distributor: M.L.A.

Volume by Shells by George Leger, 16 mm, sound and color, 8 min. Distributor: M.L.A.

Infinite Acres by Melvin Henriksen, produced by Calculus Film Project of M.A.A., 16 mm, sound and color, 10 min. Distributor: M.L.A.

What is Area? by Charles Rickart, produced by the Calculus Film Project of the M.A.A., 16 mm, sound and color, 20 min. Distributor: M.L.A.

Area Under a Curve by Charles Rickart, produced by Calculus Film Project of the M.A.A., 16 mm, sound and color, 10 min. Distributor: M.L.A.

4.6.

Shapes of the Future – Some Unsolved Problems in Geometry by Victor Klee, produced by the Committee on Individual Lectures under the direction of R. G. Long. 16 mm, sound and color, 65 min. Distributor: M.L.A.

Mathematics of the Honeycomb by Moody Institute of Science, 13 min., color, 16 mm. Distributor: Moody Inst. of Science, Educ. Film Division, Whittier, Calif. 90606.

Mathematics Peep Show, produced by C. Eames, 16 mm, sound and color, 11 min. Distributor: M.L.A.

Symmetry by J. Bregman, R. Davison, A. Holden and P. Stapp, 16 mm, sound and color. Distributor: Contemporary Films (McGraw-Hill).

4.7.

The following 12 films were produced by the College Geometry Project of the University of Minnesota. Project Director: S. Schuster.

Orthogonal Projection. Mathematician: Daniel Pedoe, 16 mm, sound and color, 12 min.

Central Similarities. Mathematician: Daniel Pedoe, 16 mm, sound and color, 10 min.

Inversion. Mathematician: Daniel Pedoe, 16 mm, sound and color, 13 min.

Dihedral Kaleidoscopes. Mathematician: H.S.M. Coxeter, 16 mm, sound and color, 12 min.

Symmetries of the Cube. Mathematicians: H.S.M. Coxeter and W.O.J. Moser, 16 mm, sound and color, 11 min.

Curves of Constant Width. Mathematician: J.D.E. Konhauser, 16 mm, sound and color, 16 min.

Equidecomposable Polygons. Mathematician: J.D.E. Konhauser, 16 mm, sound and color, 10 min.

Caroms. Mathematician: Chandler Davis, 16 mm, sound and color, 7 min.

Geometric Vectors – Addition. Mathematicians: W.O.J. Moser and S. Schuster, 16 mm, sound and color, 17 min.

Isometries. Mathematicians: W.O.J. Moser and S. Schuster, 16 mm, sound and color, 26 min.

Central Perspectivities. Mathematician: S. Schuster, 16 mm, sound and color, 13 min.

Projective Generation of Conics. Mathematician: S. Schuster, 16 mm, sound and color, 17 min.